# Yucca Mountain

## The Battle For National Energy Policy

By Stuart D. Waymire

Copyright 2017

Las Vegas

Contact: stuwaymire@aol.com
For more information see: http://www.yuccamountainexpose.com
Phone: 702-279-1385

ISBN-13:
978-1548779283

ISBN-10:
1548779288

*Stuart Dean Waymire Is a mechanical engineer, free lance writer and entrepreneur living in Las Vegas since 1980.*

# Contents

# *Preface*

The best way to introduce this book is to repeat a citizen's editorial written by this author for the Las Vegas Review Journal in 1989 which stated:

## Nevada plays 'nuclear chicken' with the feds

It is time Nevada faces the unpleasant fact that the nuclear waste repository is coming to the state. Period.

The waste is coming here for two reasons: 1) We are an underpopulated desert state with little political clout; and 2) nuclear energy is here to stay.

These two unavoidable facts of life place the no-nuclear-waste gyrations of our politicians and environmental groups in direct opposition to the well being of the majority of all Nevadans, whether you are individually for or against the repository on principal.

The conclusion is clear: the waste must go somewhere in America, and it cannot go to New York or Los Angeles (although we might like to send it there out of spite).

In short, Nevada is the only place that meets the criteria of low population, stable geology and previous experience in handling radioactive material (at the Nevada Test Site).

But can't we do without nuclear power completely and solve the waste problem that way?

Sure, if you want more tanker spills like the Exxon Valdez if you like the greenhouse effect and acid rain, and if you enjoy the carbon monoxide that pollutes the Vegas Valley.

Forget nuclear submarines for defense, too; they generate nuclear waste piled up at reactor sites around the country.

And stop dreaming - if solar power were really cost effective, every power company in America years ago would have installed sun-driven systems to cut their operating costs and boost profits.

Some will argue that Three Mile Island and Chernobyl are reasons enough to fight nuclear energy, and they even go so far as to suggest that straitjacketing nuclear waste disposal is the means to do so.

The dirty little secret that these people won't ell you, though, is that the loss of life from those disasters (zero at TMI and perhaps 1,000 over 30 years at Chernobyl) is only a small fraction of the deaths that are caused each year by coal and gas fired plants.

Hydrocarbon power plants cause thousands of deaths annually from deadly emitted carcinogens. Even when worst-case nuclear disaster scenarios are taken into account, they are not nearly so severe in their deadly cumulative effects as the smokestack plants now in operation.

So where does this leave us?

It leaves us with the realization that Nevada will be receiving nuclear waste in the near future.

The rational course of action, then, is not to continue dancing a jig to the tune of "The Sky Is Falling," but to make sure that Nevada benefits from the situation.

Constructive politics would dictate that we ask for a reasonable amount of compensation for the honor of being the nuclear waste dump for the rest of the country. A list of items of possible compensation might include:

1) We want deed to the nuclear waste so we can benefit monetarily from future recycling ventures.

2) We want substantial increases in highway funds to compensate for the wear and tear on our roads of vehicles headed to the repository, and we want highway upgrades to make those roads safe for nuclear transport.

3) We want an infusion of funds for education, both to the university system for its support in coordinating the repository and also to general education to make up for the expense of educating repository workers' children.

4) We may want various tonnage taxes on imported wastes to offset numerous state costs associated with the repository site.

You get the idea.

Our present nuclear chicken strategy, in which we have dared the feds to bring the repository here over our dead bodies, has resulted in a repository and zero benefits for Nevada as a whole.

If, on the other hand, we come to grips with the reality that nuclear waste is coming here no matter what, we might be able to salvage something from the mess the politicians and environmentalists have created for us and at least get some sort of just compensation for being the nation's nuclear dumping ground.

Maybe then we could also ask ex-governor Richard Bryan what happened to all that money that went to hotshot Washington lawyers that was supposed to save us from the repository. Did anyone in Nevada benefit from it, or was it just hogs feeding at the trough?

**Las Vegas Review Journal, Nevada Views, June 25, 1989**

In hindsight, very little would need to be changed to that editorial after nearly 30 years. Because of the above editorial, in 1990 the pro-Yucca Mountain public relations campaign of the American Nuclear Energy Council (now the Nuclear Energy Institute) run by OIZ Advertising in Las Vegas contacted me to consult as a mechanical engineer. I worked as a consultant to OIZ providing background information on technical and political issues surrounding Yucca Mountain during 1992, 1993 and early 1994. The different position papers and research I developed began to form into a book and after being released by OIZ in spring of 1994, I continued to work on this book and update it over the years and decades since.

The reason *Yucca Mountain: The Battle For National Energy Policy* is again current is that the Trump administration is now committed to going forward with reopening the Yucca Mountain program. This was of course inevitable, as my 1989 editorial predicted; nuclear waste will not be stored elsewhere for both scientific as well as political reasons. With the retirement of Senator Harry Reid in 2016, the most powerful political opponent of Yucca Mountain has now left the scene. However what is left is 300 and counting lawsuits generated by Nevada, mostly through its Nevada Nuclear Waste Project Office (NWPO), all meant to derail Yucca Mountain.

Consequently, for Yucca Mountain to go forward, the legitimacy of NWPO's contentious lawsuits will have to be addressed. However, what if NWPO's oversight and anti-Yucca Mountain advocacy was not based in science, but was the result of a radical political agenda that has nothing to do with nuclear radiation risks? This book reveals in excruciating detail the compromised science of NWPO, based on bizarre philosophies ranging from decentralist anarchism, to Maoism, Marxism, Liberation Theology, to Gaia and many in between. "Risk Perception" rather than an analysis of Risk became the marching order for NWPO. In short, Nevada's entire opposition to Yucca Mountain is based on corrupt and ideologically tainted science.

The reason this is all still so important is because the only way for America to move forward with 4th Generation Nucler Reactors that actually can burn radioactive waste, is by resolving the problem of existing nuclear waste stored at 62 plus sites. Thus the history contained in this book is a critical cog in guaranteeing our energy future; it is the only way to get rid of the phiosophical dead wood that now stops progress towards a nuclear waste solution.

Stuart Dean Waymire
July 23, 2017

# PART I: YUCCA MTN POLITICS

# 1.  Nuclear Waste Dilemma

On December 2, 1942, a team of scientists at the University of Chicago, led by Enrico Fermi created man's first controlled, self-sustaining nuclear chain reaction. Today, the United States generates 20% of its electrical power from nuclear reactors and its nuclear arsenal is perhaps the largest in the world. Both peaceful and military uses of nuclear energy generate nuclear waste which is radioactive and requires isolation from the biosphere. Since the 1950s there has been an expanding effort in the U.S. (and internationally) to develop plans for disposing the accumulating high-level nuclear wastes that result from nuclear technology and isolating this material from the environment for the long term. Over the years, proposals have varied in creativity from space launchings, to burial in deep ocean trenches, to on-site storage in what are called "dry-casks". Until recently, the plan which seemed most likely to proceed in the United States was geologic burial deep beneath the earth's surface.

Military waste is now disposed of in a geologic salt bed at the Waste Isolation Pilot Project (WIPP) facility constructed near Carlsbad, New Mexico. The United States Department of Energy began planning for the facility in 1974 and after more than 20 years of scientific study, public input, and regulatory struggles, WIPP began operations on March 26, 1999. Disposal operations are expected to continue until 2070 with active monitoring for a further hundred years. By 2010, the facility had already processed 9,000 shipments of waste. In 2010, the USDOE withdrew previous plans to develop Yucca Mountain nuclear waste repository in Nevada. WIPP was identified as a candidate for a facility to store waste for nuclear weapons defense related waste. Various mishaps at the plant in 2014 brought focus to the problem of what to do with this growing backlog of waste and whether or not WIPP would be a safe repository. The 2014 incidents involved a waste explosion and airborne release of radiological material that exposed 21 plant workers to internal doses of plutonium, which can lead to cancer of the lungs, liver, and bones.

Disposal of the much larger amounts of civilian reactor waste (75,000 plus tons) has been studied at a site called Yucca Mountain, located in the desert a hundred miles northwest from Las Vegas, Nevada. Yucca Mountain nuclear waste repository is currently identified by Congressional law and was approved in 2002 as the nation's spent nuclear waste storage facility. It was hoped that the Yucca Mountain site would be ready to accept shipments after the year 2015, but former Nevada Senator Harry Reid and the Obama Administration put the site in permanent limbo. Federal funding for the site ended in 2011 under the Obama Administration via amendment to the Department of Defense and Full-Year Continuing Appropriations Act, passed on April 14, 2011. However, licensure of the site through the Nuclear Regulatory Commission is ongoing.

## THE NEED FOR CENTRALIZED GEOLOGIC STORAGE

Most spent fuel that composes high level nuclear waste produced by nuclear power plants was originally stored in steel-lined concrete pools of water at 65 powerplant sites in more than 30 States. In the 60s and 70s when most U. S. reactors were designed, it was believed spent fuel would be stored only temporarily in at-reactor pools before being removed for reprocessing (no longer an option). The pools were therefore not designed for permanent storage. Re-racking, or rearranging, spent fuel assemblies within the pools has increased their capacity, however pool storage space cannot be maintained indefinitely.

As the Fukushima reactor disasters showed, aboveground pools are a potential hazard. Successful terrorist attacks on spent fuel pools, though difficult, are possible. If an attack leads to a propagating zirconium cladding fire, it could result in the release of large amounts of radioactive material

The Nuclear Regulatory Commission estimates many of the nuclear power plants in the United States will run out of room in their spent fuel pools and require the use of temporary storage of some kind. Spent fuel is currently also being stored in dry cask systems at a growing number of power plant sites, and at an interim facility located at the Idaho National Laboratory near Idaho Falls, Idaho. Yucca Mountain was expected to open in 2017. However on

March 5, 2009, Energy Secretary Steven Chu reiterated in a Senate hearing that the Yucca Mountain site was no longer considered an option for storing reactor waste.

A number of nuclear utilities may eventually be forced to shut down if alternative storage is not found. Consequently, the search for a nuclear waste repository for commercial spent fuel has a direct impact on whether a significant percentage of our nation's energy production facilities remain in operation. Three main arguments favor long term centralized geologic storage of nuclear waste:

*1. Geologic storage removes the high-level waste from 65 vulnerable sites around the nation (many near oceans, rivers and fault lines) and consolidates it in a sophisticated and dedicated engineered structure.*

*2. There is no guarantee later generations will have the political will or monetary resources to store the waste in the future. It would be irresponsible to force our nuclear problems on future generations*

*3. Security requirements at a central site will be diminished in comparison to funding the indefinite protection of 65 or more surface sites. The 1993 explosion at the Twin Towers at the World Trade Center in New York, and their destruction on 9/11/2001 suggests on-site storage is vulnerable to terrorist attack.*

Opponents of Yucca Mountain argue that on-site storage of nuclear waste for fifty or a hundred years at 65 sites in above-ground canisters (while an alternative disposal system is found), is preferable to Yucca Mountain, despite the instability of political, institutional and natural events. While not entirely out of the realm of possibility, these objections seem driven by a desire to preclude all nuclear technology rather than safety concerns over Yucca Mountain.

## NATIONAL ENERGY POLICY

If America's nuclear energy production capacity is to remain viable, the nuclear waste disposal problem must be resolved. The study and building of the Yucca Mountain nuclear waste repository thus has a huge impact on America's energy future. There are a number of reasons why nuclear energy is an attractive element in our national energy policy:

1) **ENERGY INDEPENDENCE:** Reserves of uranium in the U.S. makes this fuel independent of foreign suppliers. The continuing Middle-East war and potential hostilities with Iran and even Venezuela are costly battles for foreign oil reserves.

2) **ENVIRONMENTAL PROTECTION:** Hydrocarbon fuels are environmental pollutants. Nuclear power plants do not pollute the atmosphere. They also emit less radioactivity than many coal-fired plants.

3) **RESOURCE ABUNDANCE:** U.S. hydrocarbon reserves are limited. Hydro power is essentially used up. Solar has inherent limitations being geographically limited in application, resource intensive and expensive. Fusion is many years away and may never be economically competitive. Uranium and especially Thorium reserves are in contrast abundant.

4) **ECONOMIC COMPETITIVENESS:** Global markets are energy hungry. Access to nuclear energy thus becomes an important advantage in the worldwide economic foot race. China and India are well on their way to development.

5) **THIRD WORLD EMPOWERMENT:** Access to clean and inexpensive nuclear energy promises to speed economic development in the Third World and promote environmental responsibility. Burning wood in mud hearths is neither conducive to economic growth nor does it protect the environment. American nuclear technology is thus critical to ensuring Third World nuclear power is safe and nuclear waste disposal is done in a thoughtful way worldwide.

Of the available energy sources, only solar energy and nuclear stand out as being close to meeting the entire range of requirements for a sound energy policy, providing unlimited reserves and relatively benign impact on the environment. Unfortunately, little of the massive investment in alternative energies is likely to bear fruit in commercial ventures. The windmills of Altamont Pass in California stand idle. The Solar One and Solar Two solar collectors near Barstow California did not produce quantum breakthroughs. Alternative energy has thus far been cost prohibitive and may remain so.

Alternatives beyond solar are few. For example, the synfuels program during the Carter administration and current ethanol subsidies have been expensive wastes of taxpayer money. The Reagan, Bush, Clinton, Bush and Obama administrations that followed have equally lacked alternative fuel success stories and future presidencies need to be wary of venturing down alternative energy "investment" paths which are actually counterproductive to environmental goals. The one positive alternative has been natural gas production, which has been developed in spite of government help.

Another instance of the dismal energy record of the political establishment was the attempt in the 90s by former Nevada senators Harry Reid and former senator Richard Bryan to turn the Nevada Test Site into a huge hydrogen-solar energy farm. Besides being environmentally disruptive (similar to paving the Test Site with concrete) and costly (trillions of dollars said its own proponents), the hydrogen project's main purpose seemed to be a social welfare make-work project to keep the senators' Nevada constituents happy.

If we wish to keep national energy policy from degenerating into a special interest boondoggle, a philosophy of energy technology development needs to be developed that balances the competing interests of the environment, the economy and social welfare. Yucca Mountain and nuclear energy appear to be irreplaceable within this grand energy strategy.

## OPPOSITION TO YUCCA MOUNTAIN

Opposition to Yucca Mountain has often been couched as a state's rights issue pitting Nevadans against the federal government over the use of land within the state. If state's rights were the only issue, the political whirlwind surrounding the study of the nuclear waste repository would have been resolved long ago. Instead, Yucca Mountain has been used both by Nevada's political establishment and by activist environmental and social justice groups nationally and internationally to promote much more ambitious radical agendas.

Among Nevada's political elite, Yucca Mountain has long been a potent vote generator and has served to polarize the community into two lopsided fragments. Those opposing Yucca Mountain maintained a sizable majority over those who favor studying the site. A rigid anti-Yucca Mountain orthodoxy thus served as a political bludgeon to maintain political fault lines within the state.

Myriad environmental factions have also opposed the development of the repository. Greenpeace, Friends of the Earth, the Sierra Club, Public Citizens, The Safe Energy Communication Council, Citizen Alert, American Peace Test, etc. are a short list of opponents from the Green movement. The fight against Yucca Mountain by environmentalists appears to be based on a complex agenda of anti-nuclear pacifism, social reform, religious conviction and land reform more than on the potential negative health effects of nuclear waste. Consequently, anti-nuclear fanaticism at Yucca Mountain has taken on the passion of a Holy War rather than the cool logic of scientific analysis. The question this poses is whether America can afford to fashion its energy policy on the basis of radical special-interest social ideology rather than the technical merits of the nuclear fuel cycle.

## POLITICAL LANDSCAPE

The election of Bill Clinton to the presidency in 1992 caused a sea change in America's energy policy. Environmental concerns moved towards the center of energy policy, reflected in the views of Vice-president Al Gore, Interior Secretary Bruce Babbitt, EPA director Carolyn Brown, and Secretary of Energy Hazel O'Leary. The Bush administration rolled back some of the Clinton era stonewalling of energy development, but the Obama administration under the leadership of Senate Majority Leader Harry Reid attempted to foreclose Yucca Mountain and the nuclear option completely. Yet, nuclear energy seems to fit well within a combined energy/environmental policy no matter the ideology of those in power. Relative to the pollution caused by hydrocarbon technology, nuclear energy is benign. Relative to solar technology, nuclear is compact, resource efficient and cost effective.

Logically, the tradeoffs between environmental safety and cost efficiency are favorable to the nuclear industry. There is thus a strong argument that Yucca Mountain is in our national interest, isolating nuclear waste from the environment and allowing the continuation of nuclear power as an energy option.

Instead, energy policy has been driven by special interest professional anti-nuclear environmental lobbies with incestuous ties to powerful politicians (notably former Vice-President Al Gore and Senator Reid). The roots of this energy policy driven by political expediency lead to the Carter administration when Ralph Nader's Congress Watch and other environmental factions promoted the synfuels program (synthetic fuels from oil shale) and solar energy tax credits, two programs which were noted failures. One of the premier anti-nuclear coalitions in the past was The Safe Energy Communication Council, staffed largely with holdovers from this earlier movement.

Progressive energy policy is governed neither by cost efficiency nor environmental concerns, but by social considerations that form a larger Green political movement. Calls for on-site rather than geologic storage of nuclear waste made by Nevada's Nuclear Waste Project Office and Nevada's politicians, seemed derived from radical social justice equity theory rather than from engineering, environmental or economic concerns. Understanding the way the study of Yucca Mountain has been conducted and the political motivations and interests of those on both sides of the issue, is therefore critical because it has wider ramifications for all future technological energy development.

## MAKING A DECISION

The question of whether Yucca Mountain is the appropriate nuclear waste disposal technology to pursue cannot be resolved in these few pages; that is why a multiple decade and multi-billion dollar study was being conducted. What can be resolved is whether the study of Yucca Mountain has been conducted wisely and whether the political motives of those involved were based on the needs of Nevadans and Americans, or whether ulterior motives have driven the debate. A decision must be made between two extreme futures:

*1. If Yucca Mountain is as flawed and unsalvageable a technology as its opponents claim, we must prepare to shut down 20% of our nation's electrical capacity within the next thirty years. This in turn would require a massive investment (perhaps trillions of dollars) to convert the shortfall into solar capacity, or require a substantial increase in hydrocarbon imports, or result in substantial economic downsizing. Finally, it means restarting the entire multi-billion dollar repository siting process fifty years from now to dispose of already generated waste without guarantee that such efforts will be successful or that present expertise can be maintained during the delay.*

*2. Alternatively, if Yucca Mountain is found tenable (or at least salvageable), then an opportunity exists to improve America's energy independence and economic vitality through a rebirth of nuclear energy. New, higher efficiency and less polluting 4th Generation reactors could be brought on line and existing marginal nuclear facilities could be retired early. Solar and other alternative energies could still be developed, which would compete in the market place both on the basis of cost and environmental safety.*

The choices are stark. America and the world face a severe contraction of energy options if our decision on Yucca Mountain is negative. Making such a choice on the basis of special interest politics rather than after a sober analysis of risks and benefits promises multiple decades of disruption of our energy supplies.

# 2. Yucca Mountain Repository

The law that established a federally managed nuclear waste repository program is the Nuclear Waste Policy Act (NWPA) of 1982. The purpose of the act was:

**To provide for the development of repositories for the disposal of high-level radioactive waste and spent nuclear fuel, to establish a program of research, development, and demonstration regarding the disposal of high-level radioactive waste and spent nuclear fuel, and for other purposes. [1982 Nuclear Waste Policy Act, act description]**

The original Nuclear Waste Policy Act called for investigating a number of sites simultaneously, both to provide fall-back positions should any one site not prove scientifically sound and to act as controls for purpose of comparison. Unfortunately, this multi-pronged program proved to be too expensive and too politically unpalatable. There were legitimate questions of whether a multi-tracked study could be financed and considering the huge expenditures already incurred at Yucca Mountain it was unlikely three sites could have been studied concurrently. Politically, Nevada has such a small congressional delegation that it is unsurprising that larger states, when faced with accepting nuclear waste in their own backyards, chose to force the repository on Nevada.

Consequently, the Nuclear Waste Policy Amendments Act of 1987 restricted study of a repository site to Nevada:

*Sec. 160. (a) In General. -- (1) The Secretary shall provide for an orderly phase-out of site specific activities at all candidate sites other than the Yucca Mountain site. [Nuclear Waste Policy Amendments Act of 1987]*

In recognition that this was a less than equitable solution, the Amendments Act also offered a benefits package to make the medicine less bitter for Nevada:

*Sec. 171. (a) In General. -- (1) In addition to the benefits to which a State, an affected unit of local government or Indian tribe is entitled under I, the Secretary shall make payments to a State or Indian tribe that is party to a benefits agreement under section 170 in accordance with the following schedule:*

| BENEFITS | | |
|---|---|---|
| Event | **MRS** | **Repository** |
| *(A) Annual payments prior to first spent fuel receipt* | **$5mil** | **$10mil** |
| *(B) Upon first spent fuel receipt* | **$10mil** | **$20mil** |
| *(C) Annual payments after first spent fuel receipt until closure of the facility* | **$10mil** | **$20mil** |

Known not so fondly as the "Screw Nevada" bill, the 1987 amendment eliminated sites in Deaf Smith County, Texas and Hanford, Washington from consideration and shifted focus entirely towards the study of Yucca Mountain, Nevada. While this choice was significantly political and meant to shuffle the problem of siting a nuclear waste repository to Nevada (an underpopulated desert state that could ill defend itself in Congress), significant scientific research done prior to 1987 already pointed favorably towards Yucca Mountain. It is also important to remember that the amendment offered compensation to Nevada if it accepted the study. Offers to substantially sweeten the proposed benefits listed above to the $50 to $100 million level have been made since the 1987 amendment, including by no less than Sen. Bennett Johnston, D-La., author of the "Screw Nevada" bill and then Chairman of the Senate Energy Committee. Senator Lindsay Graham has made similar overtures in 2010.

**NUCLEAR ENERGY & HIGH-LEVEL WASTE**

The radioactive materials that are to be stored at the Yucca Mountain repository are for the most part the result of the commercial nuclear energy program in the United States, along with a small amount of medical and defense wastes. Nuclear energy is a crucial element of our economy and produces about twenty percent of our total electrical energy needs. The controlled release of nuclear energy became possible for commercial power became possible in the mid 1950s. In studying the interactions between neutrons and nuclei, scientists observed the following behavior of the uranium isotope Uranium-235:

$$^{235}_{92}U_{143} + n \longrightarrow {}^{138}_{56}Ba_{82} + {}^{95}_{36}Kr_{59} + 3n + \text{Energy}$$

This reaction represents the fission, or splitting, of a uranium atom after a collision with a neutron, producing a barium atom, an atom of Krypton, three neutrons and a large amount of energy through Einstein's famous equation, $E=MC^2$. The excess neutrons can cause other uranium atoms to split in a chain reaction if there are subsequent collisions, or they can cause elements to change their atomic number and state in a form of nuclear alchemy.

The nuclear reaction is accompanied by the release of an enormous amount of energy. The fission of 1 kilogram (2.2 lbs.) of Uranium-235 yields 23,000,000 kilowatt hours of energy. In comparison, one kilogram of coal only yields 9 kilowatt hours. A nuclear reactor is the means by which this tremendous energy is controlled and used to generate the steam for electricity producing turbines. The elements left in a fuel rod at the end of the uranium fuel cycle, created either as breakdown products of uranium or other fissionable material, plus elements made radioactive by collisions with neutrons, are what constitute the majority of high-level nuclear waste.

Most of the spent fuel that composes high level nuclear waste produced by nuclear power plants is currently stored either in steel-lined concrete pools of water at 65 powerplant sites in more than 30 States, or in nearby dry storage casks. Although spent fuel and high-level radioactive waste lose about fifty percent of their radioactivity after three months of storage, and about 80 percent after one year of storage, radioactivity remains for thousands of years. It is the longevity of this radiation that requires the waste to be permanently stored to isolate it from humans and the environment. The purpose of the Yucca Mountain repository is long term isolation of nuclear wastes in a stable geologic formation. The design time period under consideration is 10,000 years, the time it will take for the radioactivity of the waste to decay near the level of natural uranium.

The repository itself resembles a large mining complex with a waste-handling facility at the surface and an underground disposal facility about 1000 feet beneath the surface. The aboveground facilities at Yucca Mountain include buildings for waste handling and packaging operations, rail and truck unloading areas, water and sewage treatment plants and a storage area for excavated rock.

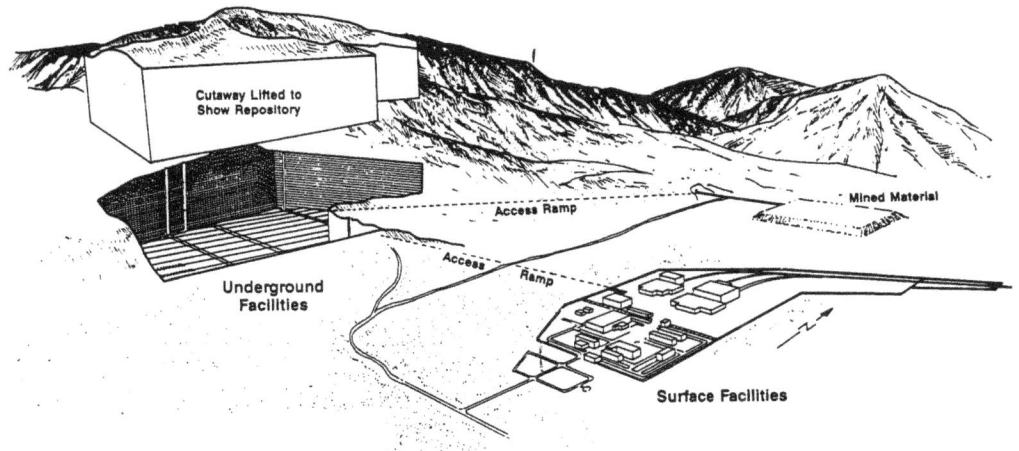

**Figure 1. Drawing of the proposed Yucca Mountain Nuclear Waste Repository**

When and if the repository is ever in actual operation, spent-fuel assemblies and high-level radioactive waste will be sealed into cylindrical canisters and transported to the site in special stainless-steel casks that have multiple containment and radiation barriers. The casks will be unloaded and inspected and the canisters enclosed in a secondary containment canister called an "overpack". This unit is then moved into the tunnel complex for final disposal. After a caretaker period of approximately twenty-five years, the Department of Energy could request approval for closure of the repository, although current thought has been given to keeping the waste retrievable for longer periods for possible reprocessing.

Yucca Mountain is designed to hold approximately 75,000 tons of nuclear waste. While this seems like a huge amount of material, in contrast the U.S. produces about 300 million tons of chemical wastes every year, much of which is not nearly as compact nor do some of these wastes decay even over large time periods. In fact, on a volumetric basis, high-level nuclear waste takes up very little space, especially as compared to the gaseous and airborne wastes of a fossil fuel like coal. According to the physicist, Bernard Cohen:

**The waste from a nuclear plant is different from coal-burning waste in a very spectacular way . . . Nuclear waste is 5 million times smaller by weight and billions of times smaller by volume. The nuclear waste from one year of operation (of a reactor) weighs about 15 tons and would occupy a volume of half a cubic yard, which means that it would fit under an ordinary card table with room to spare. Since the quantity is so small, it can be handled with a care and sophistication that is completely out of the question for the millions of tons of waste spewed annually from our analogous coal-burning plant. [Bernard Cohen, The Nuclear Energy Option, Plenum Press, p175, 1990)]**

## RADIATION CHARACTERISTICS OF HIGH-LEVEL WASTE

To understand the risks associated with Yucca Mountain, it is important to understand what radiation is, how we benefit from its use and in what ways it can harm us if not properly used. The most basic definition of radiation is that it is either a form of energy, or a particle carrying energy with it. When this radiation reacts with matter, both the radiation and the matter are altered. Radiation is all about us, whether we live near a nuclear reactor or not and it comes in many forms. The sun, light bulbs, TV sets, radios, micro-wave ovens, X-ray machines, granite blocks, our bodies and the universe itself all emit radiation, so in a sense we swim in a sea of energy and particles carrying energy.

What sets radiation from nuclear waste apart from the normal background radiation of our lives is its intensity and energy levels, not any special attribute related to its history as nuclear fuel. Cosmic rays (random energy from the universe) already shower us with the same kinds of radiation as contained in nuclear waste, and many rock formations emit similar radiation as well. Whether the radiation we face is from nuclear waste, cosmic rays or TV sets, the real danger comes from its intensity and energy, not from the uniqueness of man-made radiation versus environmental background radiation. Shielding used to diminish radiation intensity and energy to acceptable levels is effective in protecting us from man-made as well as environmental radiation.

Various common types of radiation and the particles associated with them are shown in Table 1. It is interesting to note that certain types of radiation do not pose any health risk, or do so only in limited circumstances.  For example, neutrinos are rarely if ever absorbed by mass and though many pass through our bodies at any given moment are safe at any levels. Other types, such as alpha particles, will not penetrate a piece of paper and only become dangerous when an alpha emitter (such as radon gas) is inhaled or ingested.

| Radiation Type | Particle | Energy Level | Typical Sources |
|---|---|---|---|
| Alpha | helium nucleus | 5.5 MeV | Radon Gas, Peanuts |
| Beta | electron, e⁻ e⁺ | 1.0 MeV | Co60, Tritium decay |
| Neutron | neutron | 1.5 MeV | U235 decay, A-bomb |
| Proton | proton | 1 to 1000 MeV | Atomic decay, Linear Accelerator |
| X-ray | photon | .005 MeV | TVs, X-ray Photos |
| Gamma | photon | .5 MeV | Nuclear Radiation |
| AM radio | photon | 10-14 MeV | Radio receivers and transmitters |
| Microwave | photon | 10-9 MeV | Microwave Ovens, Universe |
| Infrared | photon | 10-7 MeV | Humans, Elec. Heaters |
| Visible | photon | 10-5 MeV | Sun, Lightbulbs |
| Ultraviolet | photon | 10-4 MeV | Sun, UV lamps |
| Neutrino | massless | 1.0 MeV | Sun, Stars, Universe |
| Cosmic Rays | various | 1 to 1000 MeV | Background Radiation Universe |
| Muon | muon | 1 to 10 MeV | Cosmic Ray collisions, Accelerators |
| Kaon | kaon | 1 to 10 MeV | Cosmic Ray collisions, Accelerators |

**Table 1 Various types of radiation (a short list). Typical values given in millions of electron volts (mev), a unit of energy.**

Other points that should be understood about radiation:

- Not all radiation has the same biological effect despite similar flowrates (flux) and energies of wave-particles. The absorbed dose does not account for the severity or probability of harmful health effects since not all radiation has the same effect on body tissue even though it may dissipate the same energy in the tissue. To compensate for this discrepancy, scientists use the units of rems (roentgens equivalent man) to measure the human health hazards of radiation. Other measures use the absolute number of emissions and are given in Curies or Becquerels. We will be primarily concerned with rems as a unit in this book.

- Radiation can always be absorbed by shielding. However, it is not always appropriate to be shielded from radiation (otherwise our window panes would be opaque), and sometimes shielding is unnecessary or inappropriate (neutrinos can't be stopped by planets, dentists need to be shielded from X-rays but the dental patients need to be exposed to radiation for the film to be developed).

- The human body and all organisms have mechanisms for preventing and handling damage from radiation. Among the responses are such obvious tactics as tanning of the skin on exposure to sunlight, while at the cellular level there exist enzymes which repair or replace damaged DNA. These responses are not perfect; if they were there would be no such things as cancers or genetic mutations. On the other hand, the human body is not defenseless in the face of radiation and exposure at moderate amounts does not imply inevitable tumors and disease.

To recap, there are many forms of radiation in our environment, both from background and man-made sources. High-level nuclear waste, when properly shielded and contained, is only a small fraction (very much less than 1%) of the larger natural background radiation which already affects us, whether or not there are nuclear reactors or a high-level waste repository.

## LOW RISKS OF CONTAINED HIGH-LEVEL NUCLEAR WASTE

Ignorance of the physics and chemistry of nuclear waste has clouded the Yucca Mountain debate. It's therefore helpful to know something about the physical attributes of nuclear waste, especially the spent fuel-rod assemblies which comprise the bulk of the material to be stored at Yucca Mountain.

Within a reactor, uranium is held as small cylindrical pellets inserted in long tubes of zirconium. These rods are then clustered in packages of about 40 called fuel assemblies. Approximately 180 such assemblies form the central core of a reactor (see Figures 2 & 3).

Spent fuel rods are not considered by engineers and scientists to be the most difficult of hazardous substances to transport and store. This is not because technical people don't believe exposure to nuclear fuel rods is dangerous, but because they know the hazards are of a type that can be dealt with in a straightforward manner.

For example, unlike the gaseous methyl isocyanate that killed 3500 people at Bhopal India in 1984, nuclear waste is a solid and isn't easily dispersed in the air. In fact, uranium dioxide, the main component of the spent fuel, is a ceramic, related in its inert chemistry to coffee cups and ash trays.

Secondly, fuel rods are non-explosive. Even hammering on a nuclear pellet cannot cause a nuclear explosion, much less a chemical explosion. That is not to suggest hammering on nuclear pellets is a rational occupation, the unshielded radioactivity could kill you within the course of a day. Nevertheless, death would be radiation induced, not the result of an explosive reaction.

In 1989, Nevada residents were rudely awakened to the dangers of the solid chemical ammonium perchlorate, which exploded at the Pepcon rocket fuel plant in Henderson, Nevada and sent massive shock waves throughout the valley. In contrast, nuclear spent fuel is non-explosive, yet the public seems to have a subliminal fear that nuclear waste, which has no explosive reaction mechanism, is a volatile or explosive hazard.

The third mitigating factor about the dangers of nuclear waste is that radioactive substances can be monitored relatively easily with devices such as Geiger counters and dosimeters. Most chemical hazards are difficult to detect without costly analysis by mass spectrometers, gas chromatograms and other exotic hardware. Even in a worse case disaster you can generally find the scattered nuclear material afterward for cleanup. This is not necessarily the case for chemical spills (such as the Exxon Valdez) where toxic pollutants are dispersed on the wind or soluble in rivers and oceans.

*Illustration 1: Fuel Rod Assembly*

Finally, organisms have been successfully coping with radiation hazards from the beginning of time. Plants and animals have been exposed to naturally occurring radiation from the sun, radioactive soils, cosmic rays and other sources for literally billions of years and have adapted evolutionarily to naturally occurring background radiation levels. Consequently, this background radiation is a precalibrated biological risk standard that lets us judge whether exposures to radioactive substances are excessive and limit extra exposure (such as that from Yucca Mountain) to acceptable levels. No such natural risk calibration exists for chemicals like dioxin or DDT, making exposure to these much more prevalent chemical toxins, whose full toxicity is unknown; a greater gamble.

## POLITICIZING NUCLEAR WASTE HAZARDS

Of all the dangers faced by Americans, ranging from gasoline tankers to chlorine spills to cancerous carcinogens; nuclear waste is one of the least likely to be mishandled in a disastrous way. This isn't because nuclear waste isn't itself extremely deadly, but because the dangers are respected and the material is subject to especially protective handling procedures to ensure safety. When combined with a sophisticated, engineered disposal system at Yucca Mountain, in which the waste would be stored under a thousand feet of rock, the repository will be many times less environmentally intrusive than most other hazardous substances and less dangerous than the nuclear bomb testing that occurred in Nevadan's backyards over the last sixty years. Moreover, spent fuel disposal is dissimilar from the DOE's other nuclear waste problems because commercial nuclear waste was not developed under the secrecy and stress of Cold War national security concerns which encouraged cover-ups of contamination.

Despite the safeguards in the civilian nuclear waste disposal program, many in the political and environmental professions and many in the general population are extremely fearful of nuclear technology and the attempt to build a waste repository at Yucca Mountain. Careful analysis of these fears will shed light on whether they represent a justified wariness of technology, or are the product of political and social forces far removed from the potential dangers of nuclear waste storage.

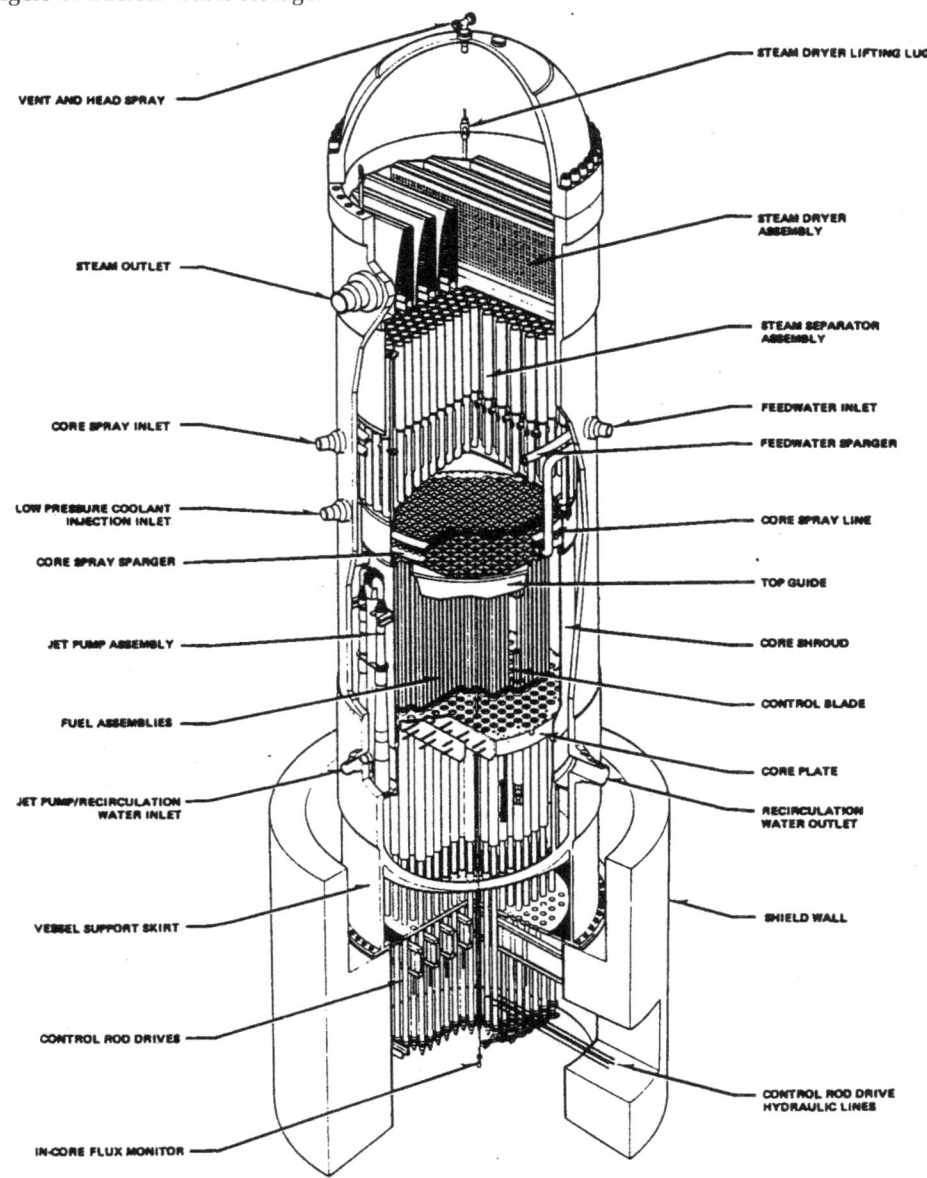

# 3.   The Nuclear Coalition

If Yucca Mountain is not merely a scientific issue, but a battle of warring ideologies vying for power, who are the warriors in this conflict? Opposition to the repository is often naively depicted as a simple grassroots environmental movement. Nuclear proponents (labeled generically as "the nuclear industry"), are in contrast seen as a monolithic entity. In reality, the structure of the two warring camps is much more complex than this and deserves explanation. We start with the nuclear proponents.

## NUCLEAR COALITION VS NUCLEAR INDUSTRY

The most important thing to understand about the nuclear industry is that there is none. There are utilities that have nuclear plants, there are companies that provide nuclear components, there are government agencies that oversee nuclear technology and there are trade organizations that promote nuclear issues. Nevertheless, as a whole these entities are not monolithic and are not engaged exclusively in nuclear energy technology, but compete in a variety of energy and industrial technologies.

Because the industry is so fragmented, nuclear energy has no dedicated champions willing to risk all in its defense. The result is that the Yucca Mountain nuclear waste repository is now being defeated politically rather than scientifically. Recognizing Nevada is perhaps a final battleground; the various nuclear entities attempted to evolve from a passive to active role in promoting the repository. The Nevada Initiative, a combined public relations and lobbying effort was created in 1991 by the American Nuclear Energy Council and other nuclear trade organizations to convince Nevadans of the inevitability and potential benefits of accepting the nuclear repository in their state. The political sparks that flew in Nevada in the 90s as a result of nuclear institutions confronting the entrenched nuclear opposition on the repository issue have now cooled as nuclear proponents have largely ceded the field.

The Department of Energy was charged with conducting the Yucca Mountain site characterization, however because of federal restrictions the DOE was not designed to be an active proponent of nuclear energy. Nuclear trade organizations like the American Nuclear Energy Council (ANEC) and more generic power industry groups like the U. S. Council on Energy Awareness (USCEA), now folded into the Nuclear Energy Institute (NEI) and Nuclear Matters, have been at the forefront of the political war to convince Nevadans of the safety and potential benefits of accepting the repository. However, these trade groups have often had conflicting allegiances and sometimes warred with themselves. For example, while ANEC strictly represented nuclear interests, the USCEA represented both nuclear and traditional utilities, some of which have little or no stake in promoting nuclear technology. Currently NEI also has little interest in promoting 4th Generation nuclear plants because the would compete against existing infrastructure.

Recognizing that they could not effectively promote nuclear energy as a disjointed industry at odds with itself, in December of 1993 four lobbying organizations combined to create the Nuclear Energy Institute (NEI). Merged into the new organization were the American Nuclear Energy Council, the Nuclear Management and Resources Council, the U.S. Council on Energy Awareness and the Edison Electric Institute. The first president and CEO of the NEI was former USCEA President Phillip Bayne. The hope was that consolidation under the NEI would allow the new coalition to mount an effective counter campaign against the environmental lobby which has for the last forty years monopolized media coverage and political debate on nuclear issues.

## BRIEF HISTORY OF NUCLEAR ENERGY

After World War II and the exploding of atomic bombs at Hiroshima and Nagasaki, research began on other military uses of nuclear energy, most notably hydrogen bomb development and the nuclear submarine reactor program. However, it was the submarine and aircraft carrier reactors developed by Admiral Hyman Rickover for the Navy in the 1950s that eventually became the basis of our commercial nuclear power industry. To counter the horrific images of Hiroshima and Nagasaki, President Harry Truman and U.S. scientists attempted to paint a positive public perception of the "friendly" atom and were for the most part successful. This was an era in which scientists were trusted almost implicitly and there was great optimism that the atom could be harnessed for peace. Over the ensuing decades, more than one hundred nuclear reactors were brought on-line in the U.S., providing as much as 21% of the nation's energy.

Although concerns for disposal of radioactive waste were lax in the earlier years, the National Academy of Science concluded as early as 1955 that geologic storage would be the most likely means of storing the accumulating radioactive wastes. Still, the first major attempt to seriously study a geologic burial site was not until a salt dome structure at a site near Lyons, Kansas was investigated in the early 1970s. This study ended when it was found that nearby brine extraction activities compromised the integrity of the site, effectively ending further study of salt domes for commercial high level waste (although not for military waste as at New Mexico's Waste Isolation Pilot Project).

Previous to the Carter presidency, an attempt was made to reduce the volume of waste through a breeder reactor program which would have recycled spent fuel into new forms of nuclear fuel. The Clinch River Breeder program was canceled by Carter shortly after construction began, in part because falling uranium prices made the separation of plutonium cost prohibitive, and because the plutonium was viewed in many circles as being part of the expanding nuclear arms race. This changed the calculus of nuclear waste disposal because it had been anticipated that spent fuel would only be stored temporarily near reactor sites in holding pools before being shipped out for reprocessing. The pools were not designed for long-term storage and therefore the nation was faced with the dilemma of finding a way to dispose of the waste.

In 1982, congress attempted to resolve the disposal issue with enactment of the Nuclear Waste Policy Act. The act committed the government to accepting the waste from commercial nuclear utilities, conditional on setting up a Nuclear Waste Fund paid through a mill levy on nuclear rate payers. Originally, nine sites were chosen for study: Yucca Mountain, Nevada; Davis Canyon and Lavender Canyon, Utah; Deaf Smith and Swisher, Texas; Vacherie Dome, Louisiana; Cypress Creek Dome and Richton, Mississippi.

In 1987, an amendment to the Nuclear Waste Policy Act (the so-called "Screw Nevada Bill") was pushed through congress which limited study to Yucca Mountain. Senator Bennett Johnston, D-La., sponsor of the amendment, also proposed legislation that would have substantially increased benefits to Nevada, but his efforts were rudely rejected by Governor Bryan and Senator Harry Reid of Nevada, an insult not forgotten by Johnston.

By 1990, it was clear that the Yucca Mountain project was in political trouble in Nevada, so much so that the entire project was in danger of being derailed. At this time, the American Nuclear Energy Council brought on Las Vegas public relations experts to bolster the pro-nuclear presence in the Nevada community. ANEC's Nevada consultants were anxious to go to a fully fledged political war against the anti-nuclear activists in Nevada, backed by promises of $10 million in funding over three years from the industry. This strategy was in sharp contrast to the passive approach the industry had taken until that time. Tentative steps towards an aggressive advocacy war brought political heat that the industry was not comfortable handling and they backed down when the Clinton administration came to power, spending only $2.5 million over four years, less than the yearly advertising budget of some Nevada car dealerships.

With the election of Bill Clinton in 1992, the nuclear industry faced the prospect of being regulated by former senator Timothy Wirth, the heir apparent to the Secretary of Energy position with strong ties to the anti-nuclear environmental movement. The industry was pleasantly surprised with Clinton's choice of Hazel O'Leary, a lawyer with Washington Beltway and nuclear utility experience who was sympathetic to the industry position on Yucca Mountain. Paradoxically, O'Leary caused severe problems for the industry's efforts in Nevada because nuclear lobbyists felt her position within the Clinton administration was vulnerable and they didn't wish to lose this lone ally.

As President Clinton's Secretary of Energy, 1993-1997, former utility company executive O'Leary, endured a stormy tenure, traveling the globe on lavish trips, and drawing criticism for failing to account for her substantial expenditures. To the chagrin of Clinton, O'Leary was also the target of a probe examining whether she agreed to meet with Chinese officials in exchange for a $25,000 donation to her favorite charity. She was cleared of wrongdoing by former Attorney General Janet Reno, but the damage was done.

Pro-repository advertising was pulled in Nevada in 1993 because Secretary O'Leary did not wish to have a confrontational atmosphere and the nuclear industry was overly eager to please; this effectively ceded the field to anti-nuclear activists who were under no such constraint. In late 1993, Secretary O'Leary began revealing past indiscretions of the Department of Energy in regard to non-consensual testing of human subjects with exposure to radiation doses. The ensuing media circus this engendered made objective discussions of nuclear risks nearly impossible during 1994, biasing the upcoming 1994 elections against those with any allegiance to the nuclear industry.

The reorganization of the "nuclear industry" lobby, the appointment of Hazel O'Leary and backroom political maneuvers substantially neutralized the Nevada Initiative in 1993 and 1994. The nuclear coalition further weakened its position by agreeing to a backroom deal, "Nuke Lite", designed to save Democrats Governor Miller and Senator Bryan from opposition in the 1994 campaign and bolster the Clinton presidency in exchange for promises of future cooperation. However, these compromises only delayed a critical future decision for nuclear interests. If nuclear facilities are to continue operation beyond 2020 (much less for new nuclear capacity to be brought on-line), the nuclear coalition must make a decision whether to go on the political offensive in Nevada.

Yucca Mountain is therefore the linchpin determining whether nuclear technology slowly strangles in decades to come, or whether it thrives and sees a new renaissance. The choice becomes whether we evolve to Small Modular Reactors, or are wedged into an extremely expensive solar future requiring a huge grid infrastructure vulnerable to attack and carrying long term costs for toxic waste removal.

# Repository Proponents

**The following list includes past and present proponents.**

## POLITICIANS
Sen. j. Bennett Johnston (D., La.)
Sen. Pet Dominici (D.,NM)
Gov. Rick Perry
State Sen. Joe Neal

## GOVERNMENT
Department Of Energy
Senate Energy Committee

## TRADE ORGANIZATIONS
Nuclear Energy Institute (NEI)
U. S. Council on Energy Awareness (USCEA)
American Nuclear Energy Council (ANEC)
Edison Electric Institute (EEI)

## RESEARCH INSTITUTIONS
Lawrence Livermore
Sandia Labs

## NEVADA PUBLIC RELATIONS
O.I.Z. Advertising
Altamira Communications

## CITIZENS GROUPS
Nuclear Waste Study Committee
Nevada CAN
Nuclear Matters

## LOBBYISTS
Ed Allison
Gov. Bob List

## CONTRACTORS
Science Applications Internationa Corporation (SAIC)
TRW
Babcock & Wilcox
Duke Engineering
E.R. Johnson Associates
Intera
Fluor Daniel

# 4. Repository Opposition

The primary individuals and groups opposing Yucca Mountain are listed in the following table. Among the wide range of individuals and organizations opposing Yucca Mountain are politicians, the Nevada Nuclear Waste Project Office and its subcontractors, and a variety of environmental and social justice movements. The length of the Yucca Mountain opposition list may come as a surprise to some, because the anti-nuclear movement is often portrayed as being an unorganized grassroots movement. In reality, the movement is a self-contained and well funded political juggernaut, relying not only on donations from individuals, but also on support from large philanthropic funds.

## Repository Opponents

*Opposition to Yucca Mountain and nuclear Energy in general is the result of an organized and well funded political juggernaut. Some individuals on the following list have been protesting nuclear technology as a profession for more than forty years. Some of the organizations are now defunct, but their membership has generally merged into newer organizations.*

### POLITICIANS
*Nationally*
Vice-President Gore
*In Nevada*
Senator Richard Bryan
Senator Harry Reid
Governor Bob Miller
Governor Sandoval
Congresswoman Shelley Berkley
Mayor Jan Laverty-Jones
Mayor Oscar Goodman
Mayor Carolyn Goodman
Congresswoman Dina Titus

### NATIONAL ACTIVISTS
Ralph Nader (PIRG, EA, PC, CW)
Amory Lovins (FOE, RMI)
Bob Loux (NWPO)
Bob Halstead (NWPO)
Steve Frishman (NWPO)
Judy T`1reichel (NNWTF, NDE, CALC, RA)
Marvin Resnikoff (SCRWC, WM, PIRG, NWPO)
Dr. Helen Caldicott (PSR)
Robert Pollard (UCS)
Dan Reicher (NRDC)
Scott Denman (SECC)
Jim Hightower (PC)
Scott Saleska (PC)
Dianne D'Arrigo (NIRS)
Jason Salzman (GP)

### STATE OF NEVADA
Sawyer Commission
    ex-Gov. Grant Sawyer
Nevada Nuclear Waste Project Office
    Bob Loux
    Bob Halstead
    Steve Frishman
Nevada Nuclear Waste Task Force
    Judy Treichel

### NWPO CONTRACTORS
Mountain Western-style
    (now Coopers Lybrand)
Center For Energy Technology & Development
    (CENTED)
Decision Research
Technology & Resource Assessment Corp.
Cygnus Scientific
Latir Enrgy
Kamer-Singer Public Relations
Mifflin and Associates
Waste Managment

### NEVADA PROTEST GROUPS
Citizen Alert
CAN-WIN
American Peace Test
Nevada Desert Experience
Lenten Desert Experience
Rural Coalition
Rural Aliance
Clergy and Laity Concerned
Western Shoshone National Council

Autonomous Women Out Loud
Rebels For Peace
Nevadans Opposing Nuclear Extinction

## WASHINGTON ANTI-NUCLEAR POLICY ORGANIZATIONS
Safe Energy Communication Council (SECC)
Public Citizens Critical Mass Energy Project (PC)
Public Information Research Group (PIRG)
Nuclear Information & Resource Service (NIRS)
Sierra Club Radioactive
          Waste Campaign (SCRWC)
Union of Concerned Scientists
Natural Resources Defense Council (NRDC)
Physicians for Social Respnsibility (PSR)

## FUNDING SOURCES
The Nuclear Waste Fund
          (tens of $millions of ratepayer
          funds through NWPO)
The Streisand Foundation
Fund for Renewable Energy and the Environment
The Tides Foundation's
The Abelard Foundation's
The Ettinger Foundation's
The beldon Fund
The CS Fund
The George Gund Fund
Alida Rockefeller

## MEDIA
Las Vegas Sun
Mary Manning (Sun Reporter)
Progressive Magazine
Media Access Project
Public Media Center

telecommunications Research and Action Center
Organizing Media Project

## ENVIRONMENTAL GROUPS
Sierra Club (SC)
Greenpeace (GP)
Greenpeace Action (GPA)
Friends of the Earth (FOE)
Rocky Mountain Institute (RMI)
Environmental Action (EA)

## NATIONAL ANTI-NUCLEAR ORGANIZATIONS
People Against Radioactive Dumping
Interfaith Action For Economic Justice
People of Faith
Hundredth Monkey
Don't Waste U. S.
STAND (Texas)
POWER (Texas)
Georgians Against Nuclear Energy
Atomic Veterans

## THINK TANKS
Institute for Energy and
          Environmental Research (IEER)
Soutwest Research and Information Center (SRIC)
American Counci for an Energy Efficient Economy
Health and Energy InstitutesInstitute for a
Sustainable Future
Environmental Policy InstitutesConcern Inc.
N.Y. Council for Economic Priorities
Foundation For Economic Trends

## PHILOSOPHICAL ROOTS

Surprisingly, opposition to Yucca Mountain and nuclear technology is not primarily motivated by environmental and health concerns over radioactive substances. Instead, social, political and religious influences dominate the debate.

One crucial issue is the question of what role technology should play in society and who controls science. Hazardous waste "dumps" are predominantly placed in the neighborhoods of poorer Americans and minorities and the "social justice" question raised by progressives is whether this is due to practical considerations or more ominously because of social biases of the technological elite. Governmental takings of land claimed by the Western Shoshone and Southern Paiute, which include Yucca Mountain, are considered by some an act of social oppression, inhibiting the native population's ability to practice traditional culture.

Other opposition at times stems from arcane political issues. One argument uses Yucca Mountain as a surrogate for the larger issue of control of Nevada's resources, but in the process links the repository with expansion of bombing ranges and secret test areas within Nevada. Others view nuclear waste as a tool of "centralizing" industrialists and part of a conspiracy to keep society under the thumb of elite capitalists, making the repository a revolutionary focal point. Some see democracy at risk at Yucca Mountain because the local community is not allowed to vote and veto the project.

A variety of religious concerns are also voiced. One is that Yucca Mountain is part of an amoral global nuclear war making apparatus because plutonium in the waste could be separated for bomb making. Others argue the repository is a pagan altar to a technology of greed on which human souls are being sacrificed. Still others believe that building a repository will destroy Mother Earth and is thus a mortal sin because we are abusing our stewardship of the planet. Apocalyptic environmentalism is in general more religious than scientific, based on paranoia that nuclear technology represents an environmental Armageddon.

Although the spectrum of anti-repository philosophies competing in Nevada seems wide and varied, they share common threads, being socially egalitarian and environmentally apocalyptic. These theories have names:

*LIBERATION THEOLOGY* - An outgrowth of Brazilian land reform struggles, its main proponents (Gutierrez and Paolo Friere) relied heavily on Marxist theory but attempted to remold this theory using Catholic concepts of "peace and justice" as outlined by the Pope. Nevada Desert Experience seems to spring from this mold, a central organizer of Nevada Test Site protest.

*RAWLSIAN ETHICS* - John Rawls of Harvard wrote his "Theory of Justice" in 1971, outlining a theoretical formulation of egalitarian thought that found acceptance in academia, though not without heated debate. It emphasizes accounting for the "least advantaged man" in all social calculations. This theory is a favorite of the academics doing socio-economic studies for Nevada.

*DEEP ECOLOGY* - An envirocentric philosophy theology formulated by David Brower (Sierra Club activist and founder of Friends of Earth) and expanded upon by other environmentalists. This is a mixture of anarchist, far-east animist and American tribal philosophies in which nuclear technology is viewed as the ultimate threat to the world ecology. Evident in many of the environmental protest movements works, from Sierra Club to Greenpeace.

*INDIAN TRIBAL SPIRITUALISM* - Many environmental movements look towards Indian philosophies of coexistence with natural resources as models. Along with this comes a reverence for supposedly egalitarian tribal lifestyles. In Nevada, the Western Shoshone have been used in this role and their geographical links to the Nevada Test Site and Yucca Mountain have tied tribal spiritualism to the debate.

*DECENTRALISM / ANARCHISM* - The organizing philosophy behind the "Green Revolution", this anarchist philosophy makes itself known in both populist environmental tracts and academia. Ralph Nader's anti-nuclear activists have been proponents of decentralism, arguing that nuclear energy centralizes political power while solar energy disperses it.

*THERMODYNAMIC NIHILISTS* - An undercurrent to Vice President Al Gore's philosophy, derived in part from the writings of environmental lawyer Jeremy Rifkin. This theory suggests man's technology (especially nuclear energy) dooms the world environment through entropic decay according to the Second Law of Thermodynamics. There are no noted thermodynamicists who believe this theory.

*MAOIST / MARXISTS* - Remnants of class warfare theory pervade many of the previously mentioned environmental theories. While Maoism and Marxism are no longer mentioned by name in environmental tracks, following the writings of environmental authors back thirty years often reveals a time when their ideas were directly attributed to these revolutionaries.

## BRIEF HISTORY OF PROTEST

These different worldviews of the dangers of nuclear technology have mixed and merged like paint in a pot over the years so that it is often hard to distinguish where one philosophy begins and another ends. Practitioners of the numerous varieties of anti-nuclearism cross the lines of debate frequently. Protest against nuclear weapons, nuclear reactors and now against the waste repository have thus interwoven into a larger tapestry. The true rationale behind the debate over the repository only becomes visible if one understands the broad historical panorama of anti-nuclear protest and its philosophical driving forces.

The history of antinuclear protest begins in the early seventies as the Vietnam War ended and the peace movement began a search for replacement causes. One event which seemed to stir up antagonism for nuclear technology among the protest factions were the musings of Alvin Weinberg, the father of nuclear reactors, who in 1972 postulated the need for a "Technological Priesthood" to watch nuclear waste for many millennia. Activists like the physicist Marvin Resnikoff began to target nuclear projects during this birthing period of the anti-nuclear movement. Resnikoff, a professor at the short lived Rachel Carson University, became critical of the West Valley nuclear reprocessing plant in New York, thus beginning what became a career of protest. Others followed similar courses during this critical period in the early 70s, becoming involved in a lifelong anti-nuclear crusade.

The growth of the Clamshell Alliance and protest at Seabrook in the early 70's led to the parallel formation of Ralph Nader's Public Citizen Critical Mass Energy Project. The Critical Mass rallies of 1974 and 1978 helped solidify and legitimize the opposition. The movement became cohesive and hit its stride with Nader's writing of "The Atomic Menace", first published in 1977, which portrayed nuclear technology as a time bomb waiting to happen.

Besides the various alliances (Clamshell, Palmetto, Cactus, Abalone, etc.) and Nader groups (Critical Mass, Public Interest Research Groups, Environment Watch, etc.), other anti-nuclear protest groups began to make their presence known. Dr. Helen Caldicott's Physicians for Social Responsibility took on the tenor of a revival crusade. Richard Pollard of the Union of Concerned Scientists and Dan Reicher the Natural Resources Defense Council also made themselves effective as anti-nuclear advocates. Amory Lovins, the British representative of Friends of the Earth, helped stop development of the Gorleben waste repository in Germany in 1979, an effort which would serve as a template for later actions. Lovins later joined forces with Marvin Resnikoff in New York to protest shipment of plutonium through the state. The accident at Three Mile Island in 1979 created a professional core of nuclear opposition which later became critical components in opposing Yucca Mountain.

Nader, Caldicott, Resnikoff, Reicher, and others formed a permanent core of anti-nuclear protesters inside the Washington Beltway with a network of offices and staff. In the academic sphere, psychologists and political geographers from a small kernel of institutions, (in particular Mountain West, Decision Research, and Clark University's CENTED) began to dominate nuclear risk perception studies based on their key participation in the socioeconomic analysis of the Three Mile Island accident for President Carter's Kemeny Commission. The ideology of these academics would later play a pivotal role in the Yucca Mountain debate.

The premier anti-nuclear activist group in Nevada has been Citizen Alert, though it lost most of its focus when Yucca Mountain was purportedly canceled by Obama/Reid in 2011. Citizen Alert started as a protest against nuclear waste disposal in Nevada in the late 70's, but cut its teeth on the MX Missile debate in the early 80's. In the same time frame, Nevada Test Site protest became the seed for creation of the Nevada Desert Experience (NDE) by Franciscans and nuns. Later, the national anti-nuclear Freeze movement merged its efforts with NDE, eventually spinning off as American Peace Test. Members of all these organizations regularly crossed boundaries and as nuclear weapons protest faded, many became regulars in Citizen Alert, the local focus of Yucca Mountain opposition. Professionals from the earlier national protest movement regularly resurrected themselves in Nevada, as for example Marvin Resnikoff who helped found the Sierra Club Radioactive Waste Campaign but traveled extensively to Nevada and later worked for the state Nuclear Waste Project Office.

In Washington, the Safe Energy Communications Council had been created in the early 80's as an umbrella organization for anti-nuclear / pro-solar environmental groups. As an offshoot of Ralph Nader's Public Citizen Critical Mass Energy Project, the SECC centralized control of nuclear protest. Anti-nuclear warriors moved from movement to movement, exemplified by Diane D'Arrigo (from the Sierra Club Nuclear Waste Campaign to the Nuclear Resource Information Center), and Caroline Petti (from Public Citizen to PIRG to Southwest Information and Resource Service to the Environmental Protection Agency).

In Nevada, the election of Richard Bryan to the governorship in 1982 ensured perpetual no-compromise opposition from the state to the Yucca Mountain repository. The state agency designed to oversee the technical study of Yucca Mountain, the Nevada Nuclear Waste Project Office (NWPO), became captive to environmental special interests under its director, Bob Loux. An unyielding opposition to the repository developed which paralleled the philosophy of the anti-nuclear movement. NWPO's science, public information and socioeconomic mandates were compromised by radical political agendas. For example, in 1985, the consulting firm Mountain West was chosen by NWPO as prime contractor to do socioeconomic studies for the state. NWPO funneled $15 million through the Mountain West research accounts, much of which was devoted to anti-nuclear advocacy.

NWPO also found itself influenced by the progressive politics of some of its key consultants. In the mid 80s the study of a potential repository in Deaf Smith County, Texas, was effectively opposed by environmentalists from the local organizations STAND and POWER. Steve Frishman of that Texas Nuclear Waste Program Office, as well as Texas Agriculture commissioner Jim Hightower, were critical to the success of this opposition. Frishman later found employment at NWPO in 1987, and Hightower joined the board of Ralph Nader's Public Citizen.

In 1987, Senator Bennett Johnston's (D-La) amendment to the Nuclear Waste Policy Act, labeled the "Screw Nevada" amendment by the opposition, designated Yucca Mountain as the sole site for study, causing an uproar in the nuclear protest movement. That year, NWPO Director Bob Loux hired Steve Frishman to serve as an opposition consultant. Also in 1987, Public Citizen produced "Shutdown Strategies", a secret plan to shutdown nuclear energy in America. The Nevada Nuclear Waste Task Force (NNWTF) was formed in late 1988 by Citizen Alert and Judy Treichel in conjunction with national anti-nuclear activists. The NNWTF was promptly given a sweetheart contract by Bob Loux through NWPO. Citizen Alert pressed the battle against Yucca Mountain in the early 90's, counting among its allies the Safe Energy Communication Council, Greenpeace, Southwest Research and Information Center, IEER, Don't Waste U.S. and many others. In short, a well organized national movement evolved to obstruct the study of Yucca Mountain because they recognized repository is the bottleneck in the continued use and development of the nuclear energy option.

When the nuclear coalition headed by the now disbanded American Nuclear Energy Council began in 1991 to present an alternative viewpoint to Nevadans opposed to that presented by the professional anti-nuclear lobby, a political war broke out. Among the anti-nuclear warriors who visited the state were Ralph Nader, Helen Caldicott, Marvin Resnikoff, Dan Reicher, Robert Pollard, Scott Denman, Rosalie Bertell and many others. However, the battle was halted before the environmentalists were sufficiently challenged. The nuclear industry reneged on its commitment to a $10 million information campaign, instead pursuing backroom deals in Washington, leaving Nevada in the hands of anti-repository environmental and political special interests.

# 5.  Fact vs. Fiction

Having introduced the proponents and opponents of Yucca Mountain, we can begin to examine whether it is science or ideology which separates the battling factions. Pro-nuclear forces claim the engineering and science at the repository, while complex, is not inherently beyond human capabilities. In contrast, charges leveled against the Department of Energy's study of Yucca Mountain by environmentalists and Nevada political entities tend to paint the study as hopelessly doomed by human incompetence and the unique dangers of radiation. While there are many legitimate scientific questions still unresolved at Yucca Mountain, our interest is in the subset of objections professional opponents have made to the site and whether these are legitimate questions or manufactured concerns.

The hailstorm of objections raised by opponents of the repository have been voiced in a variety of forums. The environmental community position has been represented by publications and media releases of the Safe Energy Communication Council, Citizen Alert, the Nuclear Information Resource Service, and an array of activist environmental organizations. The state of Nevada published its Nuclear Waste News and at one point produced a radio show called *Fact vs. Fiction* which highlighted the NWPO's differences with the DOE's study of the repository. The media have taken an activist role in supporting opponents of the repository with favorable coverage. There was, however, a certain convergence of the negative charges echoing through the environmental community, the media and NWPO, implying a certain synchronization of these concerns.

Perhaps most representative of the entire range of opposition arguments was a position paper produced in 1993 by NWPO titled *Why Nevada Is Opposed To Yucca Mountain*, which is in reality an amalgam of all the environmental positions. The issues raised by this position paper help separate the scientific statements of the repository opposition from their political statements.

## YUCCA MOUNTAIN BATTLE-LINES: NWPO EXAMPLES

*NWPO STATEMENT: "Waste considered for Yucca Mountain would contain tons of plutonium, an extremely toxic radioactive byproduct. One billionth of an ounce, if ingested, can cause cancer or genetic defects."*

Given the containment of plutonium in a nuclear waste matrix and its enclosure in multiple layers of shipping cask, overpack and geologic overburden this threat is virtually meaningless. However, environmental activists have promoted plutonium anxiety because they link it to possible bomb making by terrorists, also extremely improbable. Plutonium paranoia is an example of a NWPO and environmental concern which is actually more political and fear inducing than scientific.

*NWPO STATEMENT: "Numerous studies, both by federal government scientists and independent contractors, suggest that Yucca Mountain is scientifically unsuited for holding most dangerous nuclear material and keeping it out of the environment for the extraordinarily long time required."*

Actually, considering the complexity of site characterization there are relatively few studies which question Yucca Mountain's suitability, none of which have passed rigorous peer review. An often cited example of opposing science was former DOE geologist Jerry Szymanski's theory that groundwater could well up at Yucca Mountain. Utterly refuted after a two year study by a special review panel (the National Research Council) from the National Academy of Science, the panel rejected Szymanski's theory by a vote of 17-0.

*NWPO STATEMENT: "The tiniest amount of radiation to the reproductive cells can cause mutations. . . . The National Academy of Sciences has also concluded that exposure to any level of radiation is harmful and may lead to ill health effects."*

This again appears to be fear-mongering rather than useful science. If no level of radiation exposure is safe, then sunlight, microwave ovens, TVs, airplane flights, sunlight, etc. would all be threats to our health. The real question at Yucca Mountain is whether unreasonable levels of radiation would be emitted, a question that is still being resolved. This points, however, to the possibility that many in the opposition to Yucca Mountain may have crossed over from reasonable concern over health risks posed by a repository to a kind of nuclear paranoia.

*NWPO STATEMENT: "The site is affected by recent volcanic faulting, and nearby young volcanoes, evidence of a young and active geologic setting."*

While there are young volcanoes nearby, the real question is whether they are of a type that could disrupt the repository itself. The volcanoes are small cinder cones, not of a violent explosive nature, and do not appear to be a threat.

*NWPO STATEMENT: "The area has a history of earthquakes, including one in 1932 . . . the same magnitude as the San Francisco earthquake of 1989. A 5.6 magnitude earthquake struck an area less than 12 miles south of Yucca Mountain in 1992 . . ."*

This is a manufactured concern. Earthquakes have substantially diminished effects below the earth's surface (miners rarely feel earthquakes). Tunnels originally dug for the aborted MX Missile program at nearby Skull Mountain, nearly on top of the 1992 earthquake, were unaffected. This is another example of a quasi-scientific statement used for its political effect.

*NWPO STATEMENT: "Yucca Mountain is situated within a world-class precious metal mining district. Millions of dollars of gold and silver may be located in the area.*

This is a half-truth. Yucca Mountain itself is composed of mineralogically worthless volcanic tuff rock thrown out by a huge, ancient volcanic caldera. It is the distant edges of the caldera that contain precious metals. The nearest mineralized bedrock is many thousands of feet below Yucca Mountain or many miles distant on the surface and no commercial companies have filed claims.

*NWPO STATEMENT: "Economic studies reveal that, if built, the repository and its operational activities . . . could negatively affect future investment in Las Vegas, discourage businesses relocating to the area, and cause tourists not to visit Nevada."*

There is no empirical evidence of this being true. Las Vegas has one of the fastest growing communities in the nation, even though Yucca Mountain has long been near the top of its political and media agenda. Previous above ground nuclear testing had no measurable effect on Las Vegas' economic growth and no appreciable affect on radiation levels..

*NWPO STATEMENT: "Nevada believes that the waste should be stored in dry casks at existing sites for several decades . . . Such an arrangement would simplify eventual disposal by allowing the waste to cool thermally and radioactively before being shipped, handled and disposed of . . . This would allow spending at least 40-60 years actually searching for a scientifically sound, publicly acceptable approach to the solution of the nuclear waste problem."*

This is the environmental lobby's standard answer to the nuclear waste problem - delay the solution indefinitely. Activists in their more candid moments have admitted that this is a ruse to cause a bottleneck at nuclear reactor sites forced to store their waste and lead to the shutdown of the nuclear industry [e.g., a conversation with Diane D'Arrigo, Nuclear Information Service]. NWPO has never done its own studies of on-site storage and relies on DOE and the nuclear industry for its data, though it does not trust their findings regarding Yucca Mountain. Noted anti-nuclear activist Marvin Resnikoff now admits that on-site storage may make later transportation of casks from reactor sites problematic because of the breakdown of the fuel rods over time. In a 2013 YouTube lecture Resnikoff points to multiple dry storage dangers from bad practice and mechanical failure.

*NWPO STATEMENT: "A major concern is that radiation from nuclear waste will work its way from the proposed dump into water supplies and the air, which could expose large segments of Nevada's population.*

Time scales and geologic factors at Yucca Mountain make this exposure scenario extremely unlikely and of so limited effect as to be inconsequential. Water from Yucca Mountain drains into a natural dead end in Death Valley. Emissions to the air are expected to be near that of a single operating reactor - exceptionally small. Natural barriers to the movement of radioactive nuclides are why this site is being studied as a site for nuclear waste disposal in the first place.

*NWPO STATEMENT: "The radiation that emanates from this type of nuclear waste which would be buried at Yucca Mountain is so intense that anyone with direct contact would receive a fatal dose of radiation instantly."*

This same material is now handled daily at 110 reactor sites nationally with zero fatalities, pointing to the fear-mongering aspects of arguments posed by opposition to Yucca Mountain. The danger of radiation exposure is the entire reason for doing deep geologic burial.

*NWPO STATEMENT: The Department of Energy's credibility is so low, especially with respect to waste issues, that it is probably not capable of carrying out a program like the repository."*

If the DOE is incapable of carrying out this mandate, then there is likely no government or private entity (including NWPO nor the anti-nuclear think tanks) which can complete the task, much less study the alternative dry cask storage. Actually, the persistent paranoia about DOE which runs throughout environmental tracts is more the result of anti-war political concerns and a penchant for conspiracy theory than due to DOE's lack of credibility. While DOE does have problems, in large part the result of the complexity of its task and the legacy of the Cold War, these problems have been systematically addressed over the years.

*NWPO STATEMENT: The DOE pledged a "good faith effort" to help New Mexico acquire funding for highway bypasses, compensation for lost mineral royalties, and money for emergency management and preparedness. To date, the DOE has requested no such appropriations in its budgets."*

DOE is not a lawmaking entity and highway bypasses are not a normal part of their budgets. However, in the fall of 1992 a very favorable bill was passed in Washington D.C. that provided compensation ($431 million over 15 years) to New Mexico for their acceptance of the Waste Isolation Pilot Project (WIPP). DOE did not oppose this legislation and in fact supported it. Ironically, it was Senator Richard Bryan of Nevada who filibustered the WIPP compensation package.

Dismissing the criticism voiced by NWPO and the environmental movement as we have done here does not imply a lack of valid technological questions at the repository. Instead, our point is that the concerns voiced by DOE and its independent contractors over the suitability of Yucca Mountain are nearly inverted from those of NWPO and the environmental movement. This suggests there is an entirely hidden level of meaning to be found reading between the lines of the political and scientific battle over the nation's proposed nuclear waste repository. While we will return in depth to technical issues later, the hidden levels of political ideology expressed in opposition statements is what concerns us most in the chapters to come.

# PART II: NEVADA OVERSIGHT

# 6. Nevada Nuclear Waste Project Office

**"The State of Nevada's involvement in the issue of storing and disposing of high-level radioactive waste began in 1974 when the Atomic Energy Commission (a predecessor agency to the U.S. Department of Energy) notified then-Governor O'Callaghan that Nevada was being considered as a possible location for a "retrievable surface storage facility" for spent nuclear reactor fuel." [NEVADA NUCLEAR WASTE NEWSLETTER "Who's Who in Nevada Nuclear Waste Project Office (1989)]**

The possibility of storing nuclear waste in Nevada near the Nevada Test Site (NTS) has always been an obvious choice:

1. The surrounding Nye County population density is extremely low and likely to remain so.
2. Local experience handling radioactive substances comes from decades of bomb testing at the Nevada Test Site. Tunneling and geological expertise already exist.
3. The area is isolated desert not suitable for most industrial, agricultural or recreational uses.
4. The local geology is stable and the water table low with drainage towards Death Valley.

Opposition to the idea that Nevada might serve as a site for a nuclear waste dump was not always so highly charged. In the early seventies the state still had a fairly good working relationship with the many federal agencies that operate within its borders. Many Nevadans had watched nuclear bomb tests from their backyards while growing up and nuclear hysteria hadn't permeated the popular culture.

The Nevada Department of Energy (NDOE) began monitoring federal nuclear waste disposal plans in the late 1970s. When the NDOE was abolished by the 1983 State Legislature in response to the Nuclear Waste Policy Act of 1982, the department's responsibilities were moved to the Governor's Office where a special Nuclear Waste Project Office (NWPO) was established.

Congress tried to mitigate some of the unfairness of forced nuclear waste siting by encouraging oversight by the states under consideration to receive that waste. According to The Nuclear Waste Policy Act of 1982:

**Sec 117 B CONSULTATION AND COOPERATION. -- In performing any study of an area within a State for the purpose of determining the suitability of such area for a repository pursuant to section 112(c), and in subsequently developing and loading any repository within such State, the Secretary shall consult and cooperate with the Governor and legislature of such State and the governing body of any affected Indian tribe in an effort to resolve the concerns of such State and any affected Indian tribe regarding the public health and safety, environmental, and economic impacts of any such repository. In carrying out his duties under this subtitle, the Secretary shall take such concerns into account to the maximum extent feasible and as specified in written agreements entered into under subsection (c).**

This short section in the 1982 law was meant to protect the states of Washington, Texas and Nevada from being railroaded by outside special interests during the siting of a nuclear waste repository. Ironically, the law instead promoted just such a power grab, although not by the Nuclear Industry or Big Business as one might have suspected. Instead, the environmental movement took advantage of the oversight and consultation powers given the state, understanding that state veto power was the bottleneck in closing the nuclear fuel cycle. The 1987 Amendment to the Nuclear Waste Policy Act, the so-called "Screw Nevada" bill which set Nevada aside as the only site for characterization, also delineated oversight powers for Nevada and provided the monetary muscle for the Nevada Nuclear Waste Project Office:

**SEC. 5032 PARTICIPATION OF STATES . . . . "(B) The Secretary shall make grants to the State of Nevada and any affected unit of local government for purposes of enabling such State or affected unit of local government -**
**(i) to review activities taken under this subtitle with respect to the Yucca Mountain site for purposes of determining any potential economic, social, public health and safety, and environmental impacts of a repository on such State, or affected unit of local government and its residents.**
**(ii) to develop a request for impact assistance under paragraph (2);**

**(iii) to engage in any monitoring, testing, or evaluation activities with respect to site characterization programs with regard to such site;**
**(iv) to provide information to Nevada residents regarding any activities of such State, the Secretary, or the Commission with respect to such site; and**
**(v) to request information from, and make comments and recommendations to, the Secretary regarding any activities taken under this subtitle with respect to such site.**

The oversight latitude given the Nevada Nuclear Waste Project Office became the Achilles heel of nuclear power. Groups like Friends of the Earth, Public Citizens' Critical Mass, Citizen Alert and a long list we'll recount in later chapters, had long been convinced that nuclear materials of any sort would lead to environmental Armageddon, either through pollution or nuclear weapons proliferation. However, the long history of attempts by environmental groups to shut down the defense and civilian nuclear programs through site-by-site protest had proven ineffectual. The Nuclear Waste Project Office provided the environmentalists an opportunity to turn the tide.

If environmental elites couldn't raise popular opposition to nuclear energy in each of the states that depended on nuclear reactors, they eventually realized they could subvert the entire nuclear energy policy process by hijacking the State of Nevada's oversight of nuclear waste disposal. The State of Nevada originally had a degree of veto power over the repository and this provided a window of opportunity. If the environmentalists moved quickly enough, the entire nuclear industry could be brought to its knees by influencing a few key state politicians and agency heads in Nevada instead of battling both the entire U.S. Congress and the powerful nuclear lobbies in Washington.

Thus, professional environmental protesters became self-appointed arbiters of our national energy future through their manipulation of the Nevada Nuclear Waste Project Office. Unfortunately, since the environmentalists were predominantly from the social sciences and legal professions, their knowledge of alternative energy engineering was also lacking. This has created a situation in which national energy policy revolves around an anti-nuclear / pro-solar utopianism that offers little in the way of practical engineering remedies to national energy problems short of shutting off the nation's electrical outlets.

Most intriguing to the opposition groups was the prospect that the American nuclear power industry would pay for the noose to hang itself. Environmental groups like Greenpeace, Sierra Club, Public Citizens' Critical Mass Energy Project, etc. are by no means poor. However, The Nuclear Waste Policy Act provided an even larger reservoir of money that the radicals knew they had to tap in order to strangle the nuclear industry:

**NUCLEAR WASTE FUND Sec. 302 (a) CONTRACTS -- (1) In the performance of his functions under this Act, the Secretary is authorized to enter into contracts with any person who generates or holds title to high-level radioactive waste, or spent nuclear fuel, of domestic origin for the acceptance of title, subsequent transportation, and disposal of such waste or spent fuel. Such contracts shall provide for the payment to the Secretary of fees pursuant to paragraphs (2) and (3) sufficient to offset expenditures described in section (d). [Nuclear Waste Policy Act, 1982]**

The Nuclear Waste Fund has since generated thirty-seven billion dollars in payments (and growing), not all of which will be needed to fulfill the need for scientific studies, engineering and construction of the Yucca Mountain repository. Thirteen billion dollars has already been spent on the project. The main thing to note here is that these are ratepayer funds from nuclear utilities, not taxpayer dollars, although the federal government has control over the distribution of those funds. Senator Lindsay Graham of South Carolina authored legislation in 2009 to return those funds to ratepayers should the DOE renege on its commitment to completing a repository.

Nevada's Nuclear Waste Project Office was created using money set aside from the Nuclear Waste Fund. Under long time former director, Bob Loux, NWPO consumed over sixty million dollars between 1982 and the present, much of it employed in opposition to nuclear energy. Bob Loux was removed from his post in 2008 for misappropriation of funds for salaries for himself and his staff and while not held accountable for political reasons, the malfeasance was clear. The bitter irony of this for the nuclear industry is that much of the money spent by Nevada went to some of the most sophisticated opponents of nuclear energy in the world.

*"As federal interest in a possible Nevada waste site increased, the level of NWPO oversight also increased. The office, which was - and is - funded by federal grants (Nuclear Waste Fund) made available to states under the Nuclear Waste Policy Act of 1982, grew from a two-person staff in 1983 to a $3 million program by 1985. ["NEVADA NUCLEAR WASTE NEWSLETTER "Who's Who in Nevada Nuclear Waste Project Office (1989?)]*

One of the people on the original two-man staff left over from 1982 was Robert Loux, who became notorious in Nevada as a one-man anti-nuclear wrecking ball. A high school teacher with a major in history and minor in psychology from the University of Nevada, Reno, Loux had been involved in state energy and nuclear waste programming since 1976. In fact, except for a few years of teaching high school, this appears to have been the only career he ever pursued. Since becoming executive director of NWPO, Loux's lack of scientific expertise and technical credentials became a raw wound in the Nevada technical community which saw him as a political manipulator and engineering dilettante. This didn't stop Loux from gaining carte blanche over what grew for many years to more than $5 million dollars per year in funds, in large part distributed to foes of the nuclear industry.

As a result of action by the 1985 Nevada Legislature, NWPO became, officially, the Agency for Nuclear Projects - a statutorily established entity responsible for monitoring and overseeing U.S. Department of Energy activities related to the Yucca Mountain nuclear waste site. In the hands of then-Governor Richard Bryan, and later Senator Harry Reid, it also became part of a political strategy designed to bludgeon political opposition into submission - notably former Senator Chic Hecht in the 1988 senatorial campaign eventually won by Bryan.

Under the troika of Senator Bryan, director Robert Loux and former governor Grant Sawyer (who was enlisted to head the Nevada Commission on Nuclear Projects), the Nuclear Waste Project Office became from the incept an anti-nuclear propaganda machine. Oversight by the Sawyer Commission transformed into show trials masquerading as fact finding. Science conducted by NWPO's technical and planning division was corrupted by political considerations. The social scientists of the planning division, given lucrative contracts worth $15 million, used their expertise to generate anti-nuclear hysteria in Nevada. Less abusive but no less disturbing was that some of the technical studies were designed to support the party line rather than investigate real technical questions at Yucca Mountain

But how did what was originally a two man environmental office devoted to disposing of waste oil become so powerful a foe of nuclear energy and a thorn in the side of the Department of Energy? Director Loux simply spent the fifty million dollars in funds received for oversight as he saw fit, building a political propaganda machine in the process. Nevada's politicians, notably Senator Bryan the late ex-governor Sawyer, and Senator Harry Reid looked the other way as Bob Loux awarded millions of dollars of contracts without Requests For Proposals and without competitive bids. Even more problematic was that the Department of Energy, which was supposed to oversee the spending of NWPO, caved in to the political pressure and allowed the state to violate federal laws rather than risk making political waves. This abdication of management control over NWPO, by both the State of Nevada political establishment and the Department of Energy, allowed Loux's agency to become a loose cannon.

For example, NWPO openly violated the Federal Acquisition Regulations (FAR) against using funds to run public relations and lobbying campaigns. Whenever questioned about the legality of these public relations activities, Bob Loux simply claimed the regulations didn't apply, or that his agency was in compliance because its activities were strictly "informational". The pertinent regulation regarding limits on public relations and lobbying by agencies accepting Federal grants is FAR 31.205-22:

**FAR 31.205-22 Legislative lobbying costs.**
   **(a) Costs associated with the following activities are unallowable:**
   **(1) Attempts to influence the outcomes of any Federal, State, or local election, referendum, initiative or similar procedure, through in kind or cash contributions, endorsements, publicity, or similar activities.**

*(NOTE: NWPO was careful not to publicly endorse anti-nuclear candidates. However, they skirted this restriction by using closely affiliated environmental organizations like Citizen Alert and the Safe Energy Communication Council to lobby the local and national legislature on the Yucca Mountain issue. Other ways of skirting the law are numerous; for example, the NWPO publication, Nuclear Waste News is a political diatribe against Yucca Mountain sent to 20,000 readers, but is called "informational".)*

**(2) Establishing, administering, contributing to, or paying the expenses of a political party, campaign, political action committee, or other organization established for the purpose of influencing elections.**

*(NOTE: Citizens Against Nuclear Waste In Nevada [CAN-WIN] was specifically organized by Grant Sawyer who headed the Nevada Commission on Nuclear Projects (NCNP). NCNP oversaw NWPO and consequently NWPO was in essence a prime organizer of CAN-WIN. Other such political organizations also received covert support from NWPO.)*

**(3) Any attempt to influence (i) the introduction of Federal or state legislation, or (ii) the enactment or modification of any pending Federal or state legislation through communication with any member or employee of Congress or state legislature (including efforts to influence state or local officials to engage in similar lobbying activity), or with any government official or employee in connection with a decision to sign or veto enrolled legislation;**

*(NOTE: NWPO specifically hired lobbyists in Washington and elsewhere [Murphy, M.R. & Davenport, J.H., Sutherland, Asbill & Brennan - $60,000; Tillson, D - $144,000; Peck, D.R. - $30,000; Phillips, F. J. - $97,000; Afton &Associates - $65,000] to influence legislation. The General Accounting Office eventually balked at these expenses.)*

**(4) Any attempt to influence (i) the introduction of Federal or state legislation by preparing, distributing or using publicity or propaganda, or by urging members of the general public or any segment thereof to contribute to or participate in any mass demonstration, march, rally, fund raising drive, lobbying campaign or letter writing or telephone campaign;**

*(NOTE: Judy Treichel of the Nuclear Waste Task Force, the public relations subcontractor to NWPO, was specifically chosen in the 90s because of her affiliations with Citizen Alert, Clergy and Laity, Rural Alliance, Nevada Desert Experience and other radical organizations which regularly attempt to influence state and Federal legislation by conducting marches, letter writing, fund raising, etc. )*

**or/**

**(5) Legislative liaison activities, including attendance at legislative sessions or committee hearings, gathering information regarding legislation, and analyzing the effect of legislation, when such activities are carried on in support of or in knowing preparation for an effort to engage in unallowable activities.**

*(NOTE: Bob Loux, executive director of NWPO, extensively lobbied the 1993 Nevada legislature on matters relating to Yucca Mountain. Steve Frishman, a NWPO consultant, had a job description specifically tailored to engage in this activity. Other consultants, such as Arjun Makhijani of IEER, were hired specifically because they are active in lobbying efforts.)*

Perhaps more important than ignoring the Federal Acquisition Regulations, NWPO regularly broke the letter and spirit of 10 CFR 600-436, the procurement section of the Code of Federal Regulations. The worst shortfall was in regard to a number of clauses found in the standards of competition:

**Competition:**

**(1) All procurement transactions will be conducted in a manner providing full and open competition consistent with the standards of 600.436. Some of the situations considered to be restrictive of competition include but are not limited to:**

**(i) Placing reasonable requirements on firms in order for them to qualify to do business.** *(NOTE: NWPO does not hire companies who do not profess bias against the Yucca Mountain project. A typical example was the public relations firm Kamer, Singer which was awarded a sole-source contract specifically to oppose Yucca Mountain.)*

**(ii) Requiring unnecessary experience and excessive bonding.**

*(NOTE: A prime example was the $1.3 million contract awarded Technology $ Resource Assessment Corp. T.R.A.C's agreement with the rising groundwater theories of geologist Jerry Szymanski [theories rejected 17 - 0 by*

*the National Research Council] was a prerequisite for the contract, a requirement no other company in the U.S. could fill.)*

**(iii) Noncompetitive pricing practices between firms or between affiliated companies.**

*(NOTE This law doesn't appear to have been broken only because there was no need to circumvent what were generally sole source contracts. Why collude when your company is the only source?)*

**(iv) Noncompetitive awards to consultants that are on retainer contracts.**

*(NOTE: This applies to consultants like Judy Treichel of the Nevada Nuclear Waste Task Force, Sam Singer of Kaman and Singer Public Relations, Bob Halstead for his transportation consulting, Steve Frishman, whose consulting seems to have been more political than technical and many others.)*

**(v) Organizational conflicts of interest.**

*(NOTE: NWPO as a state agency was meant to represent the state of Nevada's interests, but contracted on the basis of the interests of Citizen Alert and national environmental groups. In fact, they employed members of these environmental special interests.)*

**(vii) Any arbitrary action in the procurement process.**

*(NOTE: NWPO's entire procurement process was arbitrary, lacking any pretense of competitive awards, rarely issuing requests for proposals, often operating in defiance of laws restricting lobbying and public relations.)*

**(2) Grantees and subgrantees will conduct procurements in a manner that prohibits the use of statutorily or administratively imposed in-State or local geographical preferences in the evaluation of bids or proposals, except in those cases where applicable Federal statutes expressly mandate or encourage geographic preference. . . . .**

*(NOTE: In fact, regional preference was encouraged for NWPO, whose mandate was after all to oversee the best interest of Nevadans in regard to the nuclear waste repository. NWPO spent, however, 60% of its research funds on out-of-state researchers specifically to avoid accountability within Nevada.)*

**(3) Grantees will have written selection procedures for procurement transactions. These procedures will ensure that all solicitations:**

**(i) incorporate a clear and accurate description of the technical requirements for the material, product or service to be procured. Such description shall not, in competitive procurements, contain features which unduly restrict competition.**

*(NOTE: NWPO regularly wrote contracts for the benefit of sole source bidders and only reluctantly distributed work descriptions through the Nevada technical and academic community.)*

The variety and extent of idiosyncrasies within the operations of the Nevada Nuclear Waste Project Office have only been hinted at in this chapter. The point we hope we have made is that there has been much more going on at Yucca Mountain than a question of state's rights. Why would Bob Loux, as executive director of NWPO, have felt compelled to skirt the competitive bidding process on the Yucca Mountain issue and ignore the Code of Federal Regulations? Why would Loux and NWPO` blatantly ignore the Federal Acquisition Regulations against lobbying and public relations efforts? Was all this activity for personal gain (some certainly was, as shown by Loux's appropriating hundreds of thousands of dollars for his staff wages). Or is there a greater underlying dynamic?

Had laws been broken merely in the time honored fashion of corrupt state politics, with money and greed as the sole motive, it would not have been a dangerous precedent but simply an irritating one. Instead, ideology has driven the Yucca Mountain war, with an icing of greed at the top. Yucca Mountain is not just a local state's rights issue run by some Nevada Bubbas trying to line their pockets, but a philosophical war of national importance.

Something much larger than local politics is unfolding in Nevada, something big enough to necessitate the violation of Federal law, the redirection of funds and the raising of political lynch mobs to oppose anyone who believes Yucca Mountain is a technology deserving study.

# 7. Technical & Socioeconomic Studies

There were two broad groups of oversight studies done by Nevada's Nuclear Waste Project Office regarding Yucca Mountain:

1. Environmental and engineering sciences conducted by various private contractors, and in part by state institutions like the Desert Research Institute, the University of Nevada at Reno and the University of Nevada Las Vegas.

2. Socioeconomic studies originally directed almost exclusively by a prime contractor called Mountain West which oversaw various socioeconomic subcontractors.

Technical studies done in-state by the Nevada academic community are fairly reliable because they didn't want to compromise their standards and this data was hard to manipulate in favor of the anti-nuclear position. However, a number of "ringers" brought in by NWPO to buttress their anti-repository party line made a number of their technical reports suspect. The socioeconomic studies in the early years were handled almost exclusively by out-of-state interests in an attempt to steer conclusions against Yucca Mountain and are tainted by the ideological agenda of key outside researchers who saw advocacy as more important than objectivity.

## TECHNICAL DIVISION

Technical studies for the state have been relatively unbiased because the state universities and the Desert Research Institute took a firm position early on that their research had to remain academically independent. This required a war of wills with NWPO director Bob Loux who at times withheld state funding from those unwilling to tailor their results to his expectations.

For example, hydrology studies were channeled to produce the results expected by Loux and the state politicians. A $1.3 million contract (without request for proposals or competitive bids) was awarded to Technical Resource and Assessment Corporation in Colorado specifically to support a rising groundwater theory. This was despite the existence of a supercomputer center at UNLV, the existence of expert hydrologists at the Desert Research Institute and the universities. Linda Lehman and Associates of Michigan was used as the state's main hydrological consultant for a period and received $500,000 in contracts, despite being distant from the problem.

Another example was Carl Johnson, who headed the technical studies section and supported NWPO's position that earthquakes are an eminent threat at Yucca Mountain, though earthquakes are generally minor events at subsurface levels. Most DOE and independent scientists believe heat-pipe condensation and criticality issues are much more significant problems at Yucca Mountain, areas neglected by state researchers.

So determined was the state to avoid reaching favorable results at Yucca Mountain, that they even refused to build monitoring stations on-site or maintain employees near the repository site, preferring to conduct science from the safety of Carson City, hundreds of miles north. Moreover, the state often blackballed the department of engineering at the University of Nevada, Las Vegas only 90 miles from the site, because of a justified fear that some UNLV professors believe Yucca Mountain an acceptable solution.

Consequently, while research conducted by Nevada academic institutions was generally of good quality, the state's technical studies were obviously biased at an institutional level. Some employees for the state, notably geologist Steve Frishman, rarely did science and spent most of their time doing political opposition work. Other anomalies in the conduct of technical oversight will be discussed later during the course of this book.

## SOCIOECONOMIC DIVISION

While Joe Strolin headed the socioeconomic side of NWPO over much of its history, the direction of research has been driven by the course set by the original outside consultants hired to conduct Nevada's socioeconomic evaluation. The linchpin to the politicization of the Nuclear Waste Project Office socioeconomic studies stems from the choice of the original prime contractor, Mountain West, in 1986. While the selection process that chose

Mountain West was exhaustive, the resulting socioeconomic research monopoly that developed locked in the ideology of the researchers as the official Nevada position.

The name Mountain West is a catchall for a group of intricately connected companies and think tanks whose end purpose seems to have been to scuttle nuclear power in America. The nominal contractual obligation of Mountain West Research of Tempe Arizona when it came on the Nevada scene in 1985 was to act as a prime contractor for evaluation of the socioeconomic impact of the Yucca Mountain nuclear waste repository on Southern Nevada. In practice it became a front for academics whose hidden political agenda left little room for an objective evaluation of the scientific risks of nuclear waste. It was Mountain West that was the philosophical seed that changed the Nevada Nuclear Waste Project Office from a state oversight agency into a political advocacy organ.

**"On September 13, 1985, the Nevada Nuclear Waste Project Office (NWPO) formally issued a Request for Proposals (RFP) for a major socioeconomic study to identify potential impacts to the state and to local communities should a nuclear-waste repository be constructed at Yucca Mountain in Southern Nevada. . . .**

**The planned Nevada socioeconomic impact study is designed to provide State and local planners with clear and quantifiable answers to questions about the social and economic effects of a nuclear repository on the State, its local governments, and its citizens. . . .**

**In the cover letter transmitting the RFP to prospective responders, the NWPO stipulated that it is not looking for traditional, run-of-the-mill economic analyses. Rather, the State is seeking innovative approaches that pertain specifically to the unique characteristics of Nevada and its localities. . . ." [Nevada Nuclear Waste Newsletter, Vol. 1, no. 1, December 1985]**

Nevada went looking for a non-traditional socioeconomic study and that's what it got in the form of Mountain West. From the beginning, this group produced not run-of-the-mill scientific investigation of socioeconomic impacts, but sophisticated political advocacy for the anti-nuclear position. What it did not produce was "clear and quantifiable answers to questions about the social and economic effects of a nuclear repository on the State, its local governments, and its citizens." In fact, so unquantifiable were Mountain West's results that Clark County (the most affected county as gateway to Yucca Mountain) opted to ignore the Mountain West results and conduct its own socioeconomic impact assessments, relying on in-house staff. Original data from Mountain West was not available for years after being collected, rendering it useless.

There were three main subcontractors under the Mountain West umbrella: Mountain West itself, Decision Research of Eugene Oregon and Clark University's Center for the Environment, Technology and Development (CENTED). There were other subcontractors and consultants such as Planning Information of Colorado and Latir Energy of New Mexico who played roles in the socioeconomic studies, at times pulling in hefty contracts in the process. But we'll see as this story progresses that Decision Research and Clark University played the critical roles in transforming the Yucca Mountain debate into nuclear hysteria.

In the process of studying risk perception, the Mountain West consortium created their own academic discipline of nuclear risk *perception* analysis, which in the minds of these non-technical social scientists became indistinguishable from the study of nuclear risk. Since these groups came to have a near monopoly in nuclear risk perception studies (they did the groundbreaking socioeconomic research after Three Mile Island accident), their theories, biased with anti-nuclear sentiments, came to replace reality in Nevada. Negative images of Yucca Mountain were reinforced in the minds of Nevadans by the psychologists from Mountain West through a misuse of research polling for political advocacy. This created a self referencing feedback virtual reality of fears and nuclear hobgoblins which precluded hard facts and hard science.

The money devoted to Mountain West was no trifling matter, amounting to over $14 million dollars in contracts over nine years. Not only were Mountain West consultants well paid (some billed at $140 per hour), but they were encouraged to used their time to forward their own particular political agenda (Rawlsian ethics) rather than do the traditional mitigation studies that the citizens of the state expected. As social scientists, one can argue Mountain West composed a powerful team of psychological manipulators, abusing their positions as social scientists to purposefully create a climate of distrust.

Since Mountain West had no compunction about moving into scientific vacuums where other hard-science researchers were hesitant to tread, they came to fill the risk assessment positions for the state, though as sociological scientists they had few credentials to do engineering risk analysis. Since physical scientists tend to avoid politics, while Mountain West viewed itself as energy policy advocates, the power of the sociologists and psychologists grew geometrically as their funding increased.

A speech given March 1993 by James Flynn (formerly of Mountain West and later of Decision Research) and attended by Joe Strolin of the Nevada Nuclear Waste Project Office and head of the socioeconomic division, gave an opportunity to clarify why it was so important that the socioeconomic studies be done by academics from outside Nevada. Strolin was asked why Nevada's university system did not receive grants to do these studies. His reply was that *"There was no one in Nevada qualified to do the work."* With fourteen million dollars to spend, it is likely the state could have built an unrivaled socioeconomic research capability within its Nevada university system that would have overcome any such limitations.

The lack of local sociological expertise to interpret results led to bizarre results. For example, the impact of nuclear imagery on the Nevada resort and gaming industry was studied by Pennsylvanians, though Nevada is foremost in the analysis of gaming impact through the University of Nevada Las Vegas Hotel Management School. Theories that negative nuclear imagery would drive away tourists wholesale, developed particularly by Paul Slovic of Decision Research, flew in the face of thirty years of nearby experience with the Nevada Test Site which paralleled a geometrically expanding economy. Eventually, socioeconomic polling done by the state became little more than part of the state's political attack apparatus, free of the constraint of local representation.

Roger Kasperson, a political geographer who led a research group from Clark University as part of the Mountain West consortium, may have provided an even more corrosive force. Early in his career, Kasperson had become convinced that sociologists were not bound by objective science but could become advocates for their subjects. This may have fatally compromised the entire socioeconomic research program within Nevada.

## COMPROMISED SCIENCE

The end result of NWPO's lack of scientific rigor has led to negative effects which extend far beyond the state's borders:

1.   Oversight by Nevada, both socioeconomic and technical, has been compromised by the politicization of its consultants. This means there is no reliable independent oversight of Yucca Mountain dedicated to the welfare of that Nevada's citizens. This negates the validity of the entire Yucca Mountain study.

2.   A precedent has been set for using the psychologists and political geographers of socioeconomic research teams as pro-active political operatives. Large scale technical projects which require socioeconomic review will now need to create higher hurdles against manipulation of socioeconomic studies. Indeed, this also occurred in New Mexico in regard to the study of the Waste Isolation Pilot Project.

# 8.   Bob Loux & NWPO

It is a curious fact of life in Nevada that a former high school teacher, who majored in history and minored in psychology, was for thirty years the executive director of the state's Nuclear Waste Project Office through its most critical periods. This is despite the fact that the state has no lack of technical expertise in things nuclear, geological or related to the mining industry and is called the Silver State to reflect a long tradition of such activity. In fact, Nevada had the opportunity to draw on some distinguished expertise in regard to nuclear waste disposal from within its borders to head its Nuclear Waste Project Office, but instead chose its director on the basis of political expediency.

The law establishing the Nevada Nuclear Waste Project Office and its seven member governing commission, the Nevada Commission on Nuclear Projects to which it reports, is Nevada Revised Statute 459.0093:

**459.0093 Agency for nuclear projects: Creation, composition; appointment and qualifications of executive director.**

**1. The agency for nuclear projects is hereby created. It consists of the commission and:**
   **(a) The division of technical programs.**
   **(b) The division of planning.**

**2. The governor shall appoint an executive director, who serves at the pleasure of the commission, and who must:**
   **(a) Be appointed from a list of three persons submitted to the governor by the commission.**
   **(b) Possess broad management skills related to the functions of the agency and have the ability to coordinate planning and communication among the Federal Government, the state and the local governments of this state on issues related to radioactive waste.**

It is difficult to understand in purely objective terms how in 1983, then Governor Richard Bryan and the Nevada Commission on Nuclear Projects, led by former governor Grant Sawyer, could justify the appointment of Bob Loux as executive director of the Nevada Nuclear Waste Project Office. Although Loux had been in the Nevada Department of Energy (later NWPO) since 1976 and handled the state's waste oil recycling efforts, that position lacked hands-on experience in nuclear issues or "broad management skills". Loux's previous employment as a high school history teacher was not related to technical issues at all.

Bob Loux survived in his niche as director of the Nuclear Waste Project Office not through his expertise in nuclear waste engineering or management, but by being a convenient political tool for Senator Bryan and various environmental groups (Citizen Alert in Nevada, the Safe Energy Communication Council in Washington D.C., among others). Even in 2008 after Loux was found diverting money to his own and underling's salaries, his only punishment was to be retired, though with accolades from everyone from Senator Reid through former Senator Bryan..

As executive director of NWPO, Loux fought hard for his agency, but with his limited technical skills and narrow job experience his efforts were often counterproductive. Staff hired by Loux had no coherent research objective other than to create objections to geologic storage at Yucca Mountain, creating an academic bubble biased against objective science. Open discussion of scientific issues at Yucca Mountain with the Department of Energy were poisoned by the state's lack of objectivity, making rigorous oversight impossible.

Loux's role as pawn to powerful politicians and environmental activist groups stems from his unchecked statutory powers. Nevada Revised Statute 459.0095 gave Loux a wide mandate:

**459.0095 Executive director of agency for nuclear projects: Powers. The executive director may:**
**1.   Provide information relating to radioactive waste to the legislature, local governments and state agencies that may be affected by the disposal of radioactive waste in this state.**

2. **Consult departments, agencies, and institutes of the University of Nevada system or other institutions of the University of Nevada system or other institutions of higher education on matters relating to radioactive waste.**

3. **Employ, within the limitations of legislative authorization, technical consultants, specialists, investigators and other professional and clerical employees as are necessary to the performance of his duties.**

4. **Make and execute contracts and all other instruments necessary for the exercise of the duties of the office.**

5. **Obtain equipment and supplies necessary to carry out the provisions of NRS 459.009 to 459.0098, inclusive.**

These powers would not have been excessive if Loux's purpose were merely to advise Nevada on nuclear waste disposal and how to mitigate any impact on the state. However, NWPO's director used his position to crush local dissent and manipulate scientific investigations while rewarding anti-nuclear activists both in Nevada and across the country. Each power granted Bob Loux was distorted to fit the needs of the national anti-nuclear lobby:

1. **Information to government agencies and the legislature:** Since Loux's position, and the position of key politicians who maintained him in power, was that they would fight Yucca Mountain to the death, all potentially positive impacts of the repository were minimized or excluded in NWPO's reports to state agencies, the legislature and the media. Consequently, the research done by NWPO was worthless as documentation on which to base legislative decisions on appropriate compensation to Nevada in the likely case that the repository was sited there.

2. **University of Nevada interface:** the loophole here is "Consult . . . the University of Nevada system or other institutions of higher education on matters relating to radioactive waste." Loux preferred to do almost all the socioeconomic impact studies using out-of-state sources although the UNLV campus in particular was uniquely qualified to do assessments of economic impacts on the gaming industry. The engineering departments at UNLV and UNR were blackballed. New Mexico avoided this problem in regard to the WIPP site by setting up a special institute within the University of New Mexico to study its similar repository based socioeconomic problems.

3. **Employment of Staff:** Loux selected key consultants based not on their scientific expertise, but on their preconceived political opposition to Yucca Mountain. NWPO's scientific reputation was thus reduced to the point of non-existence, making the agency's work the laughing stock of the national scientific community.

4. **Make executive contracts:** There is no full accounting of where major portions of nuclear rate payer money spent on NWPO went. In a later chapter, *The Money Tree*, we show what is known about Loux's budget and while it nominally balanced, this is a game of smoke and mirrors. For example, Clark University's Center for the Environment, Technology and Development and Decision Research (loosely associated with the University of Oregon) did the bulk of the socioeconomic consulting, yet we have no idea how much was spent on some of the rather dubious papers they wrote. Lump sums in the million to two million dollar range paid to Mountain West and later Coopers and Lybrand (which acquired Mountain West) seem to have been spent with little accounting of the diverse projects this money funded. The GAO objected to some of these expenditures in 1989.

5. **Obtain equipment and supplies:** Even here there is a question of whether supplies were always used strictly for State business. For example, a graphic artist for NWPO also worked for Citizen Alert. It is likely the environmental lobby had undue access to NWPO.

Rather than honest oversight, NWPO was designed from the ground up by Nevada politicians and especially NWPO's appointed director Bob Loux to roadblock ANY progress on nuclear waste storage, not only in Nevada but across the country. Bob Loux did a victory dance interview in 2011 that reveals just how deep the bad faith effort of NWPO truly was:

So the direction to me, and to the folks we work with, was fairly clear, <u>just do what you can to stop this. And, we developed a very comprehensive and I think sophisticated strategy that involved not only trying to acquire and defeat DOE in the scientific arena, but also in the public relations arena, also in the</u>

**political arena on Capitol Hill, as well as in Nevada, and also a legal strategy, and we were determined, as directed by the governor, to be as aggressive as possible in all those areas, and we developed a real four pronged strategy at that point in time to implement all of those.** We knew DOE had much more money, much more power behind them, and we knew that we could leave no stone unturned. We couldn't afford to make a mistake, and we had to use every bit of our resources to counter them.

One of the deliberate strategies we had was to try and get out in front of them on the issue, to try and develop issues associated with some of the science issues at Yucca Mountain, and force DOE to respond to us, so that we'd take a bunch of their time away from their developing their own plans and their own methods to have to try to deal with us and counter us. And I think we did that reasonably well.

We sued the Department of Energy some, over the years, probably 20 or 30 times at least. **We invoked public relations strategies, both in the State and outside the State. we even took some of or federal oversight money and began providing funds to other states along certain corridors, transportation corridors leading to Nevada, to try and enlist their population and their elected leaders in the same fight that Nevada was engaged in.**

**So, we were very deliberate, very conscious of the kind of strategy we needed to do.** We know that we really, in a sense, could leave no stone unturned, and that we needed to be aggressive and essentially give them no quarter, take no persons, and challenged every possible hing that they were doing to the best of our ability.

**Bob Loux interview 2011 page 12, 13**

In sum, Bob Loux was able to turn the Nuclear Waste Project Office into his own little fiefdom. A precedent for this hijacking of a state nuclear waste agency was provided by the Texas Nuclear Waste Program Office (TNWPO that was used to oppose Deaf Smith in the mid 80s. Steve Frishman, director of TNWPO, apparently learned the agency coup technique from the populist commissioner of the Texas Agriculture Department, Jim Hightower, (noted for his speech at the Dukakis Democratic convention). Loux hired Frishman in 1987 to be a geologist for Nevada, although Frishman produced primarily political work.

Obviously, Loux's tenure at NWPO was blessed, owing his longevity in part to Senator Richard Bryan, who had first appointed him director of NWPO in 1983 and later to Senator Harry Reid. Later governors Bob Miller, Kenny Guinn and Jim Gibbons stood in Bryan's shadow and had little control over Loux's activities. Miller's only public rebuke of NWPO came when the agency was caught red handed furnishing documents on a proposed nuclear rocket project to the environmental group Citizen Alert before anyone else in the State had seen the report.

## PUBLIC RELATIONS AND PROPAGANDA

Bob Loux's main duties indeed appear to have been to act as propagandist for the anti-nuclear position. We use the word propaganda carefully, because Loux obviously had a statutory duty to provide information and analysis of repository issues to the public. However, Loux was a political appointee, not a scientist and many of his statements were calculated to provoke instinctual reactions rather than reasoned thought.

A July 22, 1992 report by Roger O'Neil on NBC News provides a typical example. After DOE representative and noted volcanologist Bruce Crowe explained how recent earthquakes near Yucca Mountain little affected the underground site, Bob Loux, responded not to the science, but instead claimed DOE was lying:

**Well it proves one point, that DOE is going to be willing to say or do anything.**

**It reminds me of the old story about the optimist and the pessimist. When you throw the optimist into a room full of horse manure, he's the one looking for the pony. And I suspect DOE is trying to find the pony at Yucca Mountain. [O'Neil, Roger; NBC Nightly News, July 22, 1992]**

In Loux's and NWPO's statements, DOE took on a particularly ominous role, supposedly never having told the truth on any scientific issue. However, Loux's own statements often stretched the truth:

**Nobody thinks you need to bury the waste. It's safe to store it above ground for the next 100-150 years in dry cask storage. The NRC says it's safe now at the power plants. They just adopted a rule and there is no need to move it. [KDWN Radio, January 21, 1992]**

A complete review of what the Nuclear Regulatory Commission actually said ["1990 Waste Confidence Decision Review", published in the September 18,1990 Federal Register] shows something different.

**. . . the Commission finds reasonable assurance that, if necessary, spent fuel generated in any reactor can be stored safely and without significant environmental impacts for at least 30 years beyond the licensed life for operation (which may include the term of a revised or renewed license) of that reactor at its spent fuel storage basin, or at either onsite or offsite independent spent fuel storage installations.**

The NRC further stated:

**. . . the Commission does not dispute a conclusion that dry spent fuel storage is safe and environmentally acceptable for a period of 100 years. Evidence supports safe storage for this period.**

However, the NRC went on to state:

**. . . the Commission supports the timely disposal of spent fuel and high-level waste in a geologic repository, and by this Decision does not intend to support storage of spent fuel for an indefinitely long period.**

No one debates whether individual facilities *could* safely store spent fuel aboveground, but it does not follow that the entire system of 110 reactor sites nationwide constitutes a safe on-site storage system. Loux either had not considered this possibility, or felt compelled to paint Yucca Mountain in a negative light no matter what the alternative. Another statement in the January 21, 1992 interview shows the extent to which Loux made the debate one of fear:

**It would take three million nuclear bombs at the test site to equal the amount of concentrated radioactivity they want to store at Yucca Mountain.**

Of course, it could be argued that Loux and NWPO wanted to store that same three million nuclear bombs worth of concentrated radioactivity aboveground. Still another statement which stretched the truth in the January 1993 interview:

**A repository is designed to leak by regulation. It will eventually get into the biosphere, groundwater, soil and everything else. There will be no way to know if it leaks.**

Once again, on-site storage would by this analysis also be designed to leak and be even more likely to allow radioactive isotopes to reach the biosphere. Moreover, radioactive wastes are one of the easiest substances to trace if they leak through use of a simple Geiger counter or dosimeter.

Analyzing the body of Loux's statements, one finds two main currents:

1. The science at Yucca Mountain is corrupt and escape of dangerous levels of radiation inevitable

2. The DOE, the nuclear industry and even the U.S. government are not trustworthy

While the scientific facts would eventually sort themselves out, Loux's distrust of every institution outside his cloistered NWPO was troubling. Loux continually tried to paint the DOE as a potentially fascist entity driven to harm America's citizens. This is reflected in comments made to Dr. Daniel Metlay, Executive Director of the Secretary of Energy Advisory Board:

**Despite (DOE's) effort at critical self-evaluation, the final draft of the "Final Report of the Secretary of Energy Advisory Task Force on Radioactive Waste Management" continues to miss what may well be the central issue affecting trust and confidence in the civilian High Level Waste program. That is the issue of forced facility siting. Like a cancer in the corpus of the nuclear waste management and disposal effort, the continuing single-minded determination by the DOE to proceed with the Yucca Mountain repository in the face of technical concerns and strong and consistent State of Nevada opposition eats away at the credibility of any proposals that may be made for improving trust and confidence in the federal waste program.**

**As long as the State of Nevada remains the target of a forced siting process, with Yucca Mountain as its focus and symbol, none of the recommendations for improving trust and confidence in the Task Force report are likely to have much credibility. On the one hand, the DOE cannot expect to develop increased**

trust and confidence in the integrity of its program when, on the other hand, DOE continues to be engaged in the antithesis of that approach at Yucca Mountain. [Letter to Dr. Daniel P. Metlay, March 8, 1993]

More troubling was an effort to portray the U.S. government as an enemy of Nevada's citizens:

Recent letters to the editor in Nevada newspapers make the outlandish assertion that the federal government has offered the state $200 million per year since the early 1980s just to allow DOE to study the site -- no strings attached.

This is completely false.

There has never been that type of offer, nor could the federal government be held to it even if it were made. The federal government cannot be legally bound in any sort of agreement to provide compensation to a state like Nevada, even if Nevada was inclined to negotiate.

Alternatively, should the state seek "benefits," even just to let DOE study the site (as the author suggested), the state would forfeit its legal right to ever object to the project in the future. Nevada would have legally given its consent to the project, no matter how unsafe the site, no matter if the federal government made good on such an agreement. [Letter from Bob Loux to Nevada newspapers, NWPO letterhead, July 24, 1992]

While $200 million per year in benefits may not have been on the table, it is certain that in 1988 Senator Bennett Johnston, chairman of the Senate Energy and Natural Resources Committee had drafted legislation specifically setting up a schedule of $50 to $100 million in benefits, an amendment vigorously opposed by Nevada Senators Reid and Bryan as "nuclear blackmail". So Loux misstated the situation when he said "There has never been that type of offer". Indeed, Senator Lindsay Graham of South Carolina was willing to offer similar amounts as late as 2009.

Further, Loux's implication that all contracts with the federal government may be worthless because "the federal government cannot be legally bound in any sort of agreement" is more revolutionary rhetoric than part of his mandate to provide scientific oversight. If the federal government cannot be trusted to make and abide by its own laws, created by democratically elected representatives, then apparently Loux apparently believed the only route left was through obstructionism, a controversial position to say the least.

## LOUX THE PROGRESSIVE POPULIST

The nearly secessionist views of Mr. Loux placed him at the progressive populist end of the political spectrum. Indeed, a number of the consultants and friends of NWPO also share this ideology, which generically saw big business, high technology and the federal government as enemies of the people. Later we'll see that this is part of a larger decentralist movement (a hybrid of democratic anarchism and egalitarianism) which is more concerned with Yucca Mountain as part of a larger political revolution than as a technological question.

Bob Loux's opposition to Yucca Mountain, as a born and bred Nevadan, played to the Sagebrush Revolution which had a period of ascendancy in Nevada and the West. Ironically, Loux's father was a Nevada Test Site worker, so his opposition to a nuclear facility was somewhat unexpected. Indeed, the populism Loux and NWPO forwarded appeared to be imported from the academic and environmental community and not from Western individualist traditions.

## LOUX GIVEN GOLD WATCH AFTER EMBEZZLING FUNDS

So necessary was it that Yucca Mountain be shut down, that Senator Reid, former Senator Bryan (acting on a state oversight committee) and others looked the other way as the ensconced head of the Nevada Nuclear Waste Project Office, Bob Loux, between 2005 and 2008 'appropriated' pay raises for himself and his staff without authorization – otherwise known as embezzling. Normally that kind of activity gets you sent to prison and in fact the sums are similar to those taken during Nevada's G-Sting bribery scandal by politicians Dario Herrera, Lance Malone, Erin Kenny, et. al. who were indicted, imprisoned and scapegoated. But Reid and Bryan wanted Loux's 'years of service' opposing Yucca Mountain counted as a *Get-Out–Of-Jail-Free* card:

*Loux not going to resign right away*
**Nevada officials concerned that office remains strong as Yucca fight enters next stage**

CARSON CITY – The leader of Nevada's fight against the Yucca Mountain project says he's not going to decide immediately whether to resign, as Gov. Jim Gibbons requested on Wednesday.

Robert Loux, director of the Agency for Nuclear Projects, is under fire for giving himself and his five staff members pay raises above those authorized by the Legislature.

Loux has headed the nuclear projects office since its creation in 1985.

He said he wants to first talk to the Commission for Nuclear Projects, for whom he works. The position is appointed by the governor but serves at the pleasure of the commission.

Commission Chairman former Senator-Governor Dick Bryan said the commission will discuss the case at its Sept. 23 meeting. Bryan says Loux made a serious error but doesn't think he should resign.

This afternoon, Senate Majority Leader Harry Reid said that the commission should consider Loux's years of service as it assesses the situation.

"Obviously, Bob Loux has done a great job on behalf of the state, he's devoted a large chunk of his life to fighting the dump," said Reid spokesman Jon Summers. The senator "hopes the commission takes that all into account as it weighs its decisions. And whatever decision it makes, he'll support."

Assembly Speaker Barbara Buckley, also a Democrat, said that while the alleged violations are "extremely serious," the state's ability to continue fighting Yucca Mountain should also be part of the commission's decision.

"They have to weigh these allegations along with his very distinguished record over the year in fighting Yucca Mountain, and ensure that our fight against Yucca Mountain does not lose one step," the speaker said.

But former Gov. Bob Miller, a Democrat who fought the Yucca Mountain project for most of the 1990s, said "no individual is indispensable" in the state's ongoing battle. "In times of economic constraints, if you don't play by the rules you're subject to these types of ramifications," Miller said.

A spokesman for Republican Rep. Jon Porter said whether Loux remains in office is for the state to resolve. "The governor will handle it as he sees fit," said Porter's chief of staff Phil Speight.

Gibbons' office handed out research showing Loux earning $151,542 this fiscal year or 32 percent above the authorized $114,088. Gibbons' office said Loux has been giving himself and his staff extra raises for at least the last four years.

When the nuclear project office was created in 1985, Loux was named the director by Bryan who was then governor. And he was reappointed by Govs. Bob Miller, Kenny Guinn and Gibbons.

Loux said today that it would "not be productive" to comment before he has a chance to discuss the issue with his commission.

He told the Legislative Interim Finance Commission he gave himself and his staff extra money because they were doing 10-15 percent more work because of the absence of one of his employees. His staff consisted of five workers. He admitted it was an error.

For the past four years, he has earned more than other directors of state agencies with far larger numbers of workers and programs.
By _Lisa Mascaro, Cy Ryan,_ Thursday, Sept. 11, 2008 | 11:29 am.


So, Senator Reid and former Senator Bryan were willing to look the other way in regards to Loux embezzling money as long as the NWPO head was crippling the nuclear industry. We have shown that Loux's misappropriation of funds extended to the entire NWPO structure. Loux had spent a great deal of his thirty years at NWPO cultivating some of the most extreme elements of the anti-nuclear environmental movement, apparently with their approval.

Criticism of Bob Loux by representatives of DOE, site contractors and from the University system had often been viewed by repository opponents as an attempt to silence a lone voice of dissent. In reality, the technical community's concern was that Bob Loux's improprieties compromised the integrity of the Nevada's oversight over Yucca Mountain and did a disservice to the state and country. Rather than wanting a coverup, we have the unusual situation where the industrial complex begged for competent oversight by the affected state.

Since Loux and those he hired chose to promote a political and social agenda rather than do meticulous science, now that Loux has been exposed for fraud, the entire thirty years of NWPO oversight has been thrown into question. NWPO's lack of scientific rigor was embarrassingly obvious to the PhDs and high-caliber technical consultants working on the Yucca Mountain project. NWPO became the laughing stock of the scientific community and lost its ability to do competent oversight of Yucca Mountain, and much of the blame for this situation must be placed on its director who ran the state agency as a personal fiefdom for three decades.

# 9.  Nuclear Waste Task Force

In the early 1990s,The Nevada Nuclear Waste Task Force (NNWTF) acted as a contractor to the state to be the neutral education and outreach arm of the Nevada Nuclear Waste Project Office. Instead, it acted as the propaganda wing for anti-nuclear activists whose roots stretched from NWPO across the country and encompassed a variety of anti-war, environmental and religious apocalyptics.  One can argue that the entire nuclear industry has been brought to its knees from actions begun by the NNWTF.

In late 1987, Judy Treichel, Abby Johnson and Frank Clements formed the non-profit Nevada Nuclear Waste Task Force and then waited patiently until the state was able to shoehorn them into a lucrative $163,000 first year contract.  A bid proposal gives some history:

**"The Nevada Nuclear Waste Task Force was formed in December, 1987, by people who are long time Nevadans. Abby Johnson, of Carson City, and Judy Treichel, of Las Vegas, have worked together for several years on matters surrounding nuclear industry issues in Nevada. Frank Clements, of Boulder City, came to the Task Force with a comprehensive knowledge of environmental issues and thirty-six years of experience in governmental programs through his employment in the U.S. Forest Service. The three founders, each from a different Nevada city, brought together skills in grassroots organizing, knowledge of national policy decision-making and good rapport with Nevada elected officials and existing organizations. The main office of the Task Force was established in Las Vegas, Nevada in February, 1988 with two full-time employees. Judy Treichel serves as executive director and Frank Clements as (part time) office manager." [April 1991 bid proposal from NNWTF to Nevada]**

This implies the NNWTF formed out of the blue without reference to the surrounding political battle over nuclear waste in Nevada and nationally. In fact, NNWTF was specifically set up by anti-nuclear environmental groups to fight the nuclear industry on its own turf.

The parent organization to the Nuclear Waste Task Force was the National Nuclear Waste Task Force, directed by Caroline Petti. In 1985, 1986 and 1987, anti-nuclear waste groups formed a national coalition as a part of the Nuclear Safety Project of the Southwest Research and Information Center, a New Mexico based environmental organization. The coalition hired Caroline Petti as their Washington D.C. lobbyist because of her previous work with Ralph Nader's Public Citizen's Critical Mass Energy Project as well as Friends of the Earth on nuclear issues. It was due to the efforts of coalition member Citizen Alert in Nevada, that the Nevada Nuclear Waste Task Force was formed. Interestingly, Caroline Petti moved on to work for the Environmental Protection Agency writing rules governing radioactive release standards for the New Mexico Waste Isolation Pilot Project and planned to help develop similar regulations for Yucca Mountain.

Citizen Alert had been trying since 1986 to establish Judy Treichel, then a Citizen Alert board member, as director of an information center. A letter from Bob Fulkerson, then executive director of Citizen Alert to the DOE dated June 13, 1986 confirms this:

**"This is to endorse Judy Treichel's proposal to set up an information office on the high-level nuclear waste issue in Las Vegas.**
**Clearly, a full time person must be available to provide information to the public. . .**
**Because no such office currently exists in Las Vegas, we strongly recommend funding Ms. Treichel's proposal."**

Citizen Alert's central role in starting the NNWTF was reconfirmed by the late Frank Clements of the NNWTF in a conversation at the Sept. 2, 1992 meeting of the Nevada Commission on Nuclear Projects (Sawyer Commission). Treichel herself admitted that the Task Force was the idea of herself and Bill Vincent, a previous director of Citizen Alert, to counter the pro-repository activities of groups like the Nuclear Waste Study Committee [phone interview, Aug 23, 1994].

Consequently, the Nuclear Waste Task Force was from the beginning designed by the environmental lobby to fill the role of agitator within the state on the Yucca Mountain issue. Citizen Alert, the Southwest Research and

Information Center, the Safe Energy Communication Council and other national environmental organizations were all intimately involved in the creation of the NNWTF.

The links between the environmental lobby and the creation of the NNWTF are significant because the staff of NWPO and NNWTF later claimed the Nevada Nuclear Waste Task Force was a neutral entity. The attempt to coverup the origins of the NNWTF in the environmental movement was disturbing in that state officials were repeatedly forced to deny knowledge of the NNWTF political agenda though perfectly aware of its background and personnel.

Even more disturbing was that Judy Treichel and her Nevada Nuclear Waste Task Force were awarded a two-year opening contract worth $399,000, without a formal Request for Proposals, and without any competitive bids. Why would the State of Nevada let a contract of this size to NNWTF, an organization without a previous track record, without even an office, and which was formed only a month previously? Bob Loux, executive director of the Nuclear Waste Project Office and Judy Treichel had long had cordial ties through their mutual interest in environmental issues in Nevada. Apparently, the pair's mutual desire to stop the Yucca Mountain project overrode their concerns for competition on the public information contract.

The legality of such a sweetheart deal between the State of Nevada and the personal friend of the director of NWPO was obviously questionable. However, Senator Harry Reid, then Governor Richard Bryan and Lieutenant Governor Miller were using Yucca Mountain as a highly charged political football to forward their careers and simply looked the other way when it came to questioning the way NWPO picked its help. Over the years, Loux and the Staff at NWPO succeeded in awarding not only Treichel, but a number of other anti-nuclear allies similar lucrative contracts worth millions of dollars without apparent oversight and on a sole-source basis.

A letter from Judy Treichel to Bob Loux in 1986 (two years before the NNWTF was formed) shows just how tightly linked their political agendas came to be:

*Judy Treichel*
*Las Vegas, Nevada 89108*
*7 - 4 - 86*
*Dear Bob,*

*Here's the sort of stuff Bob Dickinson tosses into the weekly publication that goes to all So. Nevada contractors. It makes me crazy every year when we renew our subscription but you really need to know what's printed in here and there's no good substitute.*

*In this article he uses strange logic. The first paragraph says that we don't want govt. handouts or interference. To achieve that we would take the nuke waste for the govt. He places the govt. where he wants it -- like it has nothing to do with our test site. Wilderness projects restrict us, but nuke sites don't!*

*Dickinson's group hasn't had any public functions for long time. I think the "fact" bank was the last correspondence. However, in the local papers pro-waste letters to the editor have been frequently appearing. I have lots of stuff I want to write to the papers but I don't dare do that while my proposal is grinding its way through Don Vieth's red tape.*

*I met with Don last week. He took the proposal and letters and I am to call on Monday and see where we are. I think it will be a long negotiating process.*

*Thanks so much for your letter. It was great and should really help. The piece you wrote for the paper was good and I'm trying to see what I can do with it, without rocking the boat on my proposal. I'll let you know.*
*Till Next Time -*
*Judy*

It's clear from this letter that the NNWTF's later contract with the state was a sweetheart deal; Loux and Treichel were discussing creating a Trojan Horse public relations position two years before the Nuclear Waste Task Force was ever created. The Don mentioned in Treichel's letter was Don Vieth, who was then head of DOE Yucca Mountain operation. What Treichel and Loux were trying to do at that time was set up a public relations machine within DOE without anyone catching on to their political biases: *"I have lots of stuff I want to write to the papers but I don't dare do that while my proposal is grinding its way through Don Vieth's red tape."* In fact, Treichel's proposal was being pushed at the same time the National Nuclear Waste Task Force was being formed by national anti-nuclear forces. Treichel's ties to non-Nevadan environmental special interests devoted to obstructing the Yucca Mountain repository points to the fact that the neutrality claimed by NNWTF in future press releases was a fabrication.

Bob Dickinson, also mentioned in Treichel's letter, was at that time the head of the Nevada Nuclear Waste Study Committee (NWSC), a separate citizen's group with ties to the United States Council for Energy Awareness (USCEA). USCEA was an energy industry trade organization, so the NWSC was not itself neutral about the Yucca Mountain project. However, Dickinson and the NWSC had also proposed an educational program in 1986, which Loux stonewalled. Later, when Treichel's Nuclear Waste Task Force was selected to run the information program for the State, Dickinson and USCEA complained about the dishonest way they were excluded from consideration, but were politically straight-armed by Bob Loux. A letter from Bob Loux to Dickinson on NWPO letterhead dated Feb 18, 1988 engaged in some stretches of the truth:

**"In selecting a contractor to implement this project, which is intended to enhance public knowledge and participation, NWPO looked for proposals which would assure the broadest possible representation of citizens relative to the nuclear waste disposal issue. Publically issuing our Request for Proposals, there was a clear intent to avoid engaging groups or organizations which represented extreme positions at either pole of the debate. We received no proposal from Citizen Alert, who you and others have branded "extremist" on nuclear issues, nor did we receive a proposal from the committee you professionally represent, the Nuclear Waste Study Committee. Your group also has been viewed as polar in its position on the nuclear waste issue since it is financially supported by the nuclear power industry through its U.S. Council for Energy Awareness, an organization that, by your own statements, is actively promoting the repository in Nevada. [Letter from Bob Loux, Feb 18, 1988]**

Loux's criticism of USCEA as being biased is no doubt true, but if bias disqualified USCEA from representing the state, why did it not also disqualify NNWTF? While NWPO did not receive a proposal directly from Citizen Alert, Judy Treichel was intimately connected with the organization and received a direct recommendation from the organization on letterhead that listed her as a member of its board of directors. Citizen Alert was a key part of the coalition that formed the National Nuclear Waste Task Force, then headed by Caroline Petti, with which the Nevada Nuclear Waste Task Force was affiliated.

Moreover, how could the Nevada Nuclear Waste Study Committee and its parent USCEA not have known about a Request for Proposals in the one area it was most interested in, public education? Particularly galling was the fact that the Study Committee had already submitted an educational proposal to the state which had been rejected. In a protest of the NNWTF award, Bob Dickinson made the following comment:

**"Two years ago (1986) the NNWSC presented to the Commission on Nuclear Projects a thorough public information and education program which was well documented. In it was a program aimed at schools, libraries, public meetings and a comprehensive list of source material on all aspects of high level nuclear waste storage and transportation. Based on NNWSC's previous interest, Loux had a clear responsibility to have notified the committee directly that his office was seeking proposals." [excerpted in the Nevada Monitor, published by Nevada Nuclear Waste Study Committee, Spring 1988, page 3, see also original Dickinson letter]**

Even the Las Vegas Review Journal was forced to comment on the bizarre nature of these goings-on:

*ANTI-NUKE ACTIVIST WAS A POOR CHOICE FOR STATE TASK FORCE*
**The state's decision to hire a committed anti-nuclear activist to head up the Nevada Nuclear Waste Task Force was clearly ill advised and will destroy -- in advance -- any shot the group has at establishing credibility. . . .**

**Consider that Judy Treichel has had her finger in just about every anti-nuclear pie that's been cooked up in recent years. She has supported the demonstrators who infiltrate the Nevada Test Site. She is on the board of directors of Citizen Alert, an environmental group that hates anything nuclear. She helped organize local elements of the Great Peace March, which tromped across the country decrying nuclear weapons. She is a member of Clergy and Laity Concerned, a group that, when not decrying U. S. policy in Central America, is also stridently anti-nuclear. . . . [Las Vegas Review Journal, Opinion, Feb 18, 1988]**

In fact, Judy Treichel has spent most of her adult life as a professional anti-nuclear protester, having connections to Peace Links, Citizen Alert, Sagebrush Alliance, Clergy and Laity Concerned, Lenten Desert Experience, Nevada Desert Experience and a number of other protest organizations.

Bob Loux resolutely claimed to not know anything about Judy Treichel's political connections and agenda. On February 24, 1993, on the "Lark and the Byrd Radio Show", in an exchange with Robert Diero of the Nuclear Waste Study Committee, Loux was caught in a blatant lie:

**DIERO: I would also like to ask you this: Judy Treichel is hired to operate a division of your office called the Nuclear Waste Task Force. That particular division of your office was mandated to contact citizens and give them both the pros and cons of the repository. Now, I've met the lady, and been to a number of debates with her, would you say that Judy Treichel is impartial?**

**LOUX: Well Bob, she hands out material which is produced by the nuclear power industry which has a bias on it, as well as providing information from environmental groups. So she's providing both sides of the issue.**

**DIERO: I see. Would you categorize her as neutral on the issue or pro or con on the repository?**

**LOUX: Uh, her personal views are something that I'm not totally familiar with.**

Loux knew very well what Treichel's personal views were and his choice of the NNWTF as a nuclear information resource was not a pure business decision. For Loux to be unaware of Treichel's political agenda would require him to admit being asleep at the wheel at NWPO. He also was well aware of Treichel's lack of technical qualifications, regarding nuclear issues. As chief nuclear information officer for the state, Judy Treichel's credentials were decidedly underwhelming, except in the area of professional anti-nuclear protest. According to a biographical sheet from the NNWTF:

**"Judy Treichel moved to Las Vegas in 1969 from Minneapolis, Minnesota. She is the mother of three children and has two grandchildren. She attended the University of Minnesota.**

**Ms. Treichel worked in Las Vegas for Computer Sciences Corp., a subcontractor to the Department of Energy Nevada Operations Office.**

**She spent many years in labor/management relations and finally focused her attention on the social aspects of Nevada's environmental issues.**

**Ms. Treichel is associated with many local and national justice and environmental organizations. She is a founding member and director of the Nevada Nuclear Waste Task Force. That organization is currently under contract to the state of Nevada to provide public information on nuclear projects. She is the executive director of the Task Force. [Resume from contract proposal of Nevada Nuclear Waste Task Force to the Nevada Nuclear Waste Project Office, Spring 1988]**

Note that while Treichel attended the University of Minnesota, she did not graduate and does not in fact have any technical background or degree. While Treichel may have no technical credentials, we find in a September 18, 1983 article by Mary Manning in the Las Vegas Sun, just how busy she was with justice and environmental movements:

**"A budding human rights coalition formed in Las Vegas last spring blossomed as Clergy and Laity Concerned.**

**The umbrella organization embraces members from all Southern Nevada groups interested in human rights, peace and justice, coordinator Judy Treichel said. . . . .**

**Treichel was the coordinator for the second Lenten Desert Experience last spring, a vigil kept at the gates of the Nevada Test Site during the Roman Catholic 40-day period of Lent before Easter Sunday.**

**She also has been active in Nevada's grassroots drive to keep the MX missile out of Nevada and Utah. . .**
**.**

**Treichel formed Clergy and Laity Concerned after the Lenten Desert Experience with the help of May Miller of Northern Nevada, a long time environmentalist and peace activist, and grants from national organizations. The organization is forming a steering committee statewide, representing minorities."**
**[Manning, Mary; Las Vegas Sun, Sept 18, 1983]**

There are a number of interesting points to glean from this column, the first being that Judy Treichel was active in Citizen Alert during the MX Missile debate in the early 80s. Secondly, that Treichel was intimately tied to the Lenten Desert Experience, later known as Nevada Desert Experience, as late as 2008 which disrupted activities at the Nevada Test Site for years. Thirdly, we see reference to "grants from national organizations"; a convenient way of describing the way Treichel's organizations formed out of thin air funded by un-named out-of-state interests. In fact, the list of organizations that Treichel has been connected with shows she was a one-woman walking protest march:

- Sagebrush Alliance
- Lenten Desert Experience
- Nevada Desert Experience
- Clergy and Laity Concerned
- Peacelinks
- Rural Alliance
- Citizen Alert

The idea that Bob Loux was unaware of Judy Treichel's political bias is thus ludicrous; she was in fact picked by the state specifically because of her history of connections with nuclear protest movements.

## NON-COMPETITIVE BIDS

It would be unfair to completely belittle Ms. Treichel's activist resume, for in its own way it speaks of accomplishment. Unfortunately, it is not the same level of competence and expertise offered by many other overly-qualified individuals in Nevada with advanced degrees; remember the Nevada Test Site was long a center for atomic bomb physicists. Any of a number of professionals and organizations would have died for the same sort of contract conditions offered Judy Treichel and the NNWTF by the Nuclear Waste Project Office. In fact, two later submissions for the NNWTF information contract beat Treichel's credentials hands down, and there no doubt would have been more bids on these contract renewals had the state not carefully limited notice of Requests for Proposals in 1988, 1989, and 1991.

Evidently, the Nuclear Waste Project Office rigged the bid for public information not once, but three times in order to give preference to NNWTF and Judy Treichel. It did so not for the best interests of the state, but to ensure the anti-nuclear ideological purity of NWPO and give it a public relations propaganda arm. In response to Bob Dickinson's protests in 1988 regarding the original sweetheart contract given in February 1988 to NNWTF, a Request for Proposals was sent out May 2, 1989 with proposals to be received at the Nevada Agency for Nuclear Projects by May 19, 1989. Responding to that request was a proposal from the Center for Management Programs of the University of Nevada Las Vegas:

**"Public universities - institutions with unique responsibilities for education, research, and service in the interest of the state - provide a fertile ground for dissemination of information, public forums, and an open environment for the objective discussion of society's most complex issues. We believe the University offers uniquely qualified broad-based community programs for effective communication and public participation about the complex issues associated with locating and operating a proposed high level nuclear repository in Nevada.**

**The Center for Management Programs, one of a number of research and service centers of the University of Nevada, Las Vegas currently offers a wide-range of seminar, training, and informational programs to the general public within a professional development and public policy framework. The Center utilizes faculty, business, and industry leaders throughout Nevada. Working within the wide circle of university and community groups on a daily basis, the Center fashions tailored outreach programs. Outreach and communication constitutes the mission of the proposed program to meet the needs for citizen participation in public hearings, scoping meetings, and in review and evaluation plans and decisions on the proposed Yucca Mountain Repository." [Public Participation Program Proposal, Ed Goodin, Center for Management Programs of the University of Nevada Las Vegas, May 18, 1989]**

The proposal goes on to describe the many community contact and university resources of the Center for Management Programs, which in hindsight seem irrelevant to the stacked deck arrayed against them by NWPO. Most interesting is the credentials of four main principals:

- Edward Goodin, Doctor's Degree, (1973) Arizona State University. Concentration: Business Education, Research and Higher Education.
- Sharolyn M. Craft M.S., (1983) Business Education. Involved in successfully establishing consulting and managerial training centers in Iowa and Nevada.
- Leonard E. Goodall - Doctor's Degree, (1962) University of Illinois - Urbana. Served as President of the University of Michigan-Dearborn and the University of Nevada Las Vegas. Listed in Who's Who in America, Who's Who in the West, and American Men and Women of Science.
- R. Keith Schwer, Doctors Degree (1973), Economics. Twenty-Five years experience in applied business and economic research.
- Robert F. Smith Doctor's Degree (1963) Economics, University of Illinois. Professor of Economics and Director, Center for Economic Education, UNLV

Clearly, something was wrong when a group with three PhDs and an MS and with extensive university resources at hand (including UNLV's engineering department) was turned down. The absolute weight of such a rejection says that it was mighty important to someone that Judy Treichel, a little old housewife with little college education, be kept in her position.

To add injury to insult, in 1991 the contract cycle brought one more Request for Proposals. This time a private concern, Kadar Management, applied for the position of providing unbiased information to the citizens of the state about Yucca Mountain. Frank Darr of Kadar had some choice words to say about what happened.

**"I felt as though we were never given a chance. . ."**
**"We were first told the bid opening was set for a certain date. Then we were told that we didn't have to be there because they weren't going to award the contract that day. When we didn't show up, they awarded the contract anyway." [Interview with Frank Darr, June, 1992]**

What qualifications did the Kadar Management team bring?

- Karen Kruse B.A. Degree, Business Administration. Twenty years in office management including staff supervision, budget preparation, and disbursement.
- Frank Darr Graduate Electrical Engineer. A.A. Degree, Industrial Supervision, Safety Engineer. Thirty five years in the engineering and construction field, including supervision of the construction of power plants (Nuclear, Coal, Gas, Diesel and Hydro-electric).
- Donald Darr Graduate and Registered Civil and Mechanical Engineer. Thirty years in the design and implementation of engineering projects. John Timmerman Public Relations and Sales Manager. B.A. Communications. Twenty-five years experience in sales and financial planning.

It isn't necessary to claim Kadar's proposal was the best under the sun to prove a very important point: the bid process was rigged. Kadar could not have been ranked much lower than they were by the State and the documentary evidence is provided above, the rating sheet of NWPO consultants Steve Frishman and Dennis Baughman. Steve Frishman, a geologist for NWPO, rated Kadar a 10 vs. a 56 for Treichel. Dennis Baughman, a NWPO public relations director, ranked Kadar a 12 to Treichel's 60. Surely, Frank Darr, who had built reactors, worked at the Nevada Test Site and was a life time engineer had enough nuclear expertise to give Kadar a better ranking than zero on "Overall, working knowledge of high-level repository program". The treatment of Kadar Management would have been the typical meat and potatoes of state politics, if it hadn't also been combined with a systematic effort to keep Judy Treichel in position at all costs and no matter how many lies had to be told.

## Public Participation Program Rating Sheet

Rank respondents 1 to 10 (the best)

| Qualifications of Proposers<br>*(RFP, page 1)* | Kadar<br>Management<br>&<br>Maintenance | Nevada<br>Nuclear<br>Waste Task<br>Force, Inc. |
|---|---|---|
| Represents broad constituency, cross-section of Nevada citizens | 1 | 8 |
| Organizational capabilities to facilitate citizen involvement | 2 | 8 |
| Overall, working knowledge of high-level repository program | 0 | 9 |
| **Quality of Program Design**<br>*(RFP Attachment I, page 2-3, Organization of the program, items A, B, C and D)* | | |
| Involving appropriate citizen interest & other groups in promoting and facilitating public participation | 1 | 7 |
| Adequate expertise in developing and presenting technical material in ways understandable to general public | 5 | 8 |
| Informing public of issues & activities of nuclear waste program; assisting public in appropriate participation and involvement in these issues and activities | 1 | 9 |
| Other program elements the proposer considers appropriate and useful in organizing and carrying out comprehensive public participation effort | 0 | 7 |
| Reviewer Steve Frishman 5/8/91        Total | 10 | 56 |

## Public Participation Program Rating Sheet

Rank respondents 1 to 10 (the best)

| Qualifications of Proposers<br>*(RFP, page 1)* | Kadar<br>Management<br>&<br>Maintenance | Nevada<br>Nuclear<br>Waste Task<br>Force, Inc. |
|---|---|---|
| Represents broad constituency, cross-section of Nevada citizens | 1 | 8 |
| Organizational capabilities to facilitate citizen involvement | 3 | 8 |
| Overall, working knowledge of high-level repository program | 1 | 10 |
| **Quality of Program Design**<br>*(RFP Attachment I, page 2-3, Organization of the program, items A, B, C and D)* | | |
| Involving appropriate citizen interest & other groups in promoting and facilitating public participation | 1 | 9 |
| Adequate expertise in developing and presenting technical material in ways understandable to general public | 3 | 8 |
| Informing public of issues & activities of nuclear waste program; assisting public in appropriate participation and involvement in these issues and activities | 2 | 9 |
| Other program elements the proposer considers appropriate and useful in organizing and carrying out comprehensive public participation effort | 1 | 8 |
| Reviewer (signature) M Bauhman 5/9/91        Total | 12 | 60 |

Two things should be kept in mind about the Southern Nevada job market. First, after gambling, Las Vegas was long a government contract town (Nellis AFB, DOE, SAIC, UNLV, Nevada Test Site, Reeco, EE&G, Holmes and Narver, Lockheed, EPA, etc. . . .). This means there are always qualified technical and managerial talent available with advanced degrees and hands-on nuclear experience. Secondly, the University of Nevada at Las Vegas was starved for funding in comparison to its northern cousin in Reno. This means UNLV would have gladly provided top notch talent to provide educational services. What all these talented Nevadans lacked, however, was Judy Treichel's association "with many local and national justice and environmental organizations", specifically rabid anti-nuclear groups.

## TREICHEL'S ANTI-NUCLEAR OUTREACH

Treichel's 1993 contract renewal brought total commitments to date to the Nuclear Waste Task Force to $1.3 million dollars. This entire sum was provided by nuclear ratepayers and the nuclear industry, all supposedly for the purpose of providing educational and informational service on the nuclear waste issue to Nevadans. In reality, the money provided a technically unqualified activist housewife with the funds to operate a national political action committee dedicated to opposing nuclear energy. In other words, the Nuclear industry has paid to sabotage its own opposition. Treichel's anti-nuclear agitation was never limited to Nevada alone and she in fact serves as a clearing house for anti-nuclear activities nationwide:


**"Well, I hear from people almost every day, that are not only from Nevada, but are from a lot of different areas. And there are nationwide groups of grass roots people, the Safe Energy Communications Council works all the time. The Nuclear Information Resource Service, these are both in Washington. There are, I don't know, 6 or 8 national groups. Don't Waste U.S., and then there are groups in a whole lot of different states who are all trying to find out information. We mail them what we can. We also ask them to mail us the sort of information they have so we can keep up on what's going on there. I don't hear from people, and I really truly believe that there are a lot of people out there that even though we are told there are 49 states out there that want to dump on Nevada, because what we're hearing from other people, what Steve's (Frishman) brought up, that they don't think it's fair. But they're also worried it's not the right answer. And there are not a lot of obstructionists out there. . . ."[Treichel Statements to Citizens Against Nuclear Waste In Nevada, 6/29/93]**


In fact, there really are obstructionists out there. Daily office records of the Nevada Nuclear Waste Task Force revealed Treichel's main occupation was contacting and organizing anti-nuclear activists who wanted to shut down all nuclear activities (see below).

Treichel's 1991 contract proposal listed contacts with numerous organizations within the State of Nevada as part of its claim to being outreach specialists. The names of Treichel's associations are important, showing her objective was never to provide Nevadans with unbiased information. The following is approximately half the contact list from NNWTF's 1991 proposal, showing the anti-repository portion. While Treichel also answered informational requests from neutral and pro-nuclear inquirers, it is clear from her involvements that her purpose was to coordinate resistance from Nevada's anti-nuclear groups with national environmental activists.


## ANTI YUCCA MOUNTAIN ASSOCIATIONS
- Humane Society
- Citizen Alert
- People of Faith
- Nevadans Against Nuclear Dumping
- Sierra Club
- Nevadans for a Nuclear Test Ban
- People Against Radioactive Dumping
- American Peace Test
- American Civil Liberties Union
- American Association of University Women
- Western Shoshone Nation
- League of Women Voters (under the presidency of Abby Johnson)

The amount of time Treichel and the NNWTF devoted to contacts with the anti-nuclear environmental movement is shown most clearly in their activity logs. Among the most prominent individuals and organizations in Treichel's logs are:

- Rosalie Bertell (Concern)
- Scott Denman (Safe Energy Communications Council)
- Don Hancock (Southwest Resource Information Service, Don't Waste U.S.)
- Bill Vincent, Chris Brown, Bob Fulkerson, Marla Painter (Citizen Alert)
- Arjun Makhijani (IEER)
- Abalone Alliance
- Marla Painter, Abby Johnson (Rural Alliance)

While Treichel certainly did not limit her contacts exclusively to anti-nuclear activists, the following log excerpts show that such contacts are the primary activity of the Task Force.

NUCLEAR WASTE TASK FORCE LOGS Citations showing political and organizational interface with national anti-nuclear networks. Abbreviated. Dates from NNWTF inception 1988 through 1992.

*3/21/88 – Review material sent us by the Safe Energy Communication Council. Debate aired on Channel 10 PBS. Feedback indicates that proponents lost their case. Concern has mounted about the danger of storing plutonium.*

*4/13/88 Citizen Alert representative an office caller.*

*4/19/88  Bill Vincent, Citizen Alert, called to compliment NNWTF for the good job on the debate/forum*

*4/21/88  Conference call -- Institute of Energy, Arjun Makhijani and Bernd Franke . . .*

*4/19/88  The Task Force takes no position on the site, either pro or con [letter to candidates]*

*4/30/88 Mr. Cleve Anderson featured on talk show: Why Plutonium Should Not Be Buried In Yucca Mountain*

*5/13/88 American Peace Test called. . . . Planning caravan to form in the East and drive to Las Vegas. . .*

*5/18/88 Thirty packages of material were sent to friends of NNWTF. Twenty Seven of the packages went out of state. Report on Rural Coalition meeting in Green Bay, Wisconsin attached.*

*6/22/88  Bob Fulkerson and Bill Vincent -- Citizen Alert -- office caller's to discuss current events.*

*6/22/88 Kathy Thorpe called to advise on meeting: -- American Peace Test and Nevadans For A Nuclear Test Ban.. .*

*7/28/88  I met with Marla Painter at the Foresta Institute.  She submitted a proposal for organizing in rural Nevada . . . preliminary work to put on the debate between Dr. Rosalie Bertell and DOE . . .*

*8/17/88  Office Manager and Bill Vincent, representative from Citizen alert, transported Dr. Rosalie Burt tell from St. George Utah.. . .*

*11/3/88 Conferred with Don Hancock, Albuquerque, New Mexico. (Southwest Information Research Service)*

*12/2/88 Executive Director met at Office Of Institute For Energy to discuss a research project.*

 *12/9/88 Met with Paiute Indian Council and Rosalie Bertell to discuss the health effects of radiation from NPS*

*1/20/89  Arrange For Southwest Research And Information Service to send information to assemblywoman Myrna Williams*

*2/10/89 Confer with Tom Polikalas [Citizens Against Nuclear Waste In Nevada]*

*2/14/89 The meeting was called to order by Vice Chairman Chris Brown [Citizen Alert]*

*3/8/89  Telephone call from Institute For Energy in Washington DC*

*4/25/89  met with Chris Brown, new southern Nevada Rep. Citizen Alert*

*5/17/89 Executive Director met with Citizen Alert Task Force On Nuclear Waste*

*6/4/89 approximately 800 people visited our [Home Improvement Show] booth. . . Many asked if we had a petition to stop the dump.. . Most people thought on-site storage was the best long-term solution.*

*6/15/89  Met with Scott Denman of the Safe Energy Communication Council - also - Dr. Arjun Makhijani*

*8/12/89  attended a public session of the National Nuclear Waste Transportation Task Force*

*10/10/89  Executive Director took Scott Denman, Safe Energy Communication Council, to interviews with former Gov. Mike O'Callaghan at the Las Vegas Sun and Caryn Shatterly at Review Journal*

*2/20/90 attended Clark County Commission meeting dealing with the appointments of a county manager assistant position who has close ties to the nuclear industry . . . Two grass root level organizations, Citizen Alert and Nevadans Against Nuclear Dumping will testify.*

*3/1/90 The Task Force cosponsored an event at UNLV along with radio station KUNV which featured Lilo Woollney, member of West German parliament. . . .*

*4/23/90 Took Dr. Marvin Resnikoff to local newspapers for interviews and channel 8 TV news.*

*4/26/90  Took Dr. Arjun Makhijani of the Institute For Energy And Environmental Research to channel 8 TV for a live interview.*

*5/4/90  monitor state Democratic Convention. A plank opposing the proposed nuclear waste dump easily passed.*

*6/1/90 issued newsletter, we understand the mailing list is now over 10,000 people.*

*8/11/90 attended public session of National Nuclear Waste Task Force meeting in Washoe Valley, Nevada.*

*9/25-9/30/90  provided background on DON'T WASTE NEW YORK to Prof. Lea Sexton. Presentations at UNLV in cooperation with KUNV and UNLV's Coalition For Peace And Justice.*

*11/28/90 P. developed and sent package of material requested by Abalone Alliance.*

*1/10/91 spoke to Boulder City Peace And Social Justice Committee.*

*1/30/91 Executive Director and Dr. Arjun Makhijani participated in Task Force sponsored debate.*

*2/1/91 to Dr. Makhijani to breakfast meeting of concerned citizens connected to the Nevada Desert Experience.*

*2/28/91 met with DON'T WASTE US delegation to discuss nuclear waste issues.*

*4/18/91 met Dr. Makhijani (IEER), representatives from Nuclear Information Research Service, Don't Waste US and others for information exchange and possible contracting needs.*

*4/25/91, Coordination within NWPO officials on workshop presentations to Nuclear Information Research Service, Greenpeace conference.*

*4/26/91 attended conference in DC sponsored by Nuclear Information Research Service, Safe Energy Communication Council and Greenpeace*

*9/14/91 attended art show sponsored by Citizen Alert*

*12/5/91 Followed activities of Safe Energy Communication Council -- broke story on ANEC's Nevada Initiative*

*12/10/91 researched data pertaining to BLM/Yucca Mountain permits for Citizen Alert, Reno, Nevada*

*12/10/91 Executive Director gives presentation to grass-roots representatives at Safe Energy Communication Council in Washington DC*

*3/16/92 presentation to Public Information Research Group (Nader), Sierra Club, National Wildlife Federation, Union Of Concerned Scientists, National Taxpayers Union and Nuclear Information Research Service in Washington DC.*

# 10. The Texas Connection

In 1982, the State of Texas took a political turn with the election of Mark White to the governorship. Two Texans who filled posts in White's administration came to have special meaning in relation to nuclear energy policy and Yucca Mountain. Steve Frishman, appointed director of the Texas Governor's Nuclear Waste Program Office in the early 80s, and Jim Hightower, elected Texas Agriculture Commissioner in 1982, were instrumental in delaying the study and the building of a national nuclear waste facility.

The most well known on the national scene of the two Texans is Jim Hightower. A nationally known liberal-progressive with populist appeal, Hightower used his position as Agriculture Commissioner during the mid eighties to create a political fiefdom within the state of Texas. Hightower was dethroned from his commissioner position in the 1990 election, but remained highly visible in Democratic politics nationally and attempted to build a national platform by running his own national radio talk show.

Steve Frishman was brought on as a consultant to the Nevada Nuclear Waste Project Office as a geologist. However, Frishman's primary duties at NWPO seem to be political advocacy against the nuclear waste repository program, having now been in opposition to both the Deaf Smith site in Texas and Yucca Mountain in Nevada. His later close personal ties to Judy Treichel also played in to his role.

The progressive politics of Frishman and Hightower were similar to that espoused by the professional anti-nuclear environmental lobbyists of the Washington Beltway. In fact, Hightower long sat on the board of Ralph Nader's Public Citizen, parent to Public Citizen Critical Mass Energy Project. Thus, the progressive politics and anti-nuclearism of Frishman and Hightower help set the stage for understanding the politics of the larger anti-nuclear movement.

## HIGHTOWER VS EVIL CONGLOMERATES

In 1975, while a lobbyist in Washington D.C., Jim Hightower authored "Eat Your Heart Out; Food Profiteering In America", which indicted big agricultural companies and food wholesalers as unscrupulous. Steve Frishman, who served as director of the Texas Governor's Nuclear Waste Program Office during Democratic Governor White's term, moved to the Nevada program in 1987. Frishman served during the early years of the Hightower transformation of the Texas Agriculture Department and must have learned immensely from the experience.

Jim Hightower's progressive populism mixed with nuclear politic in Nevada in a Viewpoint article written in 1987 for the Safe Energy Communication Council, the then premier Washington anti-nuclear coalition. Titled "Restoring Our Confidence In Our Nuclear Waste Program", the article used the same arguments later used in Nevada to attempt to convince people that storage of nuclear waste is impossibly dangerous.

**Even if DOE were out of the picture, the disposal of radioactive waste in the ground using current technology is a risky one. Chances are, over the millennia, radiation would get out of the canisters. The only question is how far it would travel through the rock, at what speed, and how much damage it would wreak. . . .**

**Even without a nuclear waste accident, the economy of the Texas Panhandle, or of any other vibrant area chosen as a dumpsite, would be destroyed. . .**

**However disturbing, the Texas experience with DOE's waste program is not unique. There is little evidence that DOE seriously considered the safety or economic impact of a waste dump in any of the potential waste repository states.**

**The Nuclear Regulatory Commission has determined that wastes can be successfully stored in water basins at the plant that produced them for at least 30 years beyond the expiration of a reactor's operating license. Dry waste storage technologies would allow waste to be stored on-site even longer. By shoring up temporary on-site storage facilities, we can buy the time necessary to examine the options for permanent**

**waste disposal. [Hightower, Jim; "Restoring Confidence In Our Nuclear Waste Program", Viewpoint, Safe Energy Communication Council, 1987]**

Hightower's argument for indefinite delay of burial of high-level nuclear waste parroted the views of the Safe Energy Communications Council's position, and curiously parallels the current position of the Nevada Nuclear Waste Project Office. Jim Hightower, as a board member of Public Citizen, the Nader organization that parented numerous anti-nuclear organizations including the Safe Energy Communication Council, was part of a coalition that in appearance dictated Nevada's Nuclear Waste Project Office policy.

Some history of the Texas nuclear repository characterization and the political fallout that occurred is helpful in putting this all in perspective. Luther Carter's account in his work *Nuclear Imperative* is again a good source on what happened in Texas and how the political machinations there mesh with events at Yucca Mountain.

**The Department of Energy's siting investigation on the Texas High Plains has been taking place in Swisher and Deaf Smith counties in the Permian Basin's Palo Duro subbasin. . . . The lifeblood of the region, he has confided to me, is his view that a critical factor in turning public attitudes against the siting effort has been the "unrelieved negativism" of state officials. "They have reinforced and confirmed the worst dreads expressed by any of the local opposition," he said. "With no holding back, very quickly a frenzy of demagoguery develops, much like the theatrics of domestic war propaganda except in this case DOE and nuclear waste are the evil. Fear and negativism are easily reinforced without a continuous voice of moderation." But in truth, this does the state officials less than justice. According to his aids, Governor Mark White, knowing that Texas may need the help of Congress to keep a repository out of the Palo Duro subbasin, was somewhat restrained for tactical reasons and chose to find fault with the siting process rather than to declare DOE's interest in Palo Duro flatly unacceptable. The Texas commissioner of agriculture, Jim Hightower, has not felt so constrained. But even his comments have centered on what he not unreasonably perceives as a major land-use conflict. For instance, Hightower told a DOE hearing that, "for me, and for the agricultural community of this state, location of the nuclear dump site in the Panhandle really is not a technical issue at all. It's a human issue, an economic, cultural and moral issue . . . you could save us a lot of money, trouble, and time if we can just agree right now that prime agricultural land and a major fresh water aquifer is not a suitable site for a nuclear waste repository. [Carter, Luther; Nuclear Imperatives and Public Trust, Resources For The Future, 1987, p153]**

Jim Hightower's anti-agribusiness populism may have led him to similarly demonize nuclear energy and nuclear waste disposal in Texas as part of a broader populist Green Revolution, rather than because the Deaf Smith site was particularly dangerous. The DOE felt there was not enough fluid pressure in the underlying strata to push radioactive material up into the Ogallala aquifer even if it were breached. Hightower apparently felt the technological safety of projects like a nuclear repository are of little importance compared to the political perceptions of risk. However, broad application of such a philosophy would make it impossible for any technology to proceed because some fraction of local residents always have a perceived risk held to every technology.

If nuclear energy, coal energy, automobile manufacturing, etc. are all deemed "a human issue, an economic, cultural and even moral issue," as Hightower's philosophy assumes, then each of these technologies could be terminated by popular movements. The claim that these technologies are uneconomic, amoral and anti-cultural is made regularly by environmental and social justice activists, though it is anyone's guess how these traits could be evaluated objectively.

The larger question is whether Hightower and others in the anti-nuclear opposition based their conclusions on the repository program on the technical merits, or whether this is purely a philosophical disagreement over social means and ends. If all large technological endeavors are sinister, whether it is agribusiness in Texas or the disposal of nuclear waste in Texas or Nevada, then the debate over Yucca Mountain is pointless. This would imply that most, if not all, of the scientific criticism leveled by Hightower, Frishman, Bob Loux, Judy Treichel etc. is dishonest. In their view, scientific integrity may well be secondary to their efforts to push a Green Revolution which rejects all technological innovation.

A final quote from Luther Carter is instructive because it shows how the Texas anti-repository movement linked to Nevada:

More recently, the Holly Sugar Corporation, a major producer of beet sugar on the Texas High Plains, has given a $10,000 donation to the Nuclear Waste Task Force, a regional umbrella group for local groups such as POWER and Stand.

Judy Treichel's Nevada Nuclear Waste Task Force was a similar outgrowth of this larger, well organized effort to derail nuclear energy.

## THE FRISHMAN CONNECTION

Bob Loux's connection to Steve Frishman began at one of the many nuclear waste conferences they attended during the early eighties; they later collaborated on various reports together. When Governor Mark White was removed from office in 1988, Loux was on hand to offer Frishman a lucrative consulting position with NWPO:

**NEVADA HOLDS TALKS WITH FORMER TEXAS NUKE CHIEF**
**The former head of Texas' nuclear waste office could eventually be awarded a $65,000 contract as a geologist for Nevada's nuclear waste program, said Bob Loux, executive director of Nevada's Nuclear Projects Agency.**
**Steve Frishman was head of the Texas Nuclear Waste Programs Office but resigned Aug. 11 because of differences with Republican Gov. Bill Clements. Frishman was originally hired by former Texas Gov. Mark White, a Democrat. . . . .**
**"He knows all the nuances of the program," Loux said. "He'd be very valuable from my perspective."**
**[Las Vegas Review Journal, 8/19/1987, p5B]**

Loux and Frishman together combined to transform the Nevada Nuclear Waste Project Office into a progressive agency remarkably similar to Hightower's Texas Agriculture Department. Frishman's experience in Texas during Governor White's administration is reflected in the way NWPO later acted as an independent political entity, devising its own strategy and public relations.

Frishman was hired by NWPO as a geologist; however technical advice never seemed to be a major part of his job duties. Indeed, Frishman still spends significant amounts of time traveling the state and nation with Judy Treichel to build opposition to Yucca Mountain. At a meeting of Citizen's against Nuclear Waste In Nevada (CAN-WIN), June 29, 1993, during the critical final days of Nevada's 1993 legislature, Frishman demonstrated his lobbying abilities:

**FRISHMAN: The sites were picked on a political basis, a political game that's only gotten worse. . . Nevada is under siege by the nuclear industry, which has spent about $10 million. What they planned to get was a revocation of the Nevada legislature's 1989 law rejecting Yucca Mountain. Now they just want a crack, that citizens will exchange money for negotiating.**
**I was born in Washington D.C., but lived the last half of my life in the West. As citizens of the West, we are feeding back that Yucca Mountain is not going to happen.**
**"This is not all about nuclear waste. It's all about, among other things, how the East treats the West."**
**You can hear the ANEC ads on the radio about every ten minutes. The American Nuclear Energy Council for some reason thinks we are for sale. It's a waste of our legislator's time to even deal with this. We must oppose being the only state to accept the site. The site was selected not for its geology, but because the Nevada Test Site is already contaminated. We're not adding a little to a little at the NTS. We're adding a lot to a lot.**
**"CAN-WIN has not been very active in this part of the state. But citizen action does make a difference, it makes a very large difference. Our legislators won't act without it."**
**"A small digression, I used to own a small weekly newspaper in a relatively small town in Texas. What incensed me was that when people would get elected to city council or county commissioner, and they started doing things we all wondered about. They'd say, 'well you elected me to do what was right'. No, we elected you to do what we want"**

Nevadans no longer support nuclear testing. If Nevadans were strong in their support, you wouldn't have the Clinton administration changing its position in the last week to 'we may test, but we won't test first'. Well, that's a lot better than it was a week ago.

"Now, if you look at Yucca Mountain, we're getting the Clinton administration to finally realize that Yucca Mountain first of all is a major health and safety problem and Yucca Mountain second is a problem for the economy of the state. . .

"Third, this is what we're slowly bringing to the surface in the White House and it's the one thing that may have merit in terms of our own lifetimes, is that Yucca Mountain is undemocratic. And the White House is beginning to get a sense of this. And the only way to get that is through the participants of democracy sitting here."

"Keep on with it. Make CAN-WIN strong. Make Citizen Alert strong. Make any citizen organization and any individual citizen who feels that the thrusting of greatness of nuclear waste on the state of Nevada is absolutely the wrong thing for the state, especially if we're not willing to participate and if the generators of the waste assume we're volunteering." [break]

[In response to a question about how to stop Yucca Mountain] "The effort the state is making, the effort that local organizations are making including CAN-WIN and Citizen Alert. The effort that national organizations are making right now to get the attention of the White House that we need to undertake an overall policy review of what we do with nuclear waste in this country. For years we've been classifying them wrong, not classifying them by their level of danger or risks but by where they came from. . . We've been politically pushing on the White House on this very high level policy review, which may say, could say, that geologic storage in the near future is not the right thing to do."

It appears Frishman's real job was influencing local and national political policy in regard to nuclear waste disposal. This was outside NWPO's mandate, showing again that the agency's prime directive was obstruction of the repository rather than oversight. Frishman is admitting to being a lobbyist with the Clinton administration, not only against Yucca Mountain, but against the Nevada Test Site as well. Frishman also admitted he similarly opposed the Deaf Smith repository:

"As far as my views on the Deaf Smith site, it was from a technical and economic standpoint, judged by me, and the governor, and most elected officials in the state as unwanted. It would have been placed beneath the aquifer, a primary water source in the high Texas plains. Deaf Smith County has the second highest agricultural production of any county in the country. [KDWN Radio, "Yucca Mountain: Fact Not Fiction", sponsored by NWPO, Jan. 4, 1994]

Whether or not Deaf Smith was the perfect repository candidate (the situation was much more complex than Frishman presented), certain things are clear:

1. One of the reasons forced siting is occurring at Yucca Mountain is because of the success of Hightower, Frishman and others in politicizing and derailing the study of the Texas site. If Yucca Mountain was chosen for political reasons, Frishman and Hightower were the seed of many of those considerations.

2. It is unlikely any repository could pass muster with Frishman. Moreover, with a $100,000 a year Nevada contract to oppose the site, Frishman had little incentive to change his position.

3. Frishman's motives for opposing Yucca Mountain were driven as much by anti-Test-Site and populist sentiments as technological reasons.

In short, the motives of Hightower, Frishman, and others within the repository opposition have been an extremely complex hybrid of self interest, and ideology in which technical considerations became secondary. This tension between progressive philosophical perspectives and objective science appears unlikely to resolve the technical dilemma of nuclear waste disposal, much less any other technical dilemma short of shutting down the entire system.

# 11. The Szymanski Theories

In any new scientific venture there is bound to be dissent and alternative interpretations of the data. At Yucca Mountain, one of the most prominent early dissenters supporting Nevada's anti-repository position was Jerry Szymanski, a geologist who worked from 1983 through 1992 for DOE. The reason Szymanski's theories are important is not because of what they reveal about the science of Yucca Mountain (his scientific arguments were exhaustively rebutted), but because of what it tells us about how NWPO and the State of Nevada worked with unceasing zealotry to try and sabotage the Yucca Mountain project.

A number of alternative theories were proposed by Szymanski that would have disqualified the Yucca Mountain site. These theories hypothesized geologic mechanisms that would cause very deep underlying water to rise to the level of the repository causing corrosion of the casks containing the nuclear waste, eventually releasing radionuclides to the groundwater and environment. The media and NWPO liked to portray Szymanski as a whistleblower whose theories weren't given proper attention. Reality is something different.

The best way to dispose of the question of whether Szymanski was an unfairly dismissed whistleblower is to quote from a report from the prestigious National Research Council. Seventeen of the most highly respected scientists in the fields capable of responding to the groundwater upwelling theory were brought together in 1990 to look specifically into Szymanski's questions. The National Research Council was organized by the National Academy of Sciences in 1916 and its duty is to resolve debates of just this nature with the best scientific minds at hand, so their conclusions are the best

**"In response to a request from the DOE, the National Academy of Sciences' National Research Council established the Panel on Coupled Hydrologic/Tectonic/Hydrothermal Systems at Yucca Mountain, Nevada, under the auspices of the Board on Radioactive Waste Management, to evaluate 1) if the water table had been raised in the geologically recent past to the level of the proposed mined geological disposal system (repository), and 2) if it is likely that it will happen in the manner described in the DOE staff hydrologist's report (Szymanski) within the 10,000 year period covered by the regulations. The report claimed that such flooding had repeatedly occurred in the past and could be expected to happen again. If that were so, the water could carry still-active radioactive isotopes into the biosphere, a possibility that would lead to serious questions concerning the acceptability of the site. The panel regarded their task as not only evaluating the staff hydrologist's (Szymanski's) thesis, but also accessing the likelihood that the groundwater level could rise to the height of the repository by any plausible geologic process, or that such a rise had occurred in the past. ["Groundwater at Yucca Mountain: How High Can It Rise", Final Report of the Panel on Coupled Hydrologic / Tectonic / Hydrothermal Systems at Yucca Mountain, National Research Council, 1992, p1]**

So the review looked not only at Szymanski's original objections, but at any conceivable geological or meteorological means by which the repository could be flooded. This is hardly the profile of a scientific coverup, but was in fact an attempt to bend over backwards to investigate Szymanski's claims. After exhaustive review, the conclusions of the N.R.C. Panel did not support Szymanski's concerns:

**"The field evidence evaluated to establish whether or not deep groundwater had been forced up through faults and fractures and onto the earth's surface to produce the mineralized veins and surface deposits fell into six categories: 1) the character of soil development and geomorphic features; 2) hydrologic evidence from active and ancient springs; 3) morphologic and textural evidence from chemically precipitated mineral deposits; 4) the stratigraphic/textural/mineralogic characters of carbonate-cemented breccias; 5) geochemical and mineralogical considerations; and 6) the isotopic composition of the groundwater and mineral deposits.**

**The panel's overall conclusion was that none of the evidence cited as proof of groundwater upwelling in and around Yucca Mountain could be attributed unequivocally to that process. While some occurrences were equivocal, and some indeterminate based on observation alone, the preponderance of features ascribed to ascending water clearly (1) were related to the much older (13-10 million years old (Ma) volcanic eruptive process that produced the rocks (ash-flow tuffs) in which the features appear, (2)**

contained contradictions or inconsistencies that made an upwelling ground-water origin geologically impossible or unreasonable, or (3) were classic examples of arid soil characteristics world-wide. [How High Can It Rise, National Research Council, p2]

This should have been the end of the Szymanski controversy. The above conclusions, and many more in the Panel's report not cited here, are an academically polite way of saying Szymanski's theories are full of hot air. After the news conference of April 13, 1992 at which the findings were presented, interviews with a number of the Panel members confirmed their feeling that Szymanski's theories had been given more respect than they deserved. Vice-chair of the Panel, George A. Thompson of Stanford University said **"It would take a combined earthquake, plus a volcano, plus a climate change over geologic time to cause the water to rise to the level of Yucca Mountain."** [personal communication]

After the news conference, it was interesting to watch the actions of the media, who swarmed Jerry Szymanski, taking notes furiously as he described how he was being forced to resign from DOE because of feelings of moral indignation. Little attention at all was paid the distinguished scholars from the National Research Council who had spent two years conscientiously reviewing Szymanski's theories, but were allowed to slip away to their cars as if all their work had been for naught. Allowing science to be done by swarms of journalists may be a less than proficient way of determining the truth in highly technical matters.

Later chapters, (*Earthquakes, Volcanoes and Meteors* and *Groundwater*) will cover technical details of groundwater flow at Yucca Mountain more closely, but it should be noted that Jerry Szymanski's theories were given a much better review than presented in the media. In fact, because Yucca Mountain had become a political circus, the Department of Energy overreacted in this and other inquiries by investigating often irrelevant scientific criticism to exhaustion.

If Szymanski's theories were so thoroughly dismissed by the National Research Council (even ridiculed for their lack of scientific rigor), was Szymanski a quack geologist or were his theories driven by another agenda? A biographical sketch done by William Broad on Szymanski for the New York Times Magazine on November 18, 1990 gives us some insight into what makes Szymanski tick.

**"Szymanski grew up in Communist Poland, bright, eager to learn and marked for advancement. He was sent to the University of Warsaw, where he graduated in 1965 with a geology degree. Two years later, he was hard at work on doctoral research and teaching when, in March 1968, he joined students clamoring for economic and social reform. "We wanted a dialogue," he says. "What we saw was brute, bloody force, storm troopers beating people." Students were arrested on charges of "anti-state activities," and hundreds of Jews were fired from newspaper and university posts as the Government blamed them for shortcomings in the Polish economy. Though not Jewish, Szymanski was fed up. The Government was eager to be rid of university agitators. So, in December 1968, he was allowed to board a plane with his wife, baby and two suitcases, bound for the United States.**

**Penniless, Szymanski went to work as a consultant to the burgeoning nuclear-power industry, helping find and evaluate sites for reactors and related facilities. Paramount in his work was the study of fault lines, breaks in rock strata that can shift and cause earthquakes. In 1972, he landed a job with Dames and Moore, a civil engineering company with a global reputation for site-selection skill. Within some five years, Szymanski became a senior geologist, advising American companies and foreign governments in places like Iran, Korea, Spain and Chile on the geologic feasibility of setting up nuclear reactors. He also worked extensively with Federal regulators in charge with reactor licensing.**

**Peers held him in high esteem. "He's brilliant," says James G. McWhorter, the senior geologist at Dames and Moore. "One of his greatest talents is the ability to see the forest and the trees."**

**In 1977, at an international conference in Stockholm, Szymanski got a severe shock. Attending a session on the geologic disposal of used nuclear fuel, he was astonished to hear that the field was in its infancy, with one speaker after another talking only in theoretical terms. Nowhere in the world was there a permanent disposal site for the growing mountain of high-level nuclear waste. "I had been convinced that there was somebody on the other end," Szymanski says. "I was wrong. The fuel cycle was not complete."**

In 1982, he switched jobs, intending to address what he considered a looming obstacle to the industry's growth. At Decision Planning Corporation, a private company, he worked on Federal contracts to evaluate the merit of different types of deep geologic disposal. But he found himself only on the periphery of the process. In early 1983, he accepted a job with the Department of Energy itself, though it meant a $20,000 cut in annual pay. "I figured I would have some influence," he says. Soon after, the DOE offered him a position at the Yucca Mountain site, which was starting to undergo intense evaluation.

In February 1984, he arrived at the project offices in Las Vegas, blocks from the swirl of casinos, lights and illusion that make up the strip. Szymanski was one of the unit's four officials. He was in charge of gathering and accessing data from teams of scientists (mainly from the U.S. Geological Survey and various contractors) whose work was to form the basis for evaluating the mountain. Equally important, he was in charge of packaging that data for the Nuclear Regulatory Commission, the Federal body that would be asked to license the nuclear waste repository." [William J. Broad "A Mountain of Trouble", New York Times Magazine, November 18, 1990]

Obviously, Jerry Szymanski was no idiot; he had two years of graduate work in Poland and a sizable amount of work experience with Dames and Moore. He also specifically chose to work at Yucca Mountain despite a cut in pay, indicating he saw this as an opportunity to make an impact.

From conversations with co-workers from other projects, the consensus seems to be that Szymanski in fact believed his theories about groundwater upwelling at Yucca Mountain were plausible. Yet, other things come out: Szymanski is quick to judge and when challenged seems to have a tendency to defend his actions by claiming other scientists are at fault in their judgment, rather than perhaps admitting his own fallibility. Also, rather than being a positive attribute, his ability to see "both the forest and the trees", suggested to some associates that he was able to develop elaborate hypothesis so full of disjointed facts and dubious connections that there was a question of where the forest ended and where reality began. This is especially important because many of Szymanski's ideas seem influenced by catastrophe theory, an area of mathematics known for producing elegant theories which are difficult to analyze in the real world because of their complexity. A telling aside is that Szymanski's rising groundwater theories are handwritten on voluminous reams of paper, a curious throwback to the days before computers. What emerges is a picture of a brilliant thinker whose attempt to rewrite the geology at Yucca Mountain became entangled in an intricate web that despite its complexity failed to hold up under scrutiny.

Yucca Mountain is not the first place controversy has surrounded some of Szymanski's theories. His "rock creep" hypothesis at the Ninemile Point #2 reactor in New York (a theory that rock under stress would creep into the foundation of the reactor over time) was met with a lawsuit that was settled out of court for millions. Szymanski's analysis required numerous revisions in the plans made by the architectural firm Stoner-Webster, causing huge cost overruns. There is still contention within the geological field whether Szymanski's theories at Ninemile Point were based on sound science.

Other stories exist that show Szymanski made dubious calls in interpreting the geology of other sites in the past. This is not to say that Szymanski is incompetent; obviously geological and hydrological analysis is a highly technical field which sometimes verges on the subjective. It does however show that he does make mistakes, a trait that he and the media have not been willing to admit.

Also interesting is the mention in the New York Times article of Szymanski's work for Decision Planning Corporation before his work with the Department of Energy. There is no Decision Planning that this researcher can find, but this reference is peculiarly close to the Decision Research that has played a large role as a subcontractor to NWPO in attempting to disrupt the Yucca Mountain project. Decision Research was also involved in the Three Mile Island socioeconomic research and may have been part of a calculated effort by certain academics to derail nuclear energy in America.

After his rebuke by the National Research Council, Szymanski left the Department of Energy in May of 1992 . Later, in November, 1992, he was back in the media battle, again claiming that groundwater periodically rises at Yucca Mountain. This time he was backed by associates who had broken away from the DOE studies

## ANOTHER DOE SCIENTIST RIPS YUCCA STUDIES.
. . . "It's perfectly obvious, very clear to us, that the report by the Academy (NRC) is a very, very bad report," Livingston, who works with Technology Resource Assessment Corp., of Boulder, Colo., using data gathered by DOE scientists. "They've ignored data and misrepresented things," Livingston said. [ Las Vegas Sun, Thursday, Nov. 12, 1992, p6A]

## CONTRACT SUMMARY

(To Accompany All Contracts Submitted for Review of Board of Examiners)

## I. DESCRIPTION OF CONTRACT

1. State Agency: <u>Nuclear Waste</u>    Contractor: <u>Technology and Resource</u>
<u>Project Office</u>                              <u>Assessment Corporation</u>
<u>ATT: C.B. Archambeau</u>

Budget Account # <u>1005</u>    Address:    <u>90 Commander Spur</u>
<u>Boulder, CO  80302</u>

Expenditure Category # <u>10</u>

2. Upon Board of Examiners Approval   Termination Date: <u>October 30, 1993</u>
or Other Effective Date: <u>November 1, 1991</u>

3. New Contract or Contract Amendment? <u>New Contract</u>

4. Type of Contract: <u>Consulting</u>

5. Purpose of Contract.  Describe work to be accomplished. <u>Investigation of</u>
<u>natural groundwater hazard at Yucca Mountain.</u>

6. a. Payment for services to take place at <u>various rates</u> to a maximum of
<u>$1,351,560.00</u> for the term of the contract.

   b. If (a) not applicable, specify other basis of payment.

   c. If amendment, explain changes.

## II. JUSTIFICATION

7. What conditions mandate that this work be done? <u>U.S. Department of</u>
<u>Energy activities to site a high-level nuclear waste repository in</u>
<u>southern Nevada.</u>

8. Explain why State employees in contracting agency or other State Agencies
are not able to do this work. (Be specific). <u>Lack of qualifications and</u>
<u>experience specific to the actual project.</u>

9. a. Were bids or proposals solicited? <u>No</u>  If so, please attach.

   b. If not, why not? <u>Contractor and research team possess very unique</u>
<u>experience and qualifications relative to the actual research effort</u>
<u>and the specific location where the research is to be carried out.</u>
<u>Contractor and researchers have specialized knowledge of the</u>
<u>particular geological phenomenon in question which is critical to</u>
<u>understanding site suitability.</u>

   c. Why was this contractor chosen in preference to others?

## III. OTHER INFORMATION

10. Is contractor employed by State of Nevada, any of its political
subdivisions or by any other government? <u>No</u>  If so, is contractor
planning to render services while on annual leave, comp time, sick leave,
or on his own time?
Please explain:

11. Has the contractor ever been engaged under contract by any State agency?
If so, specify when, for what duties, for what compensation, and for
which agency. <u>No</u>

12. What is source of funds used to pay contractor? <u>Federal grant funds from</u>
<u>the U.S. Department of Energy.</u>

13. To what State fiscal year will contract cost be charged? <u>FY'92-FY'94</u>

Signature of State Agency Head

cs

# 12. The Money Tree

If philosophical rather than technical differences were what started the Yucca Mountain brushfire, money was the fuel that turned it into a firestorm. Money, tens of millions dollars worth, devoted to anti-nuclear politics by the Nevada Nuclear Waste Project Office, made the debate rage out of control. The ironic part was that the Yucca Mountain opposition was funded with the same money that had been set aside to study and implement the repository. In essence, the nuclear industry paid its own worst critics to hang them and even bought the rope.

And it paid the opposition quite well to do so. It can be argued that the better half of fifty million dollars spent on the Nevada Nuclear Waste Project Office over the years from 1983 to 1993 was used for anti-nuclear public relations, political advocacy and environmental special interests. Under director Bob Loux's guidance, nearly every study undertaken by the state of Nevada was directed towards foreclosing the possibility that the repository would come to the Silver State. Statements issued by NWPO, no matter what the scientific underpinnings, were twisted in an attempt to scuttle the entire project.

DOE also often misallocated resources as well; yet the sins of the federal agencies and its contractors were of the usual sort. Bureaucrats and subcontractors who milk the system are endemic in many large government projects (e.g., the Texas supercollider) and Yucca Mountain has been no better or worse than most projects in this respect. The difference between NWPO and DOE has been that DOE has for the most part avoided promoting an overt ideological agenda. NWPO was and is another story.

By the mid 90s, this author had an accounting of $40,975,642 worth of NWPO's contracts. Of that amount, $23,502,022, or 57.4% went to contractors outside the state of Nevada. If the Silver State weren't a mining state and one that had helped pioneer radioactive materials handling at the Nevada Test Site, the large percentage of money let to outside contractors would be understandable and accounted for by the need for outside expertise to handle technical issues. In fact, Nevada has a stock of trained professionals, not only in geotechnical, mining and engineering fields, but also sophisticated in evaluating socioeconomic impacts in regards to the hotel, gaming and convention trades on which the state depends for its existence. Consequently, money spent out-of-state by the Nevada Nuclear Waste Project Office must be viewed suspiciously, generally favoring anti-nuclear coalitions that have little interest in Nevada.

Not all NWPO money was misspent, a number of the studies conducted within the Nevada university system on geological and engineering sciences were legitimate, as was the work of the occasional independent consultant and subcontractor. However, even legitimate studies were often manipulated to try and discredit Yucca Mountain. In fact, at one point Bob Loux threatened to withhold money from studies conducted by the University of Nevada, Las Vegas, the University of Nevada, Reno and from Desert Research Institute unless they ignored objective science and supported the state's anti-repository position.

Political powers in Nevada, especially Governor Miller, Senator Bryan and Senator Reid, had a vested interest in covering up the finances and scientific dirty-laundry of the Nuclear Waste Project Office. Thorough independent audits were rarely if ever done on the agency. Certainly there were superficial audits which in essence asked the question of whether NWPO spent the money it received, and unsurprisingly found that they had. The real multi-million dollar question is whether NWPO spent its money on legitimate oversight or as a giant political slush fund.

While Nevada officials were content to look the other way at NWPO transgressions, the question of whether the Nevada Nuclear Waste Project Office spent its funds wisely and legally was a recurrent theme for a number of years at the federal level. The General Accounting Office was asked to look into this matter in 1989 and issued a report that questioned a number of the agency's activities.

## PRINCIPLE FINDINGS

**Although Nevada properly used most grant funds, it spent some funds for activities that were not authorized/or were expressly prohibited by law, regulation, court decision, or grant provision. Specifically Nevada:**

• Spent up to $608,000 of grant funds on congressional lobbying activities during the 3 years ended in June 1989. Lobbying was expressly prohibited by a provision DOE added to the grant for the period beginning March 1987 and by DOE's 1988 and 1989 appropriations acts. Also, lobbying is not authorized by the nuclear waste act.

• Used up to $75,000 to pay expenses incurred in litigation against the DOE, until a federal court ruled in September 1987 that litigation is not one of the purposes of financial assistance grants listed in the nuclear waste act. The independent auditor's reports of 1986-88 took exception to litigation expenses for the same reason.

• Exceeded by about $96,000 a spending limit of $1.5 million, contained in the DOE's fiscal 1989 appropriations act, for socioeconomic studies. Nevada also did not always have effective internal controls over grant funds. For example, the state used $275,000 from one grant period to pay expenses incurred in the prior period and did not adequately control about $226,000 in advances of funds to contractors. One of the advances effectively resulted in an interest-free loan of $210,000.

An independent audit for the year ended June 1988 also questioned the state's use of $69,000 of grant funds for state legislative activities because federal guidelines prohibit, and there is no explicit statutory authority for, the use of grant funds for this purpose. DOE had approved this use of the grant funds. On the basis of the nuclear waste act and applicable court rulings, GAO concluded that DOE had sufficient discretion to approve the use of grant funds for this purpose. [GAO / RCED-90-173, Nevada Grant Requirements, p3.]

Normally, misspending $1 million of what was in 1989 a cumulative $32.2 million in grant funds would be enough to get an agency head sacked, if not at least called on the carpet. Bob Loux has been able to successfully dodge these bullets by claiming that expenditure irregularities don't exist, in effect staring down his inquisitors. Ironically, DOE was itself at least partially to blame for letting NWPO get out of control. The GAO goes on:

**In 1985 a federal court held that DOE is required to fund Nevada's proposed tests and studies related to DOE's investigation of Yucca Mountain if they meet certain conditions. Subsequently, because of the court's decision and other factors, DOE adopted a permissive approach to administering the state's grant that has contributed to Nevada's improper use of funds. [GAO / RCED-90-173, p4.]**

In other words, DOE was sufficiently cowed by Bob Loux's and Senators Bryan and Reid's political attacks that the agency simply looked the other way and allowed NWPO to get away with what were known to be suspicious and even illegal uses of funds. But what were the "other factors" that swayed DOE's attitude toward NWPO's use of their grant money?

Privately, DOE and subcontractor managers expressed feeling that since they could never persuade the politicians of the state of Nevada to act cooperatively in regard to the study or implementation of Yucca Mountain repository, fighting the misappropriation of NWPO funds would be a losing battle. No matter what conciliatory measures were taken, DOE would be attacked by the state, and if DOE questioned the state's expenditures of funds, there would be a political hurricane. Apparently, a subconscious decision was made to allow the state to spend its funds as it pleased, and then simply ignore their input, in essence throwing the dog a bone and hoping this would solve the problem. It appears the dog took the bone and then proceeded to bite the DOE on the leg.

GAO was on the right track in their 1989 audits of NWPO, but they simply didn't go far enough. The focus of the federal audit was not to uncover fraud, or mismanagement of funds designed for scientific oversight, but to ask broad-brush questions about whether Nevada had conformed to its Congressional mandate in the Nuclear Waste Policy Act. While in the early years NWPO was viewed as being ideologically compromised rather than necessarily fiscally corrupt, some of its contractors made out like bandits (Mountain West, Decision Research, CENTED, NWTF), some of its consultants were paid excessively (Steve Frishman, Bob Halstead, Susan Zimmerman, etc.) and the lack of strict internal controls eventually led to the firing in 2008 of Bob Loux for what was in reality embezzlement of funds. Accountants from the DOE as well as the state of Nevada should have imposed quality control and internal controls on NWPO, but without their involvement or the involvement of the GAO, the agency was allowed to run free.

Nevertheless, the 1989 GAO study must have grazed somewhere close to the mark because by late 1989 the prime socioeconomic contractor, Mountain West, had been folded into the Big-Six accounting firm, Coopers and Lybrand. Whether this was done for matters of economic efficiency or to cover a trail of less than professional activities is open to question.

Requests for detailed information about the finances of NWPO and especially the funding of the Mountain West consortium bring murky responses. Bob Loux has been quite correct in saying he has provided financial figures and that NWPO has been audited, but by firms Loux has hired, not by the state auditors. The problem with the accounting trail is that it only covers block contract entities, obscuring where the money is actually spent. For example, we know Mountain West received $14 million over the years, but there is no account of what the multiple subcontractors did with their money. By simply ignoring financial controls, not issuing Requests for Proposals and avoiding competitive bidding processes, Loux hid NWPO's financial shortcomings by pretending they didn't exist.

Myriad contracts let by NWPO don't meet objective criteria set up in federal regulations (FAR 31.205-22: and 10 CFR 600.436 earlier cited in chapter three. Significant flaws in the contracting procedure include:
- Lack of Requests For Proposals.
- Lack of competitive bidding process.
- Lack of independent state audits.
- Lack of management accountability.
- Lack of preference for in-state contractors to promote local accountability.

One question a line item audit might have answered is whether political activists in the supposedly grassroots environmental movement directly benefited from their association with NWPO. We've already discussed the amazing handling of Judy Treichel's contract for the Nevada Nuclear Waste Task Force, which was awarded out of thin air. This contract eventually amounted to over a million dollars. Rural Alliance, a subcontractor to NNWTF (staffed by Citizen Alert activists) was primarily a political action committee, even opening an office in New Mexico to oppose the Waste Isolation Pilot Project. The specter of the State of Nevada running covert lobbying campaigns in other states to foment unrest is an unsettling one.

Contrary to the media image of the anti-nuclear movement as poor and ravaged crusaders, the environmental opposition is quite well funded. Greenpeace, the Sierra Club and the network of Nader organizations (Public Citizen, Environmental Watch, etc.) enjoy multi-million dollar budgets. On the nuclear issue, national environmental groups further leveraged their financial power by forming a coalition, the Safe Energy Communications Council, which allowed them to coordinate and pool resources. Donations of media resources by people like producer Oliver Stone also added to the effort without ending up in the accountant's books.

Adding the State of Nevada's Nuclear Waste Project Office grants to the mix of anti-nuclear fund raising shows the nuclear industry was more than matched in terms of the number of public relations dollars spent trying to sway Nevada voters. In fact, it is difficult to distinguish between dollars spent by NWPO on oversight/public relations and the funds of environmental groups. Activists like Judy Treichel are neither fish nor fowl, receiving funds as consultants to the state, while in turn funneling free propaganda to the affiliated Citizen Alert and other nationwide environmental organizations. Dr. Marvin Resnikoff, a NWPO contractor, ran a consulting firm called Waste Management, but was a co-founder of the Sierra Club Nuclear Waste Campaign and did anti-nuclear work for Ralph Nader's Public Interest Research Group. It is unclear whether NWPO operated as an oversight agency, or a slush fund for anti-nuclear activists.

Loux and various other agents for NWPO bitterly complained that they were underfunded in comparison to DOE and the nuclear trade organizations. For example, Loux's initial 1993 budget request was for $26 million, although he settled for $6 million. Privately, many of the engineers associated with Yucca Mountain confide that vast sums were wasted by DOE studying NWPO's poorly formed scientific theories, adding little to the safety of the site. Unfortunately, DOE couldn't afford to leave even trivial objections untested lest NWPO cry bloody murder. This caused costs to soar, which was ironically then used by opponents like Senators Bryan and Reid to claim the project fiscally mismanaged. In reality, obstructionism by the senators and the various protest groups may have driven many of the over-expenditures as much as any ineptness on the part of DOE.

Thus, the NWPO money trail is important for two reasons:

- It first shows the agency had a total lack of fiscal accountability. Any future state oversight will require financial controls.

- It shows that the environmental network that opposes Yucca Mountain is not poverty stricken, but has access to large sums of money (including even the Nuclear Waste Fund).

The following pages expose the worst irregularities present in NWPO's contracts. This money trail points to serious problems in NWPO's ability to spend money without outside controls.

# 13. Lost Benefits

A key element in campaign against Yucca Mountain has been an effort by anti-repository forces to convince the public that there will be no compensation for taking on the burden of the repository. The theory is apparently that if Nevadans are sufficiently propagandized with the idea that no benefit will come to the state, no matter what is offered by the Nuclear Waste Negotiator or the U.S. Congress, they will be less likely to give up their opposition to the site. To carry out this reeducation campaign, which hinges on convincing the populace that they should distrust their national government, a concerted effort was required from NWPO, state politicians, the socioeconomic researchers at Mountain West, and environmental activists.

In a letter to the media in July of 1992, Bob Loux had the following to say:

**". . the idea that Nevada is losing big federal dollars by opposing the Yucca Mountain project is nothing more than nuclear industry and DOE propaganda aimed at changing public opinion about the project."**

Loux was perhaps himself engaging in a bit of propaganda, attempting to convince Nevadans of the lack of any potential benefits of hosting the repository. In reality, there were many monetary and non-monetary options open to negotiation, including perhaps a limited right to veto the site even while accepting benefits. A March 26, 1987 report in the Washington Post on a bill that would have provided monetary compensation for accepting a repository puts the intent of congress and the intent of the Nevada delegation in a different light than Loux claims.

**There is a little carrot in this bill," J. Bennett Johnson (D-La.) said in announcing the legislation.**

**The proposal provides $100 million annual incentive payments to the recipient of the permanent repository and $50 million for the temporary storage facility that would prepare waste for eventual burial.**

**Such payments "for something that's going to happen anyway . . . becomes a very attractive deal," said Johnston, Energy and Natural Resources Chairman. . .**

**"Nuclear waste involves no great harm to the areas involved. It is a perceived problem," Johnston said. "These are very generous amounts of money to deal with a perceived problem." [Washington Post, March 26, 1987]**

In fact, the greatest obstacle to monetary compensation for accepting Yucca Mountain has not been the federal government. Indeed Congress has nearly begged the state of Nevada to accept benefits. Nor was the nuclear industry opposed to benefits, and in fact their none-too-secret Nevada Initiative was based on the acceptance of benefits. Instead, it was Nevada's own politicians who fought tooth and nail to torpedo any hint that Nevada might receive compensation for Yucca Mountain. Between Senators Bryan and Reid in the Senate, and Governor Miller and Robert Loux in the state of Nevada, all talk of negotiating for benefits was proactively squashed. From the Washington Post article of 1987:

**Sen. Harry M. Reid (D-Nv.) and Rep James H. Bilbray (D-Nev.) immediately denounced the money offer as a "bribe," and Nevada Governor Richard H. Bryan (D) called it "nuclear blackmail."**

**"Nevada won't be bought," Reid said. "Efforts to negotiate a high-ticket price tag to buy us off won't change our opposition."**

**Bilbray said the money offer was "a bribe that Congress will never accept." [Washington Post, March 26, 1987]**

While this posturing played to the political galleries in Nevada, it was certainly damaging to the people of Nevada. A hundred million dollars is real money in a small state like Nevada, especially with deficits in the billions of dollars, but the public was not allowed to judge for themselves. Instead, Bennett Johnson's bill was labeled "Screw Nevada" by the state's pundits and the media played down the benefits issue.

The benefits offered to Nevada by Bennett Johnston may have been a bribe as expressed by Congressman Bilbray in 1987, but they were certainly not "nuclear blackmail". Roads, bridges, supercomputers and a thousand other benefits are all part of the federal budget benefit equation. Nevada's politicians had never been shy about lobbying for fighter-bomber wings for Nellis Air Force Base or for continued testing at the Nevada Test Site in order to bring money into the state, though those projects were arguably more dangerous for Nevadans than Yucca Mountain. Yet, money for Yucca Mountain was considered a bribe while efforts to obtain "pork" in exchange for other political favors were considered good politics.

A hundred million dollars may not seem much to a neighboring state like California or even to most of the mid-sized states of the Union, but it must be remembered that the entire Nevada state yearly budget is only a couple billion dollars. Even more poignant is the fact that Governor Sandoval was forced to cope with a billion dollar shortage in 2011 in a brutal way.

David Leroy, a former Idaho Lieutenant Governor, was appointed in 1989 as U.S. Nuclear Waste Negotiator. As negotiator, Leroy communicated with governors of 58 jurisdictions and leaders of federally recognized Indian tribes to determine if any of those entities would be willing to host a repository or Monitored Retrieval Storage facility (a transitional repository). In the March 1991 edition of the Nevada Monitor, published by the Nevada Nuclear Waste Study Committee, Leroy had the following to say about the types of benefits available to Nevada:

**"It's obvious to me, particularly from historical experience, that financial benefits alone will not drive the acceptance of these facilities. When I begin to talk benefits, I have the broadest possible concept of what that term may include. I start with the concept that safety must come first. After that, we're interested in talking about choices of technology which the local citizens might be interested in so they can feel comfortable with the facilities. I then move to concepts of shared control in the operation or ownership of those facilities, and only later do I begin to talk about financial benefits or infrastructure benefits.**

**As to the later, everything is negotiable. We can talk in terms of educational benefits, of colleges and universities creating a center of excellence. We can talk about infrastructure improvements like building bridges and roads. We can go into a community, clean up an existing environmental problem and give that community a net improvement in its environmental role." [Nevada Monitor, Nevada Nuclear Waste Study Committee, March 1991]**

While Nevada may be in the process of being "screwed" by the federal government in terms of the forced siting of Yucca Mountain, it is not being denied compensation because of federal reluctance to meet its obligations. Nevada's politicians have expressed not only opposition to compensation, but opposition to even discussing the possibility of compensation. The technical community in Nevada admits in private that the actual impact of Yucca Mountain, both economically and in terms of risks, will be much less than anticipated. Consequently, the offers first made by Bennett Johnston and the government were likely overly generous. By not negotiating, Nevada may have thrown away a windfall.

Senator Lindsay Graham also has suggested payments in the $100 million range as late as 2009, as reported in Fuel Cycle Week:

**BILL CALLS FOR REBATE OF YUCCA MOUNTAIN FUNDS**

**By Jacob Mazer, Assistant Editor, *Fuel Cycle Week***

**In another gesture of resistance to the abandonment of the Yucca Mountain nuclear waste repository project, Senator Lindsay Graham (R-South Carolina) introduced a bill to the Senate calling for the return of funds set aside for the project to utilities and customers.**

**The Nuclear Waste Trust fund was created in 1982, raising money for a waste storage center using a 0.01 cent on the kilowatt rate paid by energy consumers. The fund reached about $30 billion, with about $22.6 billion now remaining after the cash sunk into Yucca Mountain.**

**Graham's Rebating America's Deposits Act mandates that the President to either confirm Yucca Mountain as "the preferred choice" for nuclear waste or begin to return to funds within 30 days of the bill's passage. About 75% of the money would go back to customers, with the rest going toward the construction of more temporary storage facilities to hold the waste on-site at power stations until a permanent solution is established. The bill also calls for separate payments of $100 million a year to states with nuclear weapons waste stating in 2017, when Yucca was scheduled to begin holding waste. Eight other Republican senators including Arizona's John McCain co-sponsored Graham's bill.**

**"No one should be required to pay for an empty hole in the Nevada desert," Graham said.**
**http://fuelcycle.blogspot.com/2009/04/bill-calls-for-rebate-of-yucca-mountain.html**

## ROOTS OF THE ANTI-REPOSITORY STRATEGY

The preemption of all discussion of negotiated benefits from popular forums tactic may in part be rooted in the work of the state's socioeconomic consultants, especially Dr. Paul Slovic of Decision Research. In a paper titled "Images of Disaster: Perception and Acceptance of Risks From Nuclear Power", written in the late 70s, Slovic addressed methods for gaining public acceptance of projects with high perceived risks In an analogy to the Hermiston, Oregon nerve gas depot at the Umatilla Army Depot, Slovic had this to say:

**"Whereas public opinion around the state was more than 90 percent opposed, residents of Hermiston were 90 per cent in favor of the transfer (of nerve gas), despite the warning that the fuses on the gas bombs deteriorate with age, but that the gas does not. . . . Several factors seem to have been crucial to Hermiston's acceptance of nerve gas. For one, munitions and toxic chemicals had been stored there safely since 1941 (so the record was good and the presence of the hazard was familiar). Second, there were clear economic benefits to the community from continued storage at the depot of hazardous substances, in addition to the satisfaction of doing something patriotic for the country. Finally, the responsible agency, the U.S. Army, was respected and trusted." [Slovic et. al.;"Images of Disaster: Perception and Acceptance of Risks From Nuclear Power", Accident At Three Mile Island, Westview, 1979]**

Consequently, Slovic long ago hypothesized that acceptance of a site like Yucca Mountain depended on:
- Similar risk experience.
- Trust of the authorities
- Economic benefit.

Risks associated with both the proposed repository and Nevada Test Site have long been a target of groups like Greenpeace, American Peace Test, Nevada Desert Experience and other environmentalists. This is despite the fact that risks associated with the Test Site have long been understood by the public as minimal and have been incorporated into their calculations of the benefits of living in Southern Nevada. Yet, trust in the DOE has been systematically attacked over the years by NWPO and the environmentalists. While the DOE has certainly made many mistakes, it has also made a good faith effort to increase its levels of trust.

Benefits form the third leg of the Yucca Mountain acceptance triangle according to Slovic's theory. It is therefore unsurprising that NWPO and the state's politicians, spurred by their socioeconomic consultants, adopted an anti-benefits philosophy and attacked the possibility of Nevada's remuneration for Yucca Mountain with a vengeance.

If Yucca Mountain turns out to be as safe as claimed by scientists, and its impact on tourism and infrastructure nil, then Nevada will have foregone billions of dollars of benefits merely to sooth the political feathers of its leading politicians. More devastating than the loss of money may be the loss of the technological community that might be brought to Nevada with a proper structuring of benefits from Yucca Mountain. For example, the development of a local energy research center that specialized in nuclear and solar technologies would have the added benefit of stabilizing a top-notch engineering and scientific community.

What repository opponents have failed to address are the downside dangers to Nevada of not developing Yucca Mountain. Betting that the next hundred years of development in Las Vegas will be entirely generated by gambling and tourism, in a risky bet at a time when gambling is being legalized nationally and the Obama administration is attacking the mining and grazing industries as too greedy. Since the state has little technology base to speak of, federal projects such as the repository become prize economic plums. Nevada's opposition to Yucca Mountain has thus cost the Silver State dearly by throwing away the one point of leverage that might have been used to trade for prime federal technical projects.

Nevada has a dismal success rate in attracting large-scale high-technology projects and the state may now even be shunned because of the perceived anti-technology bias of its politicians. It is interesting to note that the only instance where Nevada benefited from leverage due to Yucca Mountain, the supercomputer project at UNLV, was originally opposed by Senator Richard Bryan. A short list of some of the other technical projects Nevada has lost follows:

- **SP100 NUCLEAR SPACE REACTOR** - This is a development project for moderate sized nuclear reactors designed for use in space and on the moon. A $500 million project, Nevada lost this project to Hanford, Washington.

- **NSCEE SUPERCOMPUTER** - The National Supercomputing Center for Energy and the Environment is now in use at the University of Nevada Las Vegas. However, Senator Bryan and former Governor Grant Sawyer opposed the supercomputer on the grounds that since it would be used in part for Yucca Mountain, this would be accepting benefits, which Bryan and Loux claimed would be impossible.

- **SUPER COLLIDER** - Although this project was eventually canceled in 1994, Nevada did aggressively bid for this project and would have benefited even from an aborted construction effort. Then Governor Bryan could have played the Yucca Mountain card at this point, but the super-collider was lost to Texas.

- **STEALTH BOMBER WING** - Nevada was also hit in 1992 with the relocation of the F-117 Stealth fighter wing from a remote base near Tonopah, to Holloman Air Force Base near Alamogordo, New Mexico. Air Force studies indicate the move might have cost Southern Nevada $255 million in economic activity and nearly 6,000 military and civilian jobs.

- **474TH TACTICAL FIGHTER WING** - This wing was deactivated in 1988 and 1500 military personnel were reassigned to other units and bases. Although Bryan and Reid claimed they had done everything they could to stop the deactivation, the fact is their influence on almost any matter affecting Nevada is nil.

- **RADIOACTIVE DIRT** - In 1988, Richard Bryan successful opposed the transportation of slightly radioactive dirt that had formed the foundation fill around homes in New Jersey. The Beatty low-level radioactive waste dump offered to pay the state $20 million per year if allowed to reopen, so the state could benefit greatly by being a center of disposal technology.

- **NUCLEAR ROCKET** - The Air Force attempted in 1992 to begin siting of a nuclear rocket test program and one possible site was the NTS where a previous nuclear rocket program had been conducted. Bob Loux, in collusion with Citizen Alert, wrote an opposing position paper for the state of Nevada.

- **NEVADA TEST SITE** - Testing at the Nevada Test Site was put on hold in 1993 by the Clinton administration. This in part may have been because of an initial vote by Senator Bryan against the then pending Clinton economic package. Harry Reid ran proceedings that attempted to substitute the idea of a huge solar-hydrogen research project at NTS, but this was too large a boondoggle to be taken seriously.

- **BEATTY LOW-LEVEL WASTE DUMP** - Beatty Nevada had long been the site of a low-level nuclear waste dump. Governor Miller opposed and successfully closed the dump at the end of the 1993 legislative session despite offers of large subsidies. NERVA - A number of interrelated energy projects run by DOE at the NTS were vetoed by Governor Bryan's opposition to the movement of radioactive materials through the state.

Thus, opposition to Yucca Mountain is merely the most visible symptom that the state's technological potential is drifting. The opposition of key Nevada politicians to Yucca Mountain, based on exaggerated perceptions of risk, carries over to opposition and delay of other technologies as well. Recognizing this, an independent citizen advisory group backed by the the Nevada Nuclear Waste Study Committee began developing a white paper on possible negotiation points in late 1994 in preparation to the likely reopening of the Nuclear Waste Policy. The purpose of this effort was to sidestep the political establishment and attempt to procure benefits for Nevada in a rational structure before all such opportunities evaporated. Other such efforts have been envisioned in later years but have stalled. Nevertheless, among the benefits the Study Committee reviewed were the following:

## POTENTIAL COMPENSATION ELEMENTS
- Water improvements, specifically grater allotments from the Colorado River
- L.A. Highway upgrade
- High Speed Super Train to Los Angeles
- North / South Train
- General Infrastructure

- o        Street and highway construction
- o        Widening of I-15 south to Los Angeles
- o        Public transportation systems
- o        Water delivery and storm systems
- Nuclear Studies Institute
- Transmutation Facility
- Educational Upgrade all levels
- Direct Taxpayer Rebate
- Police & Fire
- General Fund
- A "per ton" fee paid directly to state treasury
- Factories for constructing transportation and storage canisters
- Trust Investment
- Ecology Preservation
- o        Desert Tortoise
- o        PupFish
- Land Swaps
- Solar Energy Complex
- Test Site Reconfiguration

# PART III: MOUNTAIN WEST

# 14. Mountain West

Mountain West Research of Tempe Arizona was founded in 1974 by James Chalmers, PhD., CRE, who served as President and Chairman of the Board. Its focus was as a consulting firm for real estate development, urban growth planning and regional economic growth. However, the event most responsible for transforming Mountain West's socioeconomic consulting into political advocacy was the Three Mile Island nuclear reactor accident in 1979.

**At the request of the President's Commission on the Accident at Three Mile Island (the Kemeny Commission), the Social Science Research Council commissioned social scientists to write a series of papers on the human dimensions of the event. [preface; Accident At Three Mile Island: The Human Dimensions, Westview, 1979]**

Among those who wrote papers for the commission were James Chalmers and Cynthia B. Flynn of Mountain West, Paul Slovic, Baruch Fischoff and Sarah Lichtenstein of Decision Research and Roger Kasperson of Clark University. This group's experience at Three Mile Island gave them the bona fides to bid on the socioeconomic studies that later were proposed in 1985 by the state of Nevada. It also shaped their view of nuclear energy.

**On Dec. 13, 1985, the Socioeconomic Study Selection Committee (comprised of seven local governments, the state Legislative Counsel Bureau and the NWPO) met in Las Vegas to review 13 proposals received in response to the Request For Proposal. As a result of the review, two proposals were selected for future consideration, and representatives of both of the firms which authored the top-ranked proposals were invited for formal interviews on Jan. 14, 1986.**

**Following an extensive interview process, Mountain West Research - Southwest, Inc. of Tempe, Arizona was selected as the manager and prime contractor for the Nevada socioeconomic impact study. [Nuclear Waste Newsletter, NWPO, spring 1986, p5]**

When the award of the contract to Mountain West was announced in the mid 1986 edition of the Nevada Nuclear Waste Newsletter, the article began with a quotation from Roger Kasperson, director of the Clark University Center for Environment, Technology and Development (CENTED) in Massachusetts. .

**Deciding whose 'backyard' shall be chosen [for disposing nuclear waste] raises questions of whether some people should bear the risks for others, of which people should bear the risks and if and how they should be compensated by the beneficiaries [of nuclear power], and of how accountability can be achieved in societal decision-making. Inadequate attention to these issues could produce highly inequitable policies. [ -- National Academy of Sciences Researcher, Roger E. Kasperson in "Equity Issues in Radioactive Waste Management (1983).**

The key here is "how accountability can be achieved in societal decision-making", a loaded phrase that has more to do with tinkering with America's social structure than the specifics of Yucca Mountain. Kasperson and a number of others involved at Yucca Mountain saw the socioeconomic studies conducted in Nevada in a much broader perspective than what is or is not good for Nevada's citizens. Instead, they found these studies to be a chance to impose their own concepts of societal decision-making on the larger national and even world community. The socioeconomic studies were meant to be a way of assessing the impact of Yucca Mountain on Nevada and its citizens, but instead became the mechanism for advocacy of the pet political theories of academics.

Roger Kasperson holds a PhD from the University of Chicago in Political Geography, and it is hard to conceive of a person more qualified to dissect the political workings of a state agency like the Nevada Nuclear Waste Project Office. Kasperson becomes a central figure behind the scenes in Nevada because his was the driving philosophical presence behind the state's anti-repository stand. Where before, the Nevada anti-nuclear movement had been a Hodge-podge of unsophisticated activists and test site protesters, now it had a skilled political geographer who knew how to bend the Nevada political landscape.

The interesting thing about the Nuclear Waste Newsletter beginning its announcement of the awarding of the Mountain West contract with a quote from Roger Kasperson is that it shows how deeply the professor was involved in the choice of the prime socioeconomic subcontractor. Even though Kasperson's group at Clark University, the Center for the Environment Technology and Development did not become the prime contract administrator, Mountain West appears to have been staffed by comparatively peripheral intellectuals compared to those from Clark University and those from another subcontractor called Decision Research. The papers produced for NWPO by the Mountain West consortium bear the imprint of Kasperson both philosophically and in terms of the sheer volume of output. Kasperson's name appears as co-author of nine papers, ranging from equity issues to post-closure risk analysis, a broad intellectual reach for a political geographer.

The reason the Clark University group is important is because they had a previously formulated master plan for America's nuclear future as far back as 1979:

**"Recognizing nuclear energy as a transitional energy source. This view will limit the role of nuclear energy to the period required to develop and deploy long-term renewable energy sources. It rules out fuel recycling and deployment of the breeder reactor, because a byproduct of this kind of reactor is plutonium, which can be used to produce bombs.**

**Limiting the total size of the commitment. No nuclear-power plants beyond those currently on order or under construction will be built. The open-ended total scale of the nuclear enterprise is a key ingredient in the nuclear debate and is not resolved by limiting the number of sites (as opposed to plants). Considered with (the above) item, this obviates the plutonium-economy anxiety." [Kasperson, Roger et.al.; Hohenemser, C.; "Institutional Responses to Different Perceptions of Risk"; Accident at Three Mile Island, Westview, 1979, p45]**

The fact that this is a call for dismantling America's nuclear energy capacity, written five years before the Nevada socioeconomic studies were started, implies Mountain West's contract with NWPO was perhaps incapable of producing a fair and neutral socioeconomic study that would help the state make informed choices. Because Yucca Mountain is the bottleneck in the nuclear fuel cycle, it afforded Mountain West academicians the opportunity to forward their own independent energy policy.

Also important to note is that the above quotation comes from an article titled "Institutional Responses to Different Perceptions of Risk". The key word here is perceptions, as in perceptions of risk versus actual risk. In conjunction with workers from the affiliated think tank called Decision Research of Eugene Oregon, the Mountain West consortium raised risk perceptions of Yucca Mountain to new heights while undermining the Nevada community's understanding of real risks. Rather than a benign academic argument, the exchange of risk perception for risk analysis became a means to politicize all technological decisions regarding Yucca Mountain.

When it came time for proposals to evaluate the socioeconomic impact of Yucca Mountain on Nevada, it is obvious that a consortium of researchers from the Kemeny Commission would hold an inside track, but it is not clear they were the best advocates for the State of Nevada's interests.

## MOUNTAIN WEST'S NEVADA IMPACT

The Mountain West consortium, whose kernel was composed of Mountain West, Decision Research, and the Clark University Center for Technology, Environment and Development, were never hired to do nuts and bolts socioeconomic impact studies. Instead, they were hired to do 'state-of-the-art' research, which in practice came to mean they were allowed extraordinary academic latitude. It is little surprise Mountain West came to be the proponents of a philosophical paradigm shift which incorporated their pet academic theories.

The Mountain West paradigm shift was an attempt, partly subconscious and partly conscious, to fuse a number of competing social philosophies which had reached dead ends into something much greater than the simple sum of those parts. Among the theories which seem to be coalescing are the following:
- Decentralization theory
- Liberation Theology
- Marxist class warfare rhetoric
- Rawlsian equity theory

- Deep Ecology
- American Indian Tribal Spiritualism

Mountain West, by winning the monopoly contract to do socioeconomic studies for Nevada, found itself in the position of acting as catalyst to fuse these divergent theories into a whole. They had a political geographer, Roger Kasperson of CENTED, whose academic philosophy already seemed molded by Marxist, decentralist, and liberation theology to act as the seed. However, Mountain West's philosophical influence was not enough by itself to create a revolution, their philosophy also needed to mesh with the indigenous movements, many of which were intimately involved in the peace protests at the Nevada Test Site.

Local environmental groups, Citizen Alert being the most prominent, had developed a populist message that was tinged by anti-military if not leftist thinking. Elements of liberation theology were present in the Test Site protests exemplified by the yearly Lenten Desert Experience and the organizing Nevada Desert Experience. Many of these test site protests were directed in part by Judy Treichel, later of the Nevada Nuclear Waste Task Force and contractor to NWPO, who had previously been part of the Lenten Desert Experience as well as the local affiliate of Clergy and Laity, a Catholic peace and justice movement. The Western Shoshone Indian tribe had long been embroiled in land conflicts over Yucca Mountain and the Test Site and their naturalistic philosophy, dovetailed with land politics, created a synergy with the other nuclear opposition groups.

Mountain West gave these competing movements coherence by providing a sophisticated philosophical center for anti-nuclear activities. Mountain West provided the link between old-line Marxist notions and the new eco-philosophies evolving at Yucca Mountain. Roger Kasperson as a political geographer and one of the principals of Mountain West, had since the early 70s contemplated a synthesis of Marxist, decentralist and liberation theology concepts. Others within the Mountain West web provided practical elements needed to implement these theories. Paul Slovic, a psychologist from Decision Research and Kristin Shrader-Frechette, a science philosopher from Florida are two who in particular helped provide the needed legal, moral and social mechanisms to solidify the new philosophy.

Slovic, as head of Decision Research, had long been associated with Mountain West, since before Three Mile Island. Slovic's theory of the 'availability heuristic', stating that people cannot accept risks of technologies to which they are not accustomed, becomes a critical link in eco-philosophy by claiming that 'fears of technology' should to be sufficient to veto technology. Thus, science is delinked from the acceptance of technology in favor of political manipulation by popular causes, such as the "ecology".

Shrader-Frechette came on the scene later, but her thoughts are so philosophically tied with Mountain West that they are inseparable. Shrader-Frechette's revival of the Doctrine of Informed Consent, which equates the designers of Yucca Mountain to Nazis war criminals doing mass experiments on unwitting human guinea pigs, was an attempt to provide a moral and legal basis for the evolving Yucca Mountain eco-philosophy.

As Mountain West's ideas were filtered through the Nuclear Waste Project Office, through Judy Treichel of the Nuclear Waste Task Force, through Chris Brown and Bob Fulkerson of Citizen Alert and through the leaders of the Western Shoshone tribe the sociologists became the de facto organizing force behind anti-nuclear protest in Nevada.

## JUDGING THE RESULTS

While Mountain West no doubt fulfilled Nevada's political needs by providing a philosophical framework for opposing the siting of a repository, it is not clear they fulfilled their obligation to provide a basis for negotiating for possible benefits should the repository be sited in the state. The state had developed a number of objectives stated during the original Request For Proposal:

## OVERALL STUDY OBJECTIVES
. . . . The study should:
1.   Provide factual bases for informed and scientifically and legally defensible decisions and to prepare the State of Nevada for subsequent negotiation, legislation, and litigation (if necessary).

2.   Identify the full range of which the State of Nevada, local governments or individual citizens might incur and therefore serve as the basis for impact planning, impact mitigation, and compensation of claims.

3.   Identify mechanisms and strategies for obtaining mitigation and compensation so as to enable the State to obtain full and timely mitigation of impacts, and compensation where impacts are not mitigable.

4.   Assist the State in minimizing or avoiding social and economic dislocations caused by site investigation, characterization, construction, operation, retrieval, closure or decommissioning relative to repository at Yucca Mountain.

5.   Enable state and local governments to minimize or avoid social and economic costs incurred by state and local governments or private citizens as a result of repository-related activities.

6.   Enable state and local governments to maximize benefits from repository development.

7.   Provide the basis and framework for allocating costs and benefits equitably through the mitigation process.

8.   Enable state and local governments to minimize risks and consequences of possible repository-related accidents. [NWPO, Request For Proposal For A Socioeconomic Impact Assessment And Mitigation Study, 1985]

Since Mountain West's general philosophy was Rawlsian egalitarianism, they focused on the equity issue at Yucca Mountain and unsurprisingly found that the siting process wasn't fair! Instead of attempting to fulfill their original mandate to provide mitigation and negotiation options to the State and local governments to relieve these inequities, they instead proposed the state resist all attempts to negotiate, but without providing a fall-back position. Thus, Mountain West's emphasis on risk perception analysis rather than economic analysis in effect locked the state into a victimization mode.

That doesn't mean Mountain West itself didn't profit from its study of Yucca Mountain, and in fact it profited handsomely from its work. There were numerous complaints that a real estate database developed by the group was unavailable for use by Nevadans, yet Mountain West bragged that this database had increased its viability as an economic development consulting entity:

Mountain West's involvement in Las Vegas has included both real estate development analysis for regional clients and a major project for the state of Nevada to study siting and transportation impacts of hazardous waste. Through this work, Mountain West has compiled a comprehensive data base on the Nevada market, which will prove valuable to clients interested in this fast growth region. [Mountain West: The Source: For Decisionmakers; Special Edition, 4th Quarter 1988]

In other words, the database Mountain West developed for NWPO was perhaps more useful to promoting the economic interests of Mountain West than the interests of Nevada.  The 1986 Nuclear Waste Newsletter announcing the selection of Mountain West made one final comment:

"The Nuclear Waste Project Office Staff is confident that the Nevada socioeconomic impact assessment project will be a state-of-the-art study -- one that will attract national attention."

It is quite certain that that national attention was warranted, especially as we uncover what state-of-the-art really means in terms of replacing hard science with political advocacy.

# 15. Marx, Mao & Paolo Friere

The one person who epitomizes the paradigm shift that evolved at Yucca Mountain is Roger Kasperson, a political geographer from Clark University in Worcester, Massachusetts. As the director of the Center for Environment, Technology, and Development (CENTED) and a sub-contractor to Mountain West, Kasperson became the philosophical center of NWPO's socioeconomic studies.

The numerous papers co-authored by Kasperson for NWPO, interweave three main areas:

1) Equity Issues;

2) Human Reliability Risk Analysis; And

3) Comparative Nuclear Waste Management.

The overall conclusion of these papers is that Yucca Mountain is inequitable to Nevadans who would shoulder greater risks than the rest of the nation and that the technologists creating the site cannot be trusted to do zero risk engineering. As an alternative, Kasperson presents the example of Sweden's consensus building approaches on their nuclear issues.

The roots of Kasperson's analysis of Yucca Mountain, however, are much more complex than an academic interest in mitigating the dangers of nuclear waste. Kasperson, a political geographer, had theorized since the early 70s how to conduct a social revolution using 'non-traditional' social advocacy. Earlier in his career he had had come to view technologists as 'elites' who were incapable of responding to the needs of citizens.

**Taken in toto elitist theory is an apology for the failures of modern democratic states. The alacrity with which contemporary political scientists have rushed to accept, indeed to enthrone, a passive role for the citizen is remarkable. The attempt to take account of empirical findings has led to a change in the principle operating values of democracy. The conception of man presented is beastly rather than noble. It offers only a gloomy prospect for the future of man. The advocacy of a politics of leadership and expertise is exceeded only by the singular absence of any basic commitment to social justice. [Kasperson, Roger and Breitbart, Myrna; Participation, Decentralization And Advocacy Planning, Association of American Geographers, 1974, p.12]**

Translated, Kasperson believed "beastly" elitist technocrats offer only a "gloomy prospect for the future of man" and are without a "basic commitment to social justice." It is little wonder that a decade after this paper was written, the State of Nevada and NWPO, influenced by Kasperson's thinking, found themselves continually bashing the scientists and technologists involved at Yucca Mountain as elitists lacking social concern for Nevada. Thus it appears it is not the particular DOE and subcontractor managers of Yucca Mountain alone that were being indicted by NWPO as untrustworthy, but the entire nebulous class of all technological elitists. Kasperson had concluded *a priori* that technologists are untrustworthy, long before he ever became involved in the nuclear debate! If elitist technocrats are hopelessly corrupt, Kasperson offered an alternative:

## THE MARXISTS

**If a notion of social justice is absent from elitist theory, it pervades Marxist conceptions of participation. For Marx, man is the measure of things. He protests a social and economic order which cripples man and prevents the full development of his capabilities. Such a full development in capabilities requires that man be productive, healthy, and genuinely interested in the world around him. But man cannot be productive when the product of his labor is "an alien object. . . ." [Participation, Decentralization And Advocacy Planning, p12]**

The above quotations, taken from a monograph titled Participation, Decentralization, And Advocacy Planning, written in 1974 by Roger Kasperson with Myrna Breitbart of Clark University, parallel what has later become in large part the philosophy of the Nevada Nuclear Waste Project Office.

Kasperson evidently believed that "elitist" scientists and technocrats are incapable of social justice. Marxism, in contrast, treats man with a sense of "social justice". What this implies is that NWPO, under Kasperson's philosophical tutelage, may have come to believe Yucca Mountain will never pass muster, no matter what scientific evidence is presented! Marxist dogma would not allow it, otherwise NWPO would have to admit that bureaucrats from the centralized Department of Energy and its private technocrat contractors may sometimes not only tell the truth, but may also care about the social outcome of their endeavors.

Kasperson's theory is part of a tradition of political philosophy holding little in common with environmentalism. Consequently, environmental sensitivity appears to have been adopted later as a front for broader political goals. Translated twenty years later, NWPO seems to have evolved towards Kasperson's viewpoint in opposition to Yucca Mountain, simply because opposing the technocratic bureaucracy is a fashionable political statement. According to Kasperson:

**Unlike Lenin, Mao Tse-Tung feared the professionalization of the leadership. His writings reveal a consistently populist orientation to participation. The peasants, in his view, are "clear-sighted" concerning those who have authority over them. Leadership involves not only knowing the will of the masses but also their active participation and support. In his famous 1943 statement, he held that "all correct leadership is necessarily from the masses to the masses . . ." The ideas of the masses are converted into concentrated and systematic ideas, brought back to the masses for testing, and then put into action. In the process, these ideas become "more correct, more vital, and richer" each time. For Mao, the participation of the masses created revolutionary energy and spirit. Like Trotsky, he was very suspicious of any leadership which became at all removed from its revolutionary base. The cultural revolution in 1966 was his response to the growth of the bureaucratic elitism that Lenin so desired. He unleashed mass participation in an effort to purify a social and cultural superstructure which had grown apart from the masses of the peasantry. [Participation, Decentralization And Advocacy Planning, p 14]**

Roger Kasperson's attempt to kindle a "cultural revolution" in Nevada and America through his hold over NWPO is hugely troubling. Mao Tse Tung's Cultural Revolution was catastrophic for the Chinese and filled with barbarous bloodshed. While a cultural revolution can no doubt be a "response to the growth of bureaucratic elitism" and "purify a social and cultural superstructure which (has) grown apart from the masses of the peasantry", unleashing such forces on the scientists working at Yucca Mountain and the citizens of Nevada is extreme.

Obviously, the Kasperson article of 1974 is not an environmental tract, but a work promoting radical social and political advocacy without reference to the environment. Written before Yucca Mountain was even proposed, Kasperson was neither an environmentalist nor risk analyst at that time, but a political geographer intent on transforming society. What this shows is that the State of Nevada's chief philosopher already viewed technologists, such as the Department of Energy, as elitists who lack a sense of "social justice" long before he ever came in contact with the Yucca Mountain project. It is little wonder then that the DOE has been consistently condemned by NWPO over the years as untrustworthy and corrupt. After all, judgment had been passed by Kasperson based on Marxist tracts twelve years before Nevada's socioeconomic studies supposedly discovered this new fact.

Is Kasperson a Marxist? It is clear he was at one point and had well defined ideas about taking over the system through the use of political advocacy, though his political center has shifted since that time. If Kasperson were just an old-line Marxist we could end our story here, but what makes his history unique is the evolution which occurs over the next years until Yucca Mountain. Kasperson's political evolution parallels a transformation of Green political theory from defunct Marxist philosophy to a new paradigm.

The critical transitional concept here is the word decentralization, which in more boisterous times was called "Power To The People". Decentralization, broadly a shift to democratic community action, is the dogma used to radicalize the debate over Yucca Mountain. However, decentralization has a far deeper meaning than merely distributing power to the people. Kasperson, the political geographer, was well aware of this even in 1974:

**"A third general objective is more self-consciously radical than the preceding ones. Some proponents of decentralization believe it to be a means for creating territorial power bases from which to contest established elite power. This notion has its roots, of course, in the revolutionary experience of Communist China and the "city and country" model of world revolution. Decentralization, in this view, is a vehicle for**

political mobilization, a strategy for creative countervailing power, a force for democratization. Milton Kotler, for example, argues (1969, p 100):

The radical task which the left must undertake if it is to assist democratic revolution is to challenge the very concept of nationalism and central control. To be radical, intellectual and physical force must accord with the original phenomena of revolution - local insurrection and local control - within which structure democracy can flourish. The left must develop and articulate a theory of local sovereignty." [Participation, Decentralization And Advocacy Planning, p 35]

Nevadans may be surprised to find that researchers at the philosophical center of the Nevada Nuclear Waste Project Office saw the debate over Yucca Mountain as a revolutionary opportunity, a chance for "creating territorial power bases from which to contest established power elite". Nevadans may want to review whether they wish to go down a path that "has its roots, of course, in the revolutionary experience of Communist China and the 'city and country' model of world revolution" and whether this is the mandate they envision for their Nuclear Waste Project Office.

Even more revealing in a historical sense is that we see juxtaposed here two divergent philosophies: Marxism and Decentralization. It appears that in the 1990s, these two worldviews began to coalesce and solidify into what is sometimes known as the Deep Ecology movement or Green Revolution. However, in 1974, a synthesis of these philosophies was only a glimmer in the eyes of elite progressive philosophers. Roger Kasperson was in this respect ahead of his time, as a political geographer and must have been one of the first to realize that Maoist Marxism and anarchist decentralization philosophies were capable of being grafted one onto the other. Kasperson already knew quite well when he came to Nevada in 1985 how such theories could cause a firestorm:

So why all the sound and fury? Most parties, the present author would contend, have seen the push for participation for what it is - a potentially effective means for redistributing political power, for eradicating the biases that keep the periphery on the periphery, for combating social injustice. That is why participation must be kept under careful control. It is also the reason that even limited activation of the poor elicits such a crushing response. . . [Participation, Decentralization And Advocacy Planning, p 25]

Certainly it will be argued that Kasperson's political philosophy changed since 1974 and his presence in the socioeconomic studies of the Nevada Nuclear Waste Project Office were innocuous, the result of random academic pinballs falling through the pins. Nevertheless, Mountain West was chosen as prime contractor in 1985, only eleven years after the writings we quote and at a time when Marxism was not yet considered dead. Kasperson is also a political geographer, for whom politics has special meaning and it is unlikely that all his revolutionary fire had been extinguished when he and similarly minded colleagues from Clark University appeared on the Yucca Mountain scene in 1985.

If it were just Kasperson who at one time espoused a radical revolutionary theory one would sleep better, but a number of other actors involved in the Yucca Mountain debate also held and continue to hold similar views. This is not necessarily because they are sophisticated academicians of political geography whose hobbies are Marxist / Maoist / Decentralization theory, but because Yucca Mountain is a gravitational attractor that draws a number of similar synergistic revolutionary viewpoints towards its center.

Does it matter whether Kasperson is a Marxist? It does when he was hired to do objective scientific work but his political ideology precludes that objectivity. The state of Nevada would not be affected one iota if Kasperson were a Marxist, or Martian for that matter, if it could trust his scientific work. But Kasperson apparently doesn't feel obligated to uphold objective scientific standards:

". . . . Traditional planners maintain a faith in the ability of the present political system to accommodate nearly all interests in society. Essentially, they accept the assumption of equal political opportunity. This process enables the planner to act as a neutral coordinator supplying individuals with information on particular planning issues and suggesting the technical solutions to problems which would best serve general public needs.

Advocates see politics in a different light. In place of consensus, the political process becomes a mélange of competing interests vying for power and influence. This process recognizes that many groups are better equipped than others to assert and maintain their interests. Advocates feel, therefore, that it is their own responsibility to compensate for their unequal distribution of political opportunity by guaranteeing everyone in society at least the chance for a meaningful expression of position on particular planning issues. In short, advocates refuse to approach planning problems as issues requiring a choice between various technical solutions. Planners cannot hope to perform their roles as problem-solvers in an aura of relative scientific objectivity - to ignore the political implications and personal values inherent in their work." [Participation, Decentralization And Advocacy Planning, p43]

Kasperson appears to have found his own advocacy niche in Nevada's Nuclear Waste Project Office as a socioeconomic consultant. Unfortunately, when Kasperson says, "Planners cannot hope to perform their roles as problem solvers in an aura of relative scientific objectivity," he calls into question the integrity of his research at Yucca Mountain, as well as the credentials of CENTED and many of the other socioeconomic consultants working for NWPO under the Mountain West banner. One cannot be certain whether their research is science, or whether it is pure advocacy.

Neutrality is something a traditional planner is stuck with, such as the professionals at DOE (who are continually challenged on their trustworthiness by NWPO). But Kasperson apparently is not a traditional planner; he's an *Advocate* who feels it is his responsibility to compensate for an unequal distribution of political opportunity. If, according to Kasperson, scientific integrity must take a back seat to the needs of political advocacy, then science as a whole is lost in a "mélange of competing interests vying for power."

It is the perversion of the Yucca Mountain scientific evaluation process to forward the political agenda of a professor of political geography that makes the participation of Kasperson, CENTED (which he directed alongside similarly opinioned faculty) and Mountain West (which he helped build) suspicious, not whether Roger Kasperson is left wing or right wing. Thus, the political agitation that occurred in Nevada may owe its hysterical nature more to Kasperson's Marxist paranoia of technological elites than to legitimate issues of risk at Yucca Mountain:

**"Going beyond advocacy suggests that we cultivate a radical self-awareness and confidence in the minds of presently powerless citizens so that they may begin to advocate meaningfully for themselves.**

**"These ideas are reflected in the writings of several well-known humanist philosophers and educators (Mills: 1959; Fromm: 1968; Fromm: 1961; Illich: 1970; Friere: 1970; Marcuse: 1964). . . . Within the radical humanist tradition . . . Marx was one of the first to protest man's estrangement and transformation into a "dependent" being with the advent of Western industrialism. Marx did not believe that man's most basic desire was for material well-being. Rather, he stressed man's basic morality and the corruption of this morality by the industrial system (mode of production) and existing social structure." [Participation, Decentralization And Advocacy Planning, p 50]**

Kasperson had devised an anti-technological advocacy strategy for co-opting the political process long before Yucca Mountain was even an issue. The goal of this strategy was never to rationally judge technological projects on their merits, because the Marxist theory Kasperson espoused claims technologists are immoral elitists. Instead, this is a social organization theory, which only coincidentally found nuclear technology a convenient scapegoat to promote its goals.

## REPACKAGING MARXISM

Of course, it is doubtful Kasperson or anyone involved in the Yucca Mountain opposition will admit to being a Marxist; Marxism is no longer in fashion even with the die-hard intelligencia of the left. However, the collapse of Marxism in the 1980's was more than the passing of a philosophical fad, it was the disintegration of a quasi-religious movement, the collapse of a moral structure.

Marxism is a beguiling moral theory, a conviction that massive redistribution of wealth and power from the elitist "haves" to the common man is a noble mission. It is also an abysmal failure as is now obvious to everyone. To fill the moral gap left by the collapse of Marxist theory, it was necessary to find a new religion that didn't completely negate the old leftist world view that massive redistribution of wealth and power was necessary for the salvation of mankind. That new religion is environmentalism, but a brand of environmentalism far removed from that of Teddy Roosevelt.

This new brand is apocalyptic environmentalism. Only if the environment is at risk of total collapse from some demonic force (conveniently provided by nuclear elitists at Yucca Mountain) could the redistributionist world view be vindicated. Thus, Marxist ideologues transformed their tattered ideology into a theology of environmental resource allocation instead of capital allocation, and the granddaddy resource of all is energy. Roger Kasperson provided the transitional theory to bridge the gap between Marxist class-warfare and the new environmental religion:

**More than an advocacy approach to planning must be employed in order to create a society based upon principles of justice and equality. . . . Paolo Friere, an exiled Brazilian educator who has managed to escape his own impoverishment and acquire a lucid understanding of its origins and elements, has developed a theory for the education of illiterate people. The crux of the theory is the notion of conscientizacao - learning to perceive social, political and economic contradictions in one's life, and to take action against the "oppressive elements" of that reality. It is Friere's belief that every human being, no matter how uneducated or impoverished, is capable of looking critically at his world in a "dialogical" encounter with others. If provided with the proper incentive and tools, Friere believes that man can begin to perceive his personal reality, its problems and its contradictions. This perception is the first requisite for changing that reality. [Participation, Decentralization And Advocacy Planning, p50]**

Paolo Friere is a key philosopher in the creation of what is called Liberation Theology. A combination of Marxist theory and Brazilian Catholic peace and justice and land reform activism, Liberation Theology finds expression at Yucca Mountain in the form of the Nevada Desert Experience (NDE), key to Nevada Test Site protest. One of NDE's founders, the notorious Judy Treichel, is also founder of the local Clergy and Laity Concerned, another peace and justice movement with Catholic and Liberation Theology roots. Conveniently, Treichel also founded the 'educational' Nevada Nuclear Waste Task Force.

While Paolo Friere is a key transitional philosophy, Kasperson and other socioeconomic researchers for NWPO eventually settled on a secular philosopher, John Rawls, on which to base their rejection of Yucca Mountain. Decentralization, Paolo Friere's liberation theology and Rawls' egalitarianism are all steps in the redefinition of Marxist concepts in palatable forms. Thus, there has been a transformation of the terminology over the last fifteen years so that now there is discussion of 'decentralization' instead of 'power to the people'. We have 'equity issues' rather than 'local insurrection and local control' and 'advocacy' rather than 'revolutionizing', but these terms are meant to obscure meaning, not enlighten.

Consequently, a complex evolution occurred at Yucca Mountain, from old-line Marxism to new branches such as decentralization, liberation theology and eventually Rawlsian ethics. Tying these threads together is Roger Kasperson, who as a sophisticated political geographer knew full well how to merge these different philosophical lines into a synergistic whole.

Sadly, Kasperson's belief that advocates can do no wrong correcting the supposed inequities of an immoral industrial system corrupted NWPO's socioeconomic studies beyond repair. The opportunity for experimenting with non-traditional advocacy was provided by the Nevada Nuclear Waste Project Office. The 'state-of-the-art' socioeconomic studies and polling conducted by the Mountain West consortium for NWPO was the means for educating the people about the 'oppression elements' in their lives (i.e., DOE and the Yucca Mountain project) a la Paolo Friere. The blueprint for the entire experiment was provided in 1974 by Kasperson's and Breitbart's *Participation, Decentralization and Advocacy Planning*. The funding for this grand experiment in social manipulation was in the amount of $15 million dollars, provided by the nuclear industry.

# 16. Decentralization, Anarchy & Green Revolution

As a political geographer, Kasperson knew as early as 1974 the radical implications of decentralization theory, as starkly evidenced by his paper *Participation, Decentralization and Advocacy Planning* cited earlier. It is useful to quote again Kasperson's view of the power of Decentralization:

**"A third goal of decentralization is the creation of territorial power bases from which to challenge elites and centralized power. Decentralization . . . certainly has this potential. But let there be no illusions as to the ease with which this will occur. The determined resistance to community action programs and school decentralization efforts suggest the likely response. Any real progress will create the "multiple abrasion points" which some decentralization critics fear. The desirability of such a development depends on one's ideology of political change; for the Marxist, decentralization holds the promise of territorial power bases in a protracted struggle; for the pluralist, it threatens the more somber prospect of urban enclaves of neglect. "[Kasperson, Breitbart; Participation, Decentralization And Advocacy Planning, p 38]**

As a revolutionary manifesto, the paper serves as a plan for taking over the political process by the manipulation of the tensions that exist in democratic institutions. It should be noted that Kasperson had spent part of the 60's dissecting the decision matrix of aldermen and Chicago's ward politics and so knew quite well how to apply such theories when eventually hired by Nevada's Nuclear Waste Project Office.

The first symptoms of decentralist philosophy applied to the nuclear issue appear in the rise of protest group like the Clamshell Alliance in the late 1970s.

**The Clamshell Alliance, which mounted mass demonstrations of civil disobedience at the construction site of the Seabrook (nuclear) power station in New Hampshire in 1976 and 1977, represented something quite different from either Critical Mass or the Union of Concerned Scientists. Inspired by the successful site occupation that German protesters staged at Wyhl in 1975, the Clams established a style of direct action that was promptly taken up by other local groups - the Catfish, Palmetto, and Abalone Alliances, among others - who during a single year held more than 120 demonstrations and rallies. The Clams and their imitators drew their members from the "radical, counter-culture, and peace movement subcultures." They sought to stop nuclear power not only because they regarded it as unsafe, but because many sought "to change American society fundamentally in the direction of decentralization, demilitarization, and egalitarianism," believing that a necessary condition for change was "a shift from centralized, capital intensive power producing systems, of which nuclear power is the epitome, to decentralized, labor-intensive, small-scale, renewable energy systems." [Carter, Luther; Nuclear Imperatives And Public Trust, Resources for the Future, 1987, p82]**

The fact that decentralization is more than a buzzword and is in fact a code word for a relatively new political formulation is apparent. The anti-nuclear movement at Seabrook evidently was already being driven by this philosophy in 1977, and it is no surprise to find its resurgence at Yucca Mountain ten years later. Decentralization's connection to the anti-nuclear movement has in part been driven by its synergy with solar utopianism. We quote from "*A Critique of the Solar Movement*" written by Ken Bossong in 1980:

**Early supporters of solar energy were attracted to the technology because it seemed capable of promoting individual and community self-reliance as well as decentralization of energy production and control. Solar technologies appeared to be the vehicles for redistributing income and thereby benefiting low-income citizens as well as other Americans. Solar was perceived as an environmentally benign technology that could lessen or eliminate dependence upon dangerous coal and nuclear technologies as well as imported oil and gas supplies.**

**However, in the half-decade since the 1973-74 Arab Oil embargo, little or no progress towards these goals of the "solar movement" has been realized. In fact, many of the political goals that had once prompted**

the pioneers of solar technologies are being lost sight of. Too many citizens groups, individual activists, small businesses, and others have come to see solar commercialization as simply a marketing problem. Consequently, the so-called solar transition is occurring in a manner that promises a continuation of the economic, environmental and political shortcomings of other energy technologies. [Bossong, Ken; A Solar Critique, "A Critique of the Solar Movement", Citizen's Energy Project, 1980, p1]

At that time, Bossong was part of Citizens' Energy Project that later became the anti-nuclear Safe Energy Communication Council. Bossong later moved to Public Citizens Critical Mass Energy Project/SUN DAY. These are Ralph Nader organizations which played a critical role in opposition to Yucca Mountain, providing direction from inside the Washington Beltway. As late as 2007 Bossong was promoting the Sustainable Energy Network, an alliance of environmental groups working for an extremely radical energy policy:

SUSTAINABLE ENERGY BLUEPRINT
A PLAUSIBLE STRATEGY FOR ACHIEVING A NO-NUCLEAR, LOW-CARBON, HIGHLY-EFFICIENT AND SUSTAINABLE ENERGY FUTURE.
The following statement outlines an ambitious but doable strategy for dramatically reducing U.S. greenhouse gas emissions, phasing out nuclear power, and ending energy imports while simultaneously creating new domestic jobs and businesses, improving energy, homeland, and national security and the economy, and enhancing the environment and public health.
Objectives:
The three primary, longer-term objectives for the nation's energy policy should be:
1.) reduce greenhouse gas emissions to a level consistent with a world-wide goal of global climate stabilization (assumes curbing U.S. $CO_2$ emissions by 60-80% from current levels by mid-century);
2.) eliminate U.S. energy imports (i.e., oil and natural gas – now 58% and 15% respectively), while reducing overall use of oil and natural gas;
3.) phase out the current generation of nuclear power while substantially curbing the production and consumption of fossil fuels, by increasing the use of energy efficiency and making a transition to sustainable, environmentally safer renewable energy sources.

Rather than a sustainable energy blueprint, this is more a starvation blueprint, so one suspects a political agenda based more on decentralist Utopian ideas than science. Decentralism early became part of the opposition to nuclear technology in the southwest U.S. in part through the Cactus Alliance:

We actively support the alternatives of strict conservation practices, the redirection of technology to meet human needs, and the full development of alternative energy sources along with decentralization of energy systems. . . [Gyorgy, Anna; No Nukes, Everyone's Guide To Nuclear Power, South End Press, 1983, p 444]

The Cactus Alliance had connections to Nevada's Sagebrush Alliance in opposition to the MX missile system in the early 80's and served as an organizing template for some of the protest against Yucca Mountain. It is apparent that among the core leadership of the anti-nuclear movement decentralization was often perceived as the cure-all for society's ills, often overwhelming concerns they may have had for the health effects of radiation. The emphasis appears to be on a class warfare struggle against centralized government and big business, in which DOE and the nuclear industry become handy whipping posts.

In an insightful book by Mildred J Loomis of the School of Living titled *Decentralism; Where It Came From, Where It is Going*, we find some of the popular roots of the decentralist movement.

. . .dependency, delinquency, disease, degeneracy and decadence. That these five D's are ever present and on the increase in modern society calls for a Fourth Revolution - a decentralist revolution.

Decentralization is not a turning back of the clock. Through decentralization, independence would replace dependency; honesty and justice would replace delinquency. Health would prevent disease and degeneracy; creative work and folk art would replace decadent and inhuman activities.

**For these desired ends, Decentralization would organize production, control, ownership, government, communications, education, and population in smaller, more human units. [Loomis, Mildred J.; Decentralism; Where It Came From; Where Is It Going, The School of Living Press, 1980, p28]**

Decentralism is obviously a Utopian philosophy, perhaps over-confidently viewed by Loomis as a cure-all for society's ills. But where did this philosophy come from? Loomis continues:

**From about 1790 to 1930, America produced a group who believed that all human activities and all organizations should be voluntary - that even defense need not be governmental and coercive. They worked hard to free the economy of monopoly and exploitation in order that crime would be reduced, and the need for defense would fall to a minimum.**

**Persons holding these beliefs and practices sometimes call themselves "individual anarchists". Examining the root meaning of "anarchy", we find that "an" means no or none, "archy" means rulership. Thus anarchy means no rulership or enforced authority. Anarchy does not mean chaos and disorder. . . . [Loomis; Decentralism, p29]**

Not all anarchist thought is created equally, however. Libertarians such as Milton Friedman and Murray Rothbard owe some of their philosophical roots to anarchist theory, however the decentralism movement and the anarchism espoused by the growing environmental movement have characteristics that are perhaps more totalitarian than libertarian in nature. Anarchy as espoused by the Green movement can very well be synonymous with chaos because it lacks the check and balance of property rights fundamental to Libertarian beliefs. Enviro-anarchism does not believe mankind has any special claim over Mother Nature, in essence stripping humans of property rights in favor of the rights of lesser organisms.

The School of Living in turn influenced the founding of Rodale Press which helped spawn much of the back-to-basics organic counter-culture known as the "Green Revolution". The linkage between decentralism, anarchism and Green politics is further explained by Loomis:

**. . .That confederation would be a new movement. It needs a name. What should it be?**
**Suggestions came. "The School of Living Movement". "Decentralization". "The New Age".**
**A sturdy peasant-like friend stood up. "I'm Peter Maurin of The Catholic Worker, just over from France.". . .**
**"My people love life and the land. In every country, there are those who do. The only hope I see for the world is in the spirit and works like School of Living. In France, we call it "The Green Revolution." . . .**
**The term found acceptance. Some used it in Free America, in The Christian Century, The Catholic Worker, and of course, in School of Living's Interpreter. From that beginning in 1940, the "Green Revolution" was our term for the decentralized, organic culture we worked for . . . [Loomis, Decentralism, p 78]**

Of course, Kasperson and the State of Nevada Nuclear Waste Project Office argued that decentralization and equity are merely benign words meant to indicate their desire to protect the local population from being ramrodded and controlled by big-government and large corporate entities (like the nuclear power industry). Kasperson and his contingent from Clark University obviously knew differently; as political geographers they had a professional understanding of the meaning of advocacy and decentralization and promoted these theories as political philosophy and not as grass-roots community involvement.

Some adherents of decentralist anarchism imply this is merely a form of democratic action, involving the masses in an expression of self rule over the environment. For example, at Yucca Mountain the claim is made that the citizens of Nevada have in essence voted (actually, they have only been polled) and have declared their opposition to the site and that this should be enough to veto the project. What is actually being advocated is a social system that has much in common with anarchist mob rule, using Yucca Mountain a side show to this debate. From Aristotle to D'Toqueville, political philosophers have taken pains to point out the dangers inherent in purely 'democratic' system. Perhaps most poignant are the comments of Aristotle, demonstrating that the dangers of decentralist thinking were well recognized two thousand years ago:

**A fifth form of democracy . . . is that in which, not the law, but the multitude have the supreme power, and supersede the law by their decrees. This is a state of affairs brought about by the demagogues. For in**

democracies which are subject to the law the best citizens hold the first place, and there are no demagogues; but where the laws are not supreme, there demagogues spring up. For the people become a monarch and is many in one; and the many have the power in their hands, not as individuals, but collectively. . . . . At all events, this sort of democracy, which is now a monarch, and no longer under the control of the law, seeks to exercise monarchical sway, and grows to a despot; the flatterer is held in honor; this sort of democracy being relative to other democracies what tyranny is to other forms of monarchy." [Aristotle, Politics, Book IV Chapter 4]

Although the Green Revolution occurring at Yucca Mountain has some roots in Marxist class struggle, the movement has taken on a whole new dimension which stems from the entangling of Marxist and anarchist roots. Maoist revolutionary theory, Liberation Theology land reform, and myriad other Green philosophies appear to merge under the unifying perspective of decentralist/anarchist theory.

Decentralist philosophy evolves further at Yucca Mountain when reformulated as equity issues based on a theory called Rawlsian Ethics, as in Roger Kasperson's book *Equity Issues In Radioactive Waste Management*. This book was a core reason Mountain West, Decision Research and CENTED were selected by the Nuclear Waste Project Office to do Nevada's socioeconomic impact studies. In essence, NWPO became the prime disseminator of decentralist / anarchist thinking in America by selecting Mountain West.

# 17. Equity & Rawlsian Ethics

The term "equity" comes up often in relation to Yucca Mountain – we all want an 'equitable' solution. However, for the socioeconomic researchers who consulted NWPO, equity has special meaning that is not as benign as the word might seem. It is the linkage of the equity with a philosophy called Rawlsian Ethics that has consequences far beyond the simple meaning of that word.

After the Kemeny Commission released its report on the accident at Three Mile Island, the socioeconomic researchers who later became the core of Mountain West tackled other subjects related to nuclear waste management. Roger Kasperson edited a work titled "Equity Issues in Radioactive Waste Management", which showed the continuing evolution of these workers towards an egalitarian framework for analyzing such issues. Running behind the equity issue, however, was a new philosophy that had only recently appeared in academic circles.

John Rawls in his seminal work, *A Theory of Justice,* hypothesized two principles which he felt formed the basis of a new global political philosophy:

**First: each person is to have an equal right to the most extensive basic liberty compatible with a similar liberty for others.**

**Second: social and economic inequalities are to be arranged so that they are both (a) reasonably expected to be to everyone's advantage, and (b) attached to positions and offices open to all.**

**[John Rawls, A Theory Of Justice, Harvard University Press, 1971, p60]**

These seemingly innocuous lines have touched off a thunderstorm of academic debate since they were penned in 1971, but only recently has this theory begun to seep down towards the common culture. The application of these principles at Yucca Mountain by political consultants to NWPO proved to be a bold social experiment. Rawlsian ethics, just over forty years old, is relatively new and untested except as an abstract academic philosophy. Nevada evidently became an unwitting guinea pig on which to test this theory.

Fears of nuclear energy and of Yucca Mountain have no legal weight unless there exists some fundamental moral right that places these fears above other considerations. For example, most people fear taxes, but this fear is not alone sufficient to defend oneself from the IRS in court. Similarly, a fear of Yucca Mountain alone is not adequate to legally halt its construction. Rawlsian Ethics, however, in essence assumes that each individual has veto power over the actions of his fellow man, for "social and economic inequalities are to be arranged so that they are . . . reasonably expected to be to everyone's advantage . . ."

This is also, of course, the definition of decentralization, the political theory. Rawlsian ethics requires power to be dispersed, i.e. decentralized to the least common denominator, the least powerful individual, based as much on social as economic inequities. Consequently, equity issues, decentralization and Rawlsian ethics are all sister philosophies. Converting the discussion of Yucca Mountain from a question of whether a centralized nuclear waste repository makes technical sense to an "equity issue" (i.e. whether it decentralizes risk in a Rawlsian sense) is thus a critical question.

Obviously Yucca Mountain is not fair, it does not evenly or equitably distribute risks or costs among America's population. If America were a pure democracy, we could have a vote on the issue and decide on the basis of our collective fears whether Yucca Mountain (or even such rights as free speech and the pursuit of liberty) should continue. In Nevada many people believe it is their inalienable right to be able to vote whether to accept or reject Yucca Mountain.

However, America is not a pure democracy but a Republic, with a democratically elected representative government. This is more than mere semantics, political philosophers and our founding fathers well understood that pure democracies are chaotic and self destructive. We are a nation first ruled by law and then by democratic processes, not the other way around. Equity as envisioned in both Rawlsian and decentralist theories would reverse our legal priorities, creating unfettered democracy in which small minorities are given veto rights over majoritarian and institutional decisions.

A critique of Rawls by Robert Paul Wolff parallels the criticism of a number of other authors regarding Rawls Theory of Justice:

**The . . . . question we must ask, following the guideline of Rawls' principles, is whether there are other, entirely different sets of economic arrangements that, while serving the fundamental purposes of production and reproduction of goods and services, will generate inequality surpluses, the feasible distribution of which would raise the expectations of the least advantaged representative man above the level that can be achieved under our present arrangements by redistributing the existing inequality surplus. I apologize to the reader for the complexity of that sentence, but the question Rawl's theory requires us to ask is very complicatedly hypothetical. I can think of a number of ways to organize the growing of wheat or the assembly of automobiles -both of which, however, are merely small-scale or micro-examples from Rawls point of view. But when I try to form a usable notion of alternatives to our present total set of economic arrangements, my mind can do no better than to rehearse the arrangements that have actually existed in some society or other - feudalism, slave-labor farming, hunting and gathering, state capitalism, collectivist socialism, and so forth. After eliminating those arrangements (slavery, for example) that violate portions of Rawl's principles other than the difference principle, I am to imagine how each of the remaining candidates would work out under the conditions of technology, resource availability, and actual or potential labor skill level obtaining in America today. Then I must gauge the size (if any) of the inequality surplus thus generated, and estimate the effect of the most favorable feasible redistribution on the reasonable expectations of the least advantaged representative man. Finally, Rawls tells me to order all the alternative sets of arrangements under consideration according to the magnitude of the expectations of the least advantaged. I now presumably know which alternatives are most just, and which less just, than present-day America, and the last step is simply to shift to the number one candidate on the list.**

**The manifest vagueness of these calculations and estimations has a very important consequence for Rawl's theory. Inevitably, one finds oneself construing the difference principle as a pure distribution principle. One simply stops asking how the goods to be distributed actually come to be in existence. . . . . [Wolff, Robert Paul; Understanding Rawls, Princeton University Press, p199]**

The reason it is necessary to understand Rawlsian ethics is because Rawlsian principles are being applied to all of us as unknowing guinea pigs, not only at Yucca Mountain but in the national political arena as well. The impact of Rawlsian ethics on the Yucca Mountain issue is substantial:

1) The fixation of the state's socioeconomic researchers on *risk perception* rather than risk analysis is a Rawlsian perspective which assumes all actions should be determined by the least advantaged members of society, (generally those who perceived risks have made most fearful).

2) The refusal by the State of Nevada to accept benefits makes sense only in relation to Rawlsian ethics in which the good of the least individual takes precedent over the utilitarian good of the whole.

3) The equity issue posed by Kasperson, Schlesinger, et al

4) The justification for the takeover of NWPO by academicians needed to determine "the best arrangement of benefits" for Nevadans.

NWPO's analysis of Yucca Mountain never calculates the possible beneficial effect of the repository on Nevada and the nation. Instead, on-site storage of nuclear waste is proposed as a solution because it distributes the pain in a Rawlsian sense and satisfies Rawl's two principles of justice. Unfortunately, this ignores the productive good that can come from the centralization of nuclear waste in a dedicated structure. Rawlsian ethics degenerates into a solely distributive theory (it can divide equitably what already exists, even the risks of nuclear waste), but it cannot create new goods and services and it cannot evaluate the possible benefits of Yucca Mountain, only its potential harm. Rawls has no mechanism for deciding what an equitable future should look like (since no one has invented that future yet), only for dividing up risk in the present. In sum, it is a recipe for paralysis, in this case the paralysis of the entire nuclear industry.

We can now see contained in their own writings the outlines of a philosophical evolution that occurred among the academics and radical environmentalists that preceded their opposition to Yucca Mountain:

**MARXISM --> MAOISM --> LIBERATION THEOLOGY --> DECENTRALISM --> RAWLSIANISM --> EQUITY ISSUES**

# 18. Doctrine of Informed Consent

Yet another philosophy, the Doctrine of Informed Consent, has been forwarded at Yucca Mountain by Dr. Kristin Shrader-Frechette, originally of the University of South Florida. Shrader-Frechette is a highly educated academic, a science philosopher, with close philosophical links to the researchers from Mountain West. Brought in by NWPO in 1992 to shore up the validity of its philosophical opposition to Yucca Mountain, Shrader-Frechette espoused some ideas that are quite controversial in regard to the future of science and civilization. Specifically, she attempted to extend Rawlsian ethics to Yucca Mountain with a legal concept called the Doctrine of Informed Consent.

Shrader-Frechette has an undergraduate degree in physics, post-graduate work in economics, hydrology and biology, and a PhD in the philosophy of science. Our purpose here is not to question Dr. Shrader-Frechette's qualifications, but to suggest that she simply has drawn inadequate conclusions about the Yucca Mountain project and about man's relationship to technology in general.

While Shrader-Frechette has criticized nuclear energy and the study of Yucca Mountain on technical grounds, the heart of her argument is actually an ethical one. For Shrader-Frechette, the question seems to be whether the utilitarian principles on which American institutions and our constitution are based should be replaced by egalitarian principles, specifically those of John Rawls.

Utilitarian theories direct us to provide the greatest safety/welfare or least risk for the greatest number of persons. In contrast, according to Shrader-Frechette, a Rawlsian perspective requires that:

". . all persons have a duty to act according to principles of justice securing the fairest distribution of goods among all people. . . . . This means, says Rawls, that persons in the original position would reject utilitarianism, because no rational being would accept a societal theory that might sanction a loss for himself in order to bring about a greater net good for everyone. Hence, if one adopted Rawls' 'Thought experiment' of the 'original position' it is unlikely that he would sanction the disparity in consumption between U.S. citizens and other people in the world. . . ." [Shrader-Frechette, K.S.; Nuclear Power and Public Policy, D. Reidel Publishing 1980, p 123]**

Applied to Yucca Mountain, Shrader-Frechette seemed to suggest that the risks of the repository, no matter how slight, are unacceptable because in an egalitarian framework the least advantaged Nevadan would never accept any such risks. The consequences of her philosophy applied to energy policy are not limited to Yucca Mountain, but are far reaching:

**"In other words, recognizing the real costs of nuclear fission might lead us to question closely-related assumptions about the value of increasing energy consumption and the value of an expanding economy. . . .**

**"If we questioned the thesis that economic growth removes poverty, then we might discover that the poor rarely share in the growth of real wealth and that they are "isolated from economic growth". Likewise if we questioned the thesis that economic growth enriches society, then we might find that the end effects of growth outweigh the good . . . "[Shrader-Frechette, K.S.; Nuclear Power and Public Policy, p124]**

Shrader-Frechette seems to imply the poor will be made worse off by expanded nuclear power facilities and the increased wealth this brings. While a total cost accounting of the economic externalities associated with nuclear energy (and fossil fuels, for that matter) is no doubt wise, *"questioning the thesis that economic growth enriches society"* seems extreme.

The problem appears to be that having once chosen Rawlsian ethics as her philosophy, Shrader-Frechette is forced to link every social and energy policy decision to this ethic. This ignores both the fact that our nation and institutions are still utilitarian by nature and the possibility that while the technology of the nuclear waste repository may not be egalitarian, it is nevertheless safe and appropriate.

In a paper funded by NWPO titled "Expert Judgment In Assessing RADWASTE Risks: What Nevadans Should Know About Yucca Mountain", Shrader-Frechette again argued that the utilitarian philosophy subscribed to by most Americans is flawed and should be replaced by Rawlsian egalitarianism:

**Utilitarian theories direct us to provide the greatest safety/welfare for the greatest number of persons. To subscribe to utilitarian theories represents a significant value judgment because utilitarianism has a number of significant flaws (see Rawls 1971, Kasperson and Abdollahzadeh 1988, Shrader-Frechette 1983, Shrader-Frechette 1985, Shrader-Frechette 1985, Shrader-Frechette 1991). [Shrader-Frechette, K.S.; "Expert Judgment In Assessing RADWASTE Risks: What Nevadans Should Know About Yucca Mountain", (NWPO-SE-054-92), June 1992, p108]**

It should be noted that Shrader-Frechette referenced herself five times in the preceding paragraph in support of Rawlsian ethics, a rather untraditional way of providing rigor to an argument. Quoting oneself is hardly a definitive proof of a position and is equivalent to claiming those who shout loudest are always correct. In the five pages Shrader-Frechette devotes to validating her Rawlsian perspective on Yucca Mountain under the NWPO grant, she cites herself no less than eleven out of twenty-six times.

Other interesting citations made by Shrader-Frechette in support of Rawlsian ethics are:

- Kasperson, Roger (3 times) NWPO consultant.

- Kneese, A.V. (3 times) NWPO socioeconomic peer review consultant.

- Kleindorfer (1 time) consultant to NWPO.

- Rawls (3 times) namesake of Rawlsian ethics.

Citing Roger Kasperson as justification for the superiority of Rawlsian ethics over utilitarianism at Yucca Mountain is problematic, since he is not independent and also works for NWPO. As shown previously, Kasperson's motivations contain elements of Maoism, Marxism, anarchism, decentralism and a tint of Brazilian liberation theology, hardly a high recommendation for Rawlsian ethics. Others cited in Shrader-Frechette's paper also worked for NWPO, notably William Schultze and Alvin Kneese, academic associates of Roger Kasperson.

Eliminating Shrader-Frechette's self-citation, Rawl's original work and the citations of NWPO researchers, leaves Shrader-Frechette's general theory of the superiority of Rawlsian and egalitarian ethics on very shaky grounds. This observation is important because it shows Shrader-Frechette's scientific proofs are not necessarily very rigorous and may be driven by the need to validate a Rawlsian worldview. Building an entire ethical treatise on the failings of DOE on the basis of such translucent logic is worrying. Shrader-Frechette used her credentials as a logician and science-philosopher (Appeal to Authority) to argue that the DOE and its contractors are error-prone, incompetent and inept. The harsh question which must be asked is whether Shrader-Frechette's analysis is not at least as error prone as anything forthcoming from the DOE. Shrader-Frechette continues:

**"The major flaw in utilitarian value judgments is that they allow minorities to be hurt. They allow various groups of persons to be treated inequitably on the grounds of expediency. Egalitarian views, on the other hand, sanction equal treatment of all persons, rather than simply using persons, perhaps violating their rights, in order to achieve some alleged social goal. Applied to Yucca Mountain, the utilitarian versus egalitarian issue focuses on the distribution of risk. Fears about inequitable or utilitarian risk distributions are what drive NIMBY (not in my back yard) syndrome. Few persons want to be members of the minority (like Nevadans) bearing the risk for the society as a whole. Moreover, because of their emphasis on providing equal protection, equal opportunity, and equal access to due process, it is arguable, from an ethical point of view, that egalitarian theories are preferable to utilitarian theories. [Shrader-Frechette, K.S.; "Expert Judgment In Assessing RADWASTE Risks, p 110]**

It is indeed very arguable whether egalitarian theories are preferable to utilitarian theories. What Shrader-Frechette, the science philosopher, has ignored is any mention of the potential failings of egalitarianism:

1) Egalitarian theories paradoxically require centralized enforcers to carry them out. Only if everyone agrees to disperse all risks and benefits in an egalitarian way can these theories work. The former Soviet Union could disperse economic risks among its people only through dictatorship.

2) There is little empirical proof that egalitarianism works. Of the many experimental communities attempted in the United States based on egalitarian models, none have thrived and Shrader-Frechette cites no examples of advanced civilizations based on her theories.

In a speech Sponsored by NWPO and the UNLV Environmental Studies program, April 1993, Shrader-Frechette attempted to further extend egalitarian ethics at Yucca Mountain. Using a concept called "The Doctrine of Informed Consent", Shrader-Frechette equated the Department Of Energy with Nazis war criminals at the Nuremberg Trials conducted at the end of World War II.

**"Free Informed Consent is something we learned about at the Nuremberg War Trials. We learned that you don't experiment on unwilling subjects without their free informed consent. We all know on account of medical ethics that doctors don't have the right to experiment on patients without their free informed consent.**

**"So the first question I want to raise in concern to Yucca Mountain, is there a question of free informed consent in the case of Yucca Mountain. Have the traditional constraints that require the public give free informed consent been met in this case? Because if, after all, we expect our medical ethics and our international ethics to conform to the doctrine of free informed consent - it's reasonable to hold our technological efforts to this doctrine as well."**

Shrader-Frechette's underlying purpose in introducing the Doctrine of Informed Consent was an attempt to give legal weight to Rawlsian ethics at Yucca Mountain, claiming a basis in international law. Rawl's "least advantaged man" is here replaced by "unwilling subjects", but with the added images of Auschwitz survivors perhaps huddled around the entrance to Yucca Mountain. The question we need to ask is whether the Doctrine of Informed Consent is actually applicable to Yucca Mountain and technology in general. According to Shrader-Frechette in her speech at UNLV, the Doctrine of Informed Consent has four components:

1) "We have to disclose the nature of the risk to the person at risk. You can't withhold information."

2) "Understanding. You can't ask people to bear a risk unless there is understanding."

3) "There is no coercion or manipulation involved."

4) "Competence. People must be competent to understand risks."

While the Doctrine of Informed Consent is rational on the surface, its logic fails in the application. For example, at any given time there is a risk that a 747 Jumbo jet will crash into Mr. Jones' cabin in a remote area of Montana. Yet we do not require United Airlines to:

1)   Disclose to Mr. Jones and millions of people within a thousand miles of such a route of that risk.

2)   Require everyone in the flight-path to study aerodynamics and structures so they can understand the full extent of their risk.

3)   We do not consider the absence of such information either manipulative or coercive.

4)   We do not require everyone under the flight-path to be a competent adult able to decide whether they are willing to take the risk.

Most disturbing about Shrader-Frechette's speech at UNLV was that she violated all of the tenets of the Doctrine of Informed Consent she had just presented. By not informing the audience of the dangers inherent in her Rawlsian viewpoint, she ignored her own four rules of free informed consent:

**1)   DISCLOSURE:** She did not tell the audience that the Rawlsian ethics she espoused is revolutionary and perhaps dangerous.

**2)   UNDERSTANDING:** Her audience clearly did not understand Rawlsian ethics and did not know she was promoting a philosophy, rather than merely discussing scientific objections to Yucca Mountain.

**3) COERCION:** Shrader-Frechette clearly attempted to manipulate, if not coerce the audience into accepting her Rawlsian worldview. At a similar speech given Union College, May 1993 in New York, Shrader-Frechette went so far as to require recording devices to be turned off to prevent further analysis of her speech.

**4) COMPETENCE:** An audience cannot be deemed competent to analyze a speech unless there is some shared understanding of terms. Rawlsian ethics is novel, untested and has not been discussed by the public.

The Doctrine of Informed Consent requires us to always inform the least advantaged man (and perhaps least qualified judge) of every physical, economic, and social condition which might put a human at risk. The obvious danger this poses is a collective neuroses, a paralysis in which no risk could be deemed negligible because of the fears of some minority. So while we wish to avoid the horrors of the premeditated evil of Auschwitz which created the need for the Nuremberg Trials, equating DOE's benign intent with such atrocities is an affront to logic.

In her work for NWPO, Shrader-Frechette judged DOE guilty of eighteen errors of methodological value judgment in estimating and evaluating radwaste risks and an additional ten logical inference errors. Shrader-Frechette did not appear to believe anyone within the DOE or its contractors addressed any of these issues, implying a remarkable degree of incompetence. While Shrader-Frechette's criticisms of DOE's buttressed the Nevada Nuclear Waste Project Office's point of view, they provided no reference against which to be measured. The correct comparison would be to apply Shrader-Frechette's logic to NWPO and to the counter proposition:

*What if we don't build Yucca Mountain and instead choose on-site storage?*

NWPO's position that on-site storage is the preferable technology would then have served as a control against which to compare DOE's logic regarding Yucca Mountain. This would have required a critical evaluation of NWPO's own scientific rigor regarding the on-site storage question, but no such analysis was forthcoming from Shrader-Frechette. Thus the problem with Shrader-Frechette's work was not that it was insufficiently brilliant, but that it was never verified through application to a control.

By arguing that nearly all investigations conducted by DOE scientists and their contractors are logically flawed, Shrader-Frechette has laid claim to being the expert of all experts. If her work is itself systematically flawed, then it might well be that DOE's estimations are indeed correct and Yucca Mountain is an appropriate technology. However, Shrader-Frechette has left herself in a rather untenable position:

1) She has not provided unimpeachable peer support for Rawlsian ethics.

2) She has not provided empirical proof for the practicality of Rawlsian ethics.

3) She has not applied the Doctrine of Informed Consent to her own proselytizing of Rawlsian ethics.

4) She has provided no control for her logical analysis of DOE through a similar investigation of NWPO. Such an analysis might well lead to the identification of many more errors of logic on the part of NWPO than even DOE is accused of committing.

5) She has provided no control for her analysis of Yucca Mountain through a parallel investigation of the problems inherent in on-site storage.

Ironically, it appears Shrader-Frechette is guilty of as great a magnitude of error in logic in her analysis of Yucca Mountain, as the errors she accuses DOE and its subcontractors of committing. Shrader-Frechette is an expert claiming the fallacy of trusting experts! This does not invalidate the individual concerns she expressed regarding DOE's methodology, but it does call into question whether Shrader-Frechette's opposition to Yucca Mountain was based on science, or on a very theoretical Rawlsian ethics in which she had a vested interest.

Even worse, if no one at NWPO was able to critique the obvious flaws in Shrader-Frechette's work, it calls into question the entire body of oversight studies conducted by that state agency.

# 19. Fear Sells

The key to understanding the hysteria that is generated by Yucca Mountain is a psychological theory called the "availability heuristic". That's not to say this is the only or best theory about why people exaggerate their fears of unlikely events like accidents at a nuclear repository. Instead, the availability heuristic is important because it was the pet theory of the social scientists at NWPO and paradoxically became the means of imprinting nuclear fears and hysteria on the population of Nevada. As such, it is another example of the closed loop feedback system that developed at the insular NWPO.

If the availability heuristic was at one point an attempt to provide an academic theory of the underlying causes of mass fear, in the context of Yucca Mountain it was twisted to become a tool for creating the impression that educating the public about real risks would be impossible in the face of this all-powerful psychological juggernaut.

**"One mode of thought that provides insight into perceived nuclear risks is a judgmental rule or strategy known as the availability heuristic (Tversky and Kahneman, 1974; Slovic, Fishoff, Lichtenstein and Hohenemser, 1979). This rule leads people to judge an event as likely or frequent if instances of it are easy to imagine or recall. Frequently occurring events are generally easier to imagine and recall than rare events; thus, reliance on availability is typically an appropriate mental strategy. However, memorability and imaginability are also affected by numerous factors not related to likelihood. As a result, this natural way of thinking leads people to exaggerate the probabilities of events that are particularly recent, vivid, or emotionally salient. Certainly, the risks from nuclear power would seem to be a prime candidate for enhancement by the availability heuristic, because of the extensive media coverage they receive and their association with the vivid, imaginable dangers of nuclear war. As Zebroski (1976) noted, "fear sells"; the media dwell on potential catastrophes, not on the successful day-to-day operations of power plants. [Slovic, Paul; "Images of Disaster: Perception and Acceptance of Risks from Nuclear Power", Energy Risk Management, Academic Press, 1979]**

The crude interpretation of the availability heuristic, i.e. fear sells, has a corollary in the Yucca Mountain debate. If fear sells, it is then easier to sell fear than to educate the public about the true nature of the risks that confront them. It was this inverted form of the availability heuristic which became the driving dogma behind the socioeconomic studies sponsored by NWPO. That is, instead of studying how unwarranted fears of nuclear technology could be overcome, NWPO used Slovic's theories of fear, encapsulated in the availability heuristic, to enhance fear!

Another interpretation of the availability heuristic is that people are incorrigibly stupid, a theory Nevadans might resent if it were fully known they were being modeled this way. If people are afraid of that to which they are not accustomed, as they are not accustomed to Yucca Mountain and the storage of nuclear waste, then the availability heuristic implies they could never learn differently no matter what the facts of the matter. Following this line of reasoning, people who do not fly in airplanes daily or ride in high speed trains should believe these activities ought to be outlawed because their perceived fear that they might one day be forced to ride in one of these devices could never be overcome by logic or education. Fortunately, there are other theories of how humans cope with the fear of the unknown, notable the Culture Theory of Aaron Wildavsky, who assumes that members of a technologically sophisticated culture will not jump at the sight of robots, space shuttles and the Yucca Mountain repository, though it may still jump at the sight of mice.

Slovic mentions in the above citation a paper written in conjunction with Christopher Hohenemser, who also happens to be from Clark University and CENTED. This broad collegiality between Decision Research and Clark University obviously had developed even before the Three Mile Island accident of 1979, the event that formed the seed of the later Mountain West consortium. Unfortunately this organizational incest eventually led to intellectual gridlock in regard to theories like the availability heuristic. When a small group of academic peers extensively quote each other, it becomes easy to believe that the theories generated by this inbred group are the only interpretations of reality possible. What seems to have formed around Slovic and Kasperson in the small sub-specialty of nuclear-risk-perception was a self confirming group bubble. After a while, it became hard to criticize long time colleagues when they drifted outside their expertise from social science into hard science and finally into political advocacy. Ultimately, these scientific transgressions were tolerated because they seemed to confirm the group's world view.

To many people, an obscure psychological research theory like the availability heuristic would seem to be the last subject on which to base arguments for or against a project as technologically sophisticated as Yucca Mountain. The problem for NWPO, however, was that they were significantly outgunned on the hard science intellectual front; for every high school teacher, housewife, or historian they could squeeze into the mold of a qualified Yucca Mountain critic, DOE had five all-star scientists who had a specialty in the field at hand.

Thus, a symbiosis developed between Paul Slovic, (the intellectual big gun of Decision Research), Kasperson at Clark University and Director Loux of the Nuclear Waste Project Office. Slovic got NWPO funding to test his availability theory as long as it supported the view (developed by Kasperson and Clark University and adopted by NWPO) that the inordinate fears of the public about nuclear waste were unchangeable. In a vicious cycle, NWPO then generated more unreasonable nuclear fears, that Paul Slovic in turn studied and documented, so CENTED could justify their anti-nuclear equity policy formulations, which were fed back into NWPO's campaign of inflaming unreasonable nuclear fears among Nevadans. In this roundabout way, the consultants of CENTED and Decision Research began to imprint national energy policy with their own version of reality by means of their philosophical hold on NWPO.

All this may all be hard to believe, but some examples of the NWPO / Slovic symbiosis in this "fear sells" mode aren't hard to come by. In a letter on December 18, 1991 sent by Bob Loux of NWPO to the media, we see the State attempting to use a Slovic paper about nuclear fear to induce fear.

**"The enclosed article, "Perceived Risk, Stigma, and Potential Economic Impacts of a High-Level Nuclear Waste Repository in Nevada," . . . . reports on careful, scientifically sound research conducted by a leading authority in the field of social impact assessment, Dr. Paul Slovic. Dr. Slovic is internationally regarded for his ground-breaking work in understanding the nature of hazards and the processes by which hazards and risks affect individual behavior.**

**You will find this article to be in significant contrast to the type of wild speculation and unsupported claims about the laundry list of "benefits" associated with the Yucca Mountain repository being made by the nuclear industry and supporters of the repository in Nevada. . . ."**

The fear being pushed by Loux was that no benefits are possible to Nevada from the Yucca Mountain repository and that tourists will be driven away in droves by "nuclear stigma". This begs the question of who was guilty of "wild speculation and unsupported claims", the nuclear industry and supporters of the repository forces, or perhaps Bob Loux and the State of Nevada? Because Nevada is highly dependent on gaming revenue, any adverse effects of nuclear images on the gambling rake would be a legitimate concern. The problem comes when fears of revenue losses due to nuclear "stigma" are used to heighten nuclear fears which in turn heighten fears of economic loss. This is easily recognizable as a positive feedback loop in which the experimenter has become part of the system and provides the positive feedback necessary to drive the system ballistic.

As far as being "scientifically sound research", it's interesting to look at the conclusions of the "Perceived Risks, Stigma and Potential Economic Impacts" article sited by Loux:

**"The mechanisms of perceived risk, social amplification, and stigma are observable in the record of past experience with nuclear and other types of hazards. In the context of the Yucca Mountain repository, these mechanisms appear to have the potential to cause substantial losses to each of the various economic sectors at risk.**

**Judging from the Phoenix survey, the test site has worked its way into the imagery of Nevada for only a small percentage of people and is rarely associated with Las Vegas. Moreover, the operations of the test site have been restricted and unavailable to full public scrutiny. Nuclear-waste transport, the operation of the waste repository, and any controversies over the safety of these activities will likely be far more visible to the public and the media. In particular, tens of thousands of nuclear-waste-shipments by truck or rail throughout the United States will be a prominent reminder of the repository and its risks. As these shipments converge upon Las Vegas, nuclear associations with that city may be built to a far greater extent than has occurred with the secret, contained, underground explosions at the test site. Finally, there is no**

evidence that the small degree of association of the test site with the region has not actually impaired tourism and business development. Apart from the gambling industry, business development has shown little progress despite the potential attractiveness of Las Vegas for many kinds of industries. [Slovic, et.al."Perceived Risk, Stigma, and Potential Economic Impacts of a High-Level Nuclear Waste Repository in Nevada", Risk Analysis, Vol. 11, No. 4, 1991]

Some comments are in order here. Slovic admits that the Nevada Test Site has had an only minuscule affect on the images associated with Las Vegas in the minds of residents outside Nevada. Had Slovic actually been a Nevada resident instead of an out-of-state researcher, he would also have known that the underground explosions at the test site were hardly "secret", or "unavailable for public scrutiny". Experiments were publicly announced before each test, Las Vegas residents often felt ground temblors and the newspapers were filled with analysis and reports of protests at the site every time a bomb went off. Moreover, test site workers were ubiquitous, their families respected residents of Las Vegas, and happenings at the site had a multitude of leaky-sieve means of being reported to the public. Slovic goes on:

It may also be the case that the test site and the repository will interact in a synergistic way to produce nuclear imagery to an extent that is greater than the sum of the individual contributions from each facility. Little is known about the dynamics of the process by which images become salient. It is certainly true, however, that individuals have a number of images associated with any particular place. There may be some threshold of repetition that moves a weak or unstable image from the periphery into the core image of a place. If so, Nevada's link to the nuclear weapons test site may increase its potential for stigmatization from the repository relative to a state with no existing base of nuclear imagery. ["Perceived Risk, Stigma, and Potential Economic Impacts", p694]

Since the one group most responsible for exploiting fears about Yucca Mountain was the Nuclear Waste Project Office, the "repetition that moves a weak or unstable (nuclear) image from the periphery into the core image of a place", would need to come from NWPO and from Slovic's studies of the possible negative economic impact of nuclear imagery on the State. From Slovic we have a final analysis:

In sum, our analysis indicates that the development of the Yucca Mountain Repository will, in effect, force Nevadans to gamble with their future economy. The nature of that gamble cannot be specified precisely, but it appears to include credible possibilities (with unknown probabilities) of substantial loses to the visitor economy, the migrant economy, and the business economy. ["Perceived Risk, Stigma, and Potential Economic Impacts", p694]

Was Slovic's article, as Loux claimed, scientifically sound research? The number of 'will likely', 'It may be', 'There may be', 'If so', and 'it appears' qualifiers laced in the small sections of Slovic's writing quoted above, should make even laymen pause to consider whether Slovic is doing science or editorializing. However, criticism extends into the academic community as well.

Ross Hemphill of Argonne National Laboratory and Gilbert Bassett of the Department of Economics at the University of Illinois wrote a scathing rebuttal to the Slovic article which also appeared in the journal Risk Analysis. It is not usual for a journal to run a full length rebuttal to a previous author's article, usually such comments are limited to letters of discussion. It's worth quoting Bassett and Hemphil's main arguments:

Slovic et al. consider the potential for a high-level nuclear waste repository to have "adverse economic effects" on the city of Las Vegas and the state of Nevada. The conclusion seems to be that the potential is great. But no quantitative estimates of the magnitude of impacts are presented. Further, if "potential" is supposed to connote more than "anything is possible," then, in our opinion, the conclusion does not follow from the survey data or method of analysis. We are not saying the repository will be without affects, but rather that no impacts are demonstrated. The conclusion about large impacts is the opinion of the authors and not a consequence of empirical data. [Bassett, Gilbert; Hemphill, Ross;"Comments On 'Perceived Risk, Stigma, and Potential Economic Impacts'", Risk Analysts, Vol. 11, No. 4, 1991, p697]

In a conversation with Gilbert Bassett, he used the words 'bad science' to refer to the Slovic results. In their exuberance to prove that dire things would happen if Yucca Mountain was built, research conducted under the auspices of Mountain West often drifted from science into the realm of political advocacy. According the Bassett and Hemphill:

**The scenarios about what will happen at the repository in the future, while "potentially" valid, are speculative and not based on the evidence. The validity of the assertions is arguable, but beside the point. The assertions are opinions that are not informed by the survey data on public perceptions. This jeopardizes the aim of providing a convincing demonstration of the importance of perceptions in social impact assessment. It also means the conclusions regarding the repository are not based on scientific evidence. [Bassett; Hemphill, Risk Analysts, Vol. 11, No. 4, 1991, p698]**

In fact, many of the statements issued by NWPO were not based on scientific evidence, but on the opinions of its researchers, some of whom had been working for twenty years to stop nuclear energy. It was little wonder that their social and energy policy agendas began to encroach on their science. A final word from Bassett and Hemphill:

**If anything, the evidence goes the other way -- 50 years of the test site has put "nuclear" into the Las Vegas image set for only 1% of the sample. This does not mean the same will hold for the repository. It means that the formation of images and stigma is a complicated process that will be determined by intervening events and, "by proper safety design and management . . that instill and maintain trust and that work to protect the economic base of those individuals and communities whom the facility puts at risk. [Bassett; Hemphill, Risk Analysts, Vol. 11, No. 4, 1991, p699]**

Slovic and the other social scientists under Mountain West were under pressure to make their data and conclusions regarding nuclear stigma fit the results expected by NWPO and the state's politicians. In 1979 when Slovic et al. were not in the employ of an anti-nuclear state agency fighting Yucca Mountain to the death, they'd had the leisure to be more candid about possible routes to the acceptance of nuclear energy:

**These examples (x-rays, Hermiston nerve gas repository) illustrate the slow path through which nuclear power might gain acceptance. It requires an incontrovertible long-term safety record, a responsible agency that is respected and trusted, and a clear appreciation of benefit. [Slovic, Paul, et.al.; "Psychological Aspects of Risk Perception", Accident At Three Mile Island, Westview, 1982, p13]**

Of course, the converse is also the formula to kill nuclear power: distort the safety record of nuclear energy, destroy the credibility of the agency involved at Yucca Mountain (DOE) and question at every point the possible benefits to people of Nevada. NWPO and Nevada's politicians have fully employed these negative tactics in their fight against Yucca Mountain, following in inverse the recipe so clearly stated by Slovic et al. in 1979.

Fear sells. A normal citizen led protest of Yucca Mountain might have focused on technological and safety issues, states' rights and adequate compensation as its primary concerns. Under the tutelage of Mountain West and its social scientists, Nevada's citizens were instead molded by sophisticated psychological techniques that drove anti-nuclear hysteria home. Soon, fears of a relatively benign Yucca Mountain repository were implanted in a population that had lived thirty-five years in harmony with the Nevada Test Site where hydrogen bombs were regularly exploded above ground.

According to NWPO and the Mountain West researchers, DOE wasn't to be trusted, there would be no benefits, nuclear waste would be transported through the I-15 Interchange (Spaghetti Bowl) downtown, the Las Vegas economy would collapse, volcanoes would erupt under the site, earthquakes would shake Yucca Mountain to pieces, ad infinitum In all this, there was a decided lack of credible technical evidence that Nevadans would be at any overwhelming real risk. Instead, perceived risks were raised to new heights and fear was sold to Nevadans by the Nuclear Waste Project Office and its psychologists. They knew only too well how to mass market nuclear hysteria with a theory they affectionately called the availability heuristic.

# 20. Perceived Risks

The measurement of perceived risk formed the backbone of the socioeconomic studies of Yucca Mountain for the Nevada Nuclear Waste Project Office. Both Roger Kasperson of CENTED and especially Paul Slovic of Decision Research consider themselves experts in this area, in a sense having created the field of nuclear risk perception (vs. risk analysis) from scratch to buttress their sociological interpretation of technology. There is, however, a huge difference between perceived risk and risk itself.

The difference between risk and risk perception is so great that it forms the difference between modern civilized man and his superstitious aboriginal ancestors. The New Guinean who makes a god of the airplane flying overhead is indulging in risk perception; the pilot flying the airplane is making use of science and objective risk analysis (we hope!). Earth Worship, through the assignment of mystical perceived properties to the rocks, streams, air and organisms that surround us, makes for interesting cultural anthropology, but abysmal science.

A century ago, some Indian tribes of the Southwest believed that a photograph captured part of their soul. According to Earle R. Forrest, who traveled with his camera through Navaholand in 1900:

**This was the West I had been looking for, and I decided right there to capture it with my camera.**

**When I mentioned the subject to Meadows (the Trading Post keeper), he smiled. "Well, you'll have a hard time until they get acquainted with you. . . . Only a few have ever seen a camera, and it's a very mysterious affair to them, for they cannot understand it. They think it's magic of some kind, and they're afraid of bad spirits" [Forrest, Earle R.; With A Camera In Old Navaholand, University of Oklahoma Press, 1970, p27]**

There was thus a perceived risk for certain Navajos before the 1920s that having their image transferred to paper involved a highly threatening technological. This was a perceived risk certainly as great as that posed by Yucca Mountain to Nevadans.

**Navahos are not cowards. They are courageous. But like all primitive people they are superstitious about anything they do not understand. And they could not comprehend how I could make a picture of a person by pointing a black box at him. It was magic of some kind, and they were afraid of evil spirits in the white man's witchcraft. . . ." [Forrest.; With A Camera In Old Navaholand, p33]**

But why would the Navahos be afraid of this technology? After all, the camera was just a box and the pictures it produced little different from the reflections in a pool.

**(Meadows) explained a little more of their belief in witchcraft. "A Navajo never dies from natural causes," he said. "If he gets sick it's because some enemy paid a medicine man to plant evil spirits in his body, and it's those spirits that make him sick. . . . They really believe that they take a great risk if they let you take a picture, for if an enemy hears about it, he'll try to have bad spirits break or damage the picture." [Forrest.; With A Camera In Old Navaholand, p33]**

The Yucca Mountain repository may be perceived to threaten one's life, but even its worst critics don't believe it can steal your soul. Should we have banned cameras in the Southwest because some local populations perceived it as threatening? Should Yucca Mountain be scuttled simply because of the perception of the masses (not the scientific reality) that it is dangerous?

The consultants of NWPO and Mountain West apparently interpreted their surveys and socioeconomic data as justification for a theory that perceived risk _is_ risk. Such a theory would be an employment bonanza for sociologists suddenly called upon to measure each subconscious perception the public holds on even the most minor technological issue. Before any road, building, machine or flea catcher could be constructed, a survey would need to be conducted to measure the public's "perceived risk" of the new technology. This theory would normally be rejected at face value, but in an academic world of Rawlsian ethics, sometimes the ridiculous is rewarded.

Paul Slovic of Decision Research and Roger Kasperson of Clark University in their socioeconomic work for NWPO forwarded in synergy a technological philosophy for Nevadans that in many ways mimicked the fear of witchcraft practiced by the Navaho at the introduction of the camera. Here is how it worked:

1.    Slovic's pet theory, the "availability heuristic", hypothesizes people are afraid of the unaccustomed and are unlikely to change their perceptions through education. Slovic long studied the inflated fears of the populace regarding nuclear technology despite himself admitting the actual risk levels are much below the perceived risk.

2.    Kasperson's Rawlsian equity theories propose that societal ethics should be based on the needs of the "least advantaged" member. In other words, policy should be determined by the fears of the least members, often the most naive and misinformed.

3.    Combining Slovic's and Kasperson's theories at Yucca Mountain, their logic demands that the nuclear fears of the least advantaged man (who presumably cannot be enlightened through education) should be the overriding consideration in regards to storage of nuclear waste. No matter how beneficial nuclear technology might be for the community at large, nor how safely radioactive materials were used and contained, it would be the perceptions of risk of the least advantaged man that ruled the use of that technology. Utilitarianism would be replaced by egalitarianism.

4.    Finally, Slovic and Kasperson self-validated their psychological and political theories by spreading fears through their own socioeconomic studies. Relying on directed polling of the population, they not only measured fear in a scientific sense, but used those polls through the publicity arm of NWPO to enhance nuclear fear, much as if they were Navajo medicine men invoking evil spirits.

Polls done in normal socioeconomic impact research are designed to be as non-invasive as possible. Sociologists have long recognized that polling results can feed back on themselves causing a swaying of the results. Polls done in the political arena have long taken advantage of this feedback effect, and are often specifically designed to not only gauge voter sympathies, but to also sway those sympathies (push-polls). The Mountain West consortium apparently turned their polls from scientific tools into political weapons, abdicating objectivity to promote the anti-nuclear cause. Unfortunately, the ramifications of this are not limited to Yucca Mountain, but potentially threaten to disrupt our national economy.

For example, in late 1992 NWPO's hopes that risk perceptions alone would be the wild card that would kill the Yucca Mountain project rose to a peak. The data Mountain West had collected on the impacts of perceived risks on real estate values, were in fact an attempt to set the groundwork for lawsuits to stop transportation of nuclear waste through the State of Nevada and the country as a whole. The claim was made that property values would be adversely affected by the negative images of nuclear waste and that the government would be unable to compensate residents.

A lawsuit in Santa Fe, New Mexico awarded $550,376 in condemnation proceedings to John and Lemonia Komis whose property was near a nuclear waste transportation route to the Waste Isolation Pilot Project facility. An additional $337,815 was awarded on the theory that their property had lost perceived value in public perception and should be compensated. The New Mexico Supreme Court held that property owners are entitled to compensation for the loss in value to land which is caused by public fear, whether or not that fear is reasonable. The Court said:

**. . . If loss of value can be proven, it should be compensable regardless of its source. Thus, if people will not purchase property because they fear living or working on or near a WIPP route, or if a buyer can be found, but only at a reduced price, a loss of value exists. If this loss can be proven to a jury, the landowner should be compensated. [City of Santa Fe, 845 P.2d at 756-757]**

**The New Mexico Court further held that a public opinion poll, a videotape regarding the WIPP site and expert testimony was properly admitted in support of the claim of "public-perception damages." [pages 757-759]**

The fact that the state of Nevada had pinned much of its hope for defeating the repository on the "perceived risk" transportation liability issue is clear. A letter titled "NEW MEXICO CASE RAISES POTENTIAL LIABILITY FOR NEVADA, FEAR THEORY COULD HAVE MAJOR IMPACT FOR HIGH LEVEL DUMP", from the Nevada Office of the Attorney General, dated April 7, 1993, demonstrates the impact of the Slovic and Kasperson theories on Nevada and the nation. According to Attorney General Frankie Sue Del Pappa :

**"My greatest concern is the serious implications this case has for the state. The theory of the case, if applied by Nevada courts to the inevitable claims associated with the repository, could have far-reaching fiscal impacts,"**

Senator Bryan also was quoted in the letter:

**Nevada's Attorney General today has outlined yet another potential liability for the Yucca Mountain project. Her analysis of the case should serve notice that the Department of Energy should no longer duck dealing with the transportation of high level nuclear waste, but should give that issue the attention it deserves, Bryan said. "From the nuclear power plants on the east coast, across the Appalachian states and across the Midwest, implications from the New Mexico case will raise incredible complications and obstacles to the DOE and the nuclear utility's plans to build a nuclear dump in Nevada."**

Indeed, the implications extended far beyond Yucca Mountain to the entire question of interstate commerce of any sort. For example, could perceptions of the fear of overflying airplanes devalue property near airports, requiring compensation from municipalities? Could Nevada Power be sued because citizens perceive a gas main transporting natural gas could explode, decreasing the perceived value of their property? Could you sue if you were within a half mile of a high-voltage transformer and the public's *perceived* fears of cancers were suddenly raised by the media, though real risks were minimal?

It is an interesting concept that people can sue not only for real damages, but also for perceived damages. The late Frank Clements from the Nuclear Waste Task Force, suggested in a conversation after the Sept 4, 1992 Sawyer Commission meeting that the State had high hopes this tactic would work in Nevada and all along the corridor routes to the repository. But perceived risks work both ways.

Suppose the State of Nevada inflated the perceived risks of nuclear waste beyond what was reasonable, causing an unjustified rise in nuclear risk perception? By the state's own reasoning they would be liable for the diminished value of property in Nevada caused by their anti-nuclear campaign. If the State promoted fear specifically to stop Yucca Mountain, they might well be liable for countless billions in lost property value at casinos and residences throughout the state due to revenue lost when tourists, made fearful by NWPO's public relations campaign, avoided the state.

A nation run on Rawlsian risk perception (the fears of the uninformed carrying the most weight), might well suffer from the paralysis of the entire economy. No planes could fly because of the perceived risk that they would crash into hotels, no trains could carry chemicals for fear of toxic spills, nothing could be transported or invented or built for fear that perceptions of risk would halt that endeavor. Utilitarians, in contrast, would argue that the final arbiter of technological progress must be real risk, measured as best science can allow.

Politics is the art of perceived risks; science is the objective measurement and estimation of real risks. Adding politics and its baggage of perceived risks to science as advocated by progressives like those of the State of Nevada would invalidate science and corrupt any political decisions based on what was supposed to be objective research. Unfortunately, this is a prescription for paralysis and begs the question of how much difference there is between our modern elevation of perception to reality and primitive superstitions that attributed mystical properties to inanimate objects?

In 1926, Earle Forrest returned to Navajo country, a short twenty-four years after he'd first encountered a primitive populace afraid of the magic spirits contained in his technological black-box camera:

**It was at Tohatchi that I took my first photographs of Navahos this trip. In the shade of the store was an old man and his wife with a child about three years old. . . The trader told me they had many sheep and some cattle and horses. When at my request, he asked if I could take their pictures, much to my surprise they told him they would like it. The old fear of the magic black box with the evil eye had apparently vanished with the years. [Forrest; With A Camera In Old Navaholand, p196]**

In twenty years, a primitive society living in an era of still glacially slow communications had become adjusted to a photographic technology they previously believed capable of stealing their souls. This is a rather poignant counter-example to Paul Slovic's "availability heuristic" theory and NWPO's claim that no amount of positive scientific information about Yucca Mountain can sway popular opposition. A primitive Navajo culture accepted an equally threatening technology without the benefit of socioeconomic impact studies, Rawlsian ethics, or any of the other sociological theories advanced by the Nevada Nuclear Waste Project Office and its consultants.

# 21. Selective Polling

The difference between good science and bad science is one of methodology rather than outcome. All theories are good theories as long as the methodology used to prove or disprove them is logical and objective. The socioeconomic studies run by NWPO crossed the line into bad science when their experimenters became part of the experiment and confused scientific methodology with political advocacy against Yucca Mountain. This became obvious at the interface between the socioeconomic researchers and the public, i.e. the polling techniques utilized by Mountain West not only measured popular sentiment but were designed to influence it as well.

Paul Slovic of Decision Research concentrated his studies for the last forty years in the area of risk perception, especially in regard to nuclear issues. Early on, Slovic became convinced that because of the psychological theory called the "availability heuristic" it would be nearly impossible to teach any population (and by inference Nevadans) about the actual low risks of nuclear technology and the Yucca Mountain repository.

In order to prove such a theory, a social scientist polls public sentiments on issue that are "easy to imagine or recall" but have low "availability" (as for example a nuclear repository disaster). Where science failed in Nevada was when social scientists like Slovic not only polled public sentiment, but entered the public arena to advocate the results for which they were polling.

Nevadans would likely be afraid of nuclear energy and Yucca Mountain with or without the presence of the NWPO socioeconomic researcher. The fear of mutations and cancers due to silent and deadly radiation, no matter how unlikely, is difficult to change even with education, a result which supports the availability heuristic theory. However, humans do adapt and become accustomed to many other fears, living near volcanoes, earthquake faults, flood zones, etc. Las Vegas itself is within eye distance of the nuclear test site. The Mountain West researchers seemed to dismiss this adaptation process, evidently theorizing that fears expressed in their polls were the same as fears experienced outside an experimental situation. Moreover, they became advocates of their own theory, claiming that their polls showed Nevadans were unlikely to ever learn to accept the presence of a nuclear repository, in effect acting as advocates for that very outcome.

Mountain West social scientists wrote articles in the Wall Street Journal and Science, appeared on radio and were the catalyst for numerous newspaper articles. Not only did they attempt to prove the "availability heuristic" by claiming their polls showed it to be impossible to educate the public about infrequent nuclear risks, but they also claimed public fears of Yucca Mountain were justified because scientists are incapable of calculating each and every limb of a risk tree. In other words, they became advocates for the specific outcomes they were polling (negative images towards Yucca Mountain).

This ethical lapse is similar to a chemist looking for organic compounds who knowingly stirs the pot with a finger covered with his lunch at Taco Bell. It would be no surprise for the chemist to find organic traces of burrito and taco sauce in his results. It is little surprise to find that Mountain West's results claimed Nevadans could never accept information about nuclear energy, especially since that information was tainted by the people doing the polling.

For example, David Pijawka of Arizona State and a contractor to NWPO in a July 21, 1992 interview on KDWN radio showed his biases all too clearly:

**BAUGHMAN: Dr. Pijawka, what was the survey and what were the results?**
**PIJAWKA: I was pleased to see this make front page news. It shows the repository has become a salient issue. . . . We interviewed 701 households. Not surprisingly, they said earthquake activity is important and it should be a criteria for dropping Yucca Mt. as a site. . . .**
**CALLER: I object to the DOE commercials that say Three Mile Island was a non-event. I have an aunt whose legs were burned from that fallout in the grass and know how she suffered.**
**PIJAWKA: Three Mile Island wasn't a "non-event." We may see some health manifestations show up many years from now. It cost the industry billions of dollars. . . We find that the TV ads put on by the**

industry are a failure. 41% of those interviewed are less trusting of DOE. 55% have not changed their level of trust. Only 3% are trusting now.

BAUGHMAN: Carl Gertz (Director of Yucca Mountain for DOE) said your survey is designed to create negative responses.

PIJAWKA: (laughter) That is expected rhetoric from the Department of Energy. The survey was very objective and neutral. People could put any answer they wanted on the scale they were given.

Pijawka's responses were disturbing:

1. When polls create front-page news, there is a question of whether they were designed for science or political advocacy.

2. The public's perceived risk of earthquakes causing failures at Yucca Mountain has nothing to do with the actual risks. People perceive rollercoasters as risky, but this hardly stops them from flocking to ride them.

3. Three Mile Island was not a nuclear disaster. TMI's health manifestations were trivial. and it is considered an accident with negligible effects on local radiation levels (unlike the Chernobyl disaster). As a non-engineer, non-health scientist, Pijawka's comments were political.

4. Tracking ANEC advertisements for political effectiveness implies Pijawka et.al. were running a political campaign against those ads rather than doing research. Questioning the ability of the ANEC ads to educate Nevadans is not a Yucca Mountain socioeconomic impact.

5. Dismissing the respected Carl Gertz's comments as laughable DOE rhetoric implies Pijawka himself has an agenda in trying to undermine the credibility of the DOE.

The fact that Pijawka was on radio advocating an anti-Yucca Mountain position suggests that he acted more as a public relations consultant than an objective social researcher. Indeed, KDWN's show, "Yucca Mountain: Fact Not Fiction", never interviewed a pro-Yucca Mountain guest in its two years of shows spanning 1991 through 1993, calling into doubt the objectivity of this radio program.

Pijawka quoted an earlier autumn 1991 statewide poll conducted by Decision Research which tracked the effectiveness of the ads being generated by the American Nuclear Energy Council [see Report of the Nevada Commission Nuclear Projects, 1992, Attachment II, Autumn 1991 Nevada State Poll]. Question (4) from this poll is especially interesting because of the spin put on it by both sides of the debate:

**4) Based upon these (ANEC) advertisements are you, personally: More supportive of the Yucca Mountain program Less Supportive, About the Same, Don't Know**

The state argued that these results showed the ANEC commercials, mostly dealing with transportation casks and pellets safety issues, were ineffective. The interpretation of the industry was that the 32% who were less supportive had already been lost to their efforts and that a 14.8% gain (not 3% as suggested by Pijawka) was actually highly encouraging in such a negative market after touching only a few issues.

The last question in this survey (7) asked for written feelings about the ANEC advertisements featuring spokesman Ron Vitto. The majority of the 601 comments were negative, typically:

**I feel they were a gimmick. I didn't believe it for a minute. I know about radiation. I have a medical background. I know you can't smell, hear or see it. But it's still very dangerous. They are trying to tell us it's not. I know it is. (respondent #9)**

And:

**He sorta reminds me of a snake oil salesman. If it were a real pellet, he'd glow in the dark - it's a con. (respondent #4)**

A few were positive:

**I think they're good. I think the advertisements are educational. It makes you at ease, because in the ads they tell you if the wastes fall off the truck in their barrels, they won't crack, so there's no problem. I feel real at ease. (respondent #43)**

The larger question was why a state agency felt compelled to track an industry public relations campaign and how this related to the socioeconomic status of Nevadans. In fact, it is an example of how the NWPO pollsters were tied intimately into the anti-nuclear political campaign, much as the pollsters for politicians running for office are integral parts of modern political warfare. Tracking ad responses is not a normal part of socioeconomic research which generally tries to determine mundane questions like migration patterns, income and the like.

In contrast to the state's surveys was a poll conducted by the Southwestern Social Science Research Center at the University of Nevada Las Vegas, titled Nuclear Issues Survey, - Spring 1993. The UNLV poll came to a number of surprising conclusions that had not been generally publicized about perceptions of Nevadans towards Yucca Mountain. According to the final evaluation of the UNLV poll:

## EVALUATION

**Nearly three quarters of respondents (73.0%) place their trust in scientists as opposed to less than two percent (1.8%) who place their trust in politicians about the technical study of Yucca Mountain. As in other polls, Nevadans support scientists and appear willing to let them undertake the study without political intervention.**

**Nevadans are in firm support (80.6%) of a stronger role by the scientists of the state university and college system in the scientific oversight and analysis of the study period. In combination with the high trust in scientists we are led to conclude that Nevadans have confidence in the state university system, and believe it should have a vested interest in the study period at Yucca Mountain.**

**Nearly ninety percent (87.2%) believe the evaluation of Yucca Mountain could be done better if state scientists were actually on the Yucca Mountain site versus only conducting off-site analysis. The confidence in the role of the state as an oversight actor would be enhanced through on-site opposed to off-site evaluation.**

**Over ninety percent (94.6%) believe the state should provide an itemized list of how the money it receives for the oversight of the Yucca Mountain study is spent.**

**Among the most important issues facing Nevada today, six out of sixty-three are ranked above all others. These are Water, the Economy, Jobs, Education, Population Growth, and Crime.**

**Well over fifty percent of the respondents (57.2%) do not rank Yucca Mountain as one of the two most important issues facing the state. One quarter (22.8%) feel Yucca Mountain ranks as one of the most important issues in the state. Based on support levels for benefits a significant number of these respondents see the issue as an important avenue for securing benefits for Nevada. Approximately one-fifth (19.7%) did not choose to respond or did not know how they would rank Yucca Mountain in relationship to other issues facing the state.**

**Over two-thirds (67.3%) feel nuclear energy and technology should be an important part of the nation's energy future. No stigma about nuclear energy and technology is apparent among the respondents.**

**While this poll was sponsored by the nuclear industry, neither its results nor even the results of Decision Research are much in dispute. What is in dispute is the scope, interpretation and political use of these polls. If the industry had followed NWPO's lead, it would have focused on whether Nevadans perceived the Nevada Nuclear Waste Project Office to be successful in persuading them to fear radioactive disasters at Yucca Mountain. Instead, some substantive questions were addressed.**

**Taking the results together, from various pro and anti sources, the situation appears to be this: Nevadans are overwhelming opposed to hosting the site (73%), but also overwhelmingly (68%) believe Yucca Mountain will be built. They do not rank Yucca Mountain as the most important issue, being superseded by water, the economy, jobs, education, population growth and crime, but it certainly is important. While only (3%) trust DOE, (73%) trust scientists versus (1.8%) who trust politicians.**

These seemingly paradoxical results indicate that Nevadans hold complex multidimensional views of the study of Yucca Mountain. This indicates they may not be as dumb as the availability heuristic suggests and are capable of more complex responses to the dangers posed by a radioactive waste repository than cringing paralyzed fear. Perhaps a robust discussion of not only the risks of Yucca Mountain, but also its benefits, would bring a sophisticated response from a public that weighs the benefits, risks and perceptions of the repository in a rational way and leads to informed decisions on how to deal with this national problem.

# 22. NWPO Socioeconomic Reports

The Nevada Nuclear Waste Project Office funded studies on a variety of socioeconomic issues in the early years and the citizens of Nevada should still be curious what more than $15 million initially expended paid for. Knowing what was left out the studies is perhaps even more important. While the Nevada socioeconomic studies may have been 'state-of-the-art', they did not serve Nevadans well as a means of planning their future in regard to Yucca Mountain. In fact, since they were advocacy studies rather than objective, they must all be thrown out if Yucca Mountain is restarted.

The reports issued by NWPO socioeconomic researchers unfortunately suffer from a number of flaws:

1. Except for research on the Indians tribes, the research was done almost exclusively by out-of-state academics with little feel for local culture, economic conditions or traditions.

2. Many of the studies have little bearing on Nevada and are in fact philosophical works more suited for academic research papers than the real world.

3. The overriding concern in many papers is to find a way to defeat Yucca Mountain, rather than do practical social and economic research. This is evident in some rather bizarre studies.

4. Much of the polling is politically motivated.

5. The main studies were monopolized by a few social researchers who appeared willing to encroach on the engineering sciences without proper credentials.

6. Much of the philosophy proposed is based on Rawlsian ethics. Rawlsian ethics is not part of main stream socioeconomic research because it is speculative and perhaps even revolutionary.

Some of the socioeconomic research problems are evidenced in the titles of their resulting reports and for that reason we've printed the following list of publications available from NWPO to show their philosophical trend. The implications of some of the NWPO studies are obscure, so we've attempted to provide clarification where possible to explain why many of these studies are suspect.

## NWPO SOCIOECONOMIC REPORTS

**NOTE: Those reports which either have little bearing on Nevada's socioeconomic future, are poor science, or otherwise deserve further study and validation by researchers are marked with a (\*). A bracketed comment [ ] following a report suggests possible problems.**

- NWPO-SE-001-88 Retirement Migration and Military Retirement Planning Information Corporation (June 1988)

- NWPO-SE-002-88 Characteristics of the Las Vegas / Clark County Visitor Economy. Planning Information Corporation (June 1988)

- NWPO-SE-003-88 Current Target Industry Analysis: Las Vegas Metropolitan Area. M. Ross Boyle, Growth Strategies Organization (June 1988)

- NWPO-SE-004-88 Business Profile of Metropolitan Las Vegas. M. Ross Boyle, Growth Strategies Organization (June 1988)

- NWPO-SE-005-88 Nevada State Revenue Analysis. Planning Information Corporation (June 1988)

- NWPO-SE-006-88 Nevada Local Government Revenue Analysis. Planning Information Corporation (June 1988)

- * NWPO-SE-008-88 The Effects of Human Reliability in the Transportation of Spent Nuclear Fuel. Seth Tuller, Roger E. Kasperson and Samuel Ratick, Center for Technology, Environment and Development, Clark University (June 1988). **[Psychological rather than engineering analysis. Tends to overemphasize inevitability of human error without allowing for adaptive engineering responses. Relies heavily on interpretations of Lindsay Audin of the environmental movement.]**

- * NWPO-SE-008-88 Risk Management and Organizational Systems for High-Level Radioactive Waste Disposal: Issues and Priorities. Jacque Emel, Brian Cook and Roger Kasperson with Halina Brown, Robert Goble, Jeff Himmelberger and Seth Tuller, Center for Technology, Environment and Development, Clark University (September 1988) **[Lacks engineering analysis]**

- * NWPO-SE-009-88 Distributional Equity Problems at the Proposed Yucca Mountain Repository. Roger E. Kasperson and Sassan Abdollahzadeh, Center for Technology, Environment and Development, Clark University (July 1988) **[Emphasizes egalitarianism solutions over utilitarianism. Nuclear technology apparently does not conform to egalitarian values]**

- * NWPO-SE-010-88 Potential Retrieval of Radioactive Wastes at the Proposed Yucca Mountain Repository: A Preliminary Review of Risk Issues. Robert Goble, Dominic Golding and Roger E. Kasperson, Center for Technology, Environment and Development, Clark University (June 1988) **[little engineering expertise evident. Overemphasizes human risk factors.]**

- * NWPO-SE-011-88 Postclosure Risks at the Proposed Yucca Mountain Repository: A Review of Methodological and Technical Issues. Jacques Emel, Roger E. Kasperson, Robert Goble and Ortwin Renn, Center for Technology, Environment and Development, Clark University (June 1988) **[little engineering expertise evident. Overemphasizes human risk factors.]**

- * NWPO-SE-012-88 The Accident at Gorleben: A Case Study of Risk Communication and Risk Amplification in the Federal Republic of Germany. Hans Peter Peters and Leo Hennen, Center for Technology, Environment and Development, Clark University (June 1988) **[Risk amplification was to an extent premeditated by anti-nuclear factions.]**

- * NWPO-SE-013-88 New Mexico's Waste Isolation Pilot Plant (WIPP): An Historical Overview. Ronald G. Cummings, University of New Mexico (June 1988) **[Emphasizes opposition politics.]**

- * NWPO-SE-014-88 The U.S. Department of Energy's Attempt to Site the Monitored Retrievable Storage Facility (MRS) in Tennessee, 1985 - 1987. Michael R. Fitzgerald and Amy Snyder McCabe, Energy, Environment and Resources Center, University of Tennessee (May 1988)

- * NWPO-SE-015-88 Goiana Incident Case Study. John S. Petterson, Impact Assessment Corporation (June 1988) **[Brazilian situation is not very similar to Yucca Mountain, technically or socially]**

- NWPO-SE-016-89 Assessment of the Impact of a Nuclear Waste Repository at Yucca Mountain on the Economic Development Potential of Las Vegas, Clark County, and the Surrounding Area. M. Ross Boyle, Growth Strategies Organization (January 1989) **[Boyle is headquartered in Virginia]**

- NWPO-SE-017-89 Summary of Background Fiscal Data and Analysis for Nevada Socioeconomic Study to Date. Planning Information Corporation (January 1989)

- * NWPO-SE-018-89 Assessing the State/Nation Distributional Equity Issues Associated with the Proposed Yucca Mountain Repository: A Conceptual Approach. Roger E. Kasperson and Samuel Ratick, Center for Technology, Environment and Development, Clark University (June 1988) **[Based on Rawlsian ethics, the equity question is not relevant to assessing the actual socioeconomic impact, whether fair or not, from Yucca Mountain.]**

- * NWPO-SE-019-89 A Framework for Analyzing and Responding to the Equity Problems Involved in High-Level Radioactive Waste Disposal. Roger E. Kasperson, Sam Ratick and Ortwin Renn, Center for Technology, Environment and Development, Clark University (June 1988) **[Does not address practical socioeconomic impact issues, but favors Rawlsian issues]**

- NWPO-SE-020-89 Nevada State and Local Government Comments on the U.S. Department of Energy's Report to Congress Pursuant to Section 175 of the Nuclear Waste Policy Act, as Amended. Prepared by the Nevada Agency for Nuclear Projects and Affected Local Government (March 1989)

- * NWPO-SE-021-89 The Convention Planning Process: Potential Impact of a High-Level Nuclear Waste Repository in Nevada. Howard Kunreuther, Doug Easterling and Paul Kleindorfer, Center for Risk and Decision Processes, The Wharton School, University of Pennsylvania (September 1988) **[Done out-of-state despite local convention analysis expertise.]**

- NWPO-SE-022-89 Yucca Mountain Socioeconomic Project: An Interim Report on the State of Socioeconomic Studies. Mountain West Research, Las Vegas, Nevada (June 1989) **[ Little objective peer review]**

- NWPO-SE-023-89 Perceived Risk, Stigma, and Potential Economic Impacts of a High-Level Nuclear Waste Repository in Nevada. Paul Slovic, et al. of Decision Research, and James Chalmers, et al. of Mountain West Research, Las Vegas (June 1989) **[Overemphasizes reactions of people to fears. Ignores Nevada's positive experience with the Nevada Test Site]**

- NWPO-SE-024-89 Executive Summary: An Interim Report on the State of Nevada Socioeconomic Studies. Mountain West Research, Las Vegas (June 1989) **[ Little objective peer review]**

- NWPO-SE-025-89 Yucca Mountain Socioeconomic Project Preliminary Findings: 1989 Nevada Telephone Survey. Mountain West Research (June 1989)

- NWPO-SE-026-90 Native Americans and Yucca Mountain: A Summary Report. Catherine S. Fowler, Maribeth Hamby and Mary Rusco, University of Nevada Reno/Cultural Resources Consultants, Ltd (September 1990)

- NWPO-SE-027-90 Social Amplification of Risk: An empirical Study. William Burns, Paul Slovic, University of Oregon / Decision Research, and Roger Kasperson, Jeanne Kasperson, Ortwin Renn and Srinivas Emani, Center for Technology, Environment and Development, Clark University (September 1990) **[Driven by need to justify amplification theory despite contrary local evidence.]**

- * NWPO_SE-028-90 What Comes to Mind When You Hear the Words "Nuclear Waste Repository"? A Study of 10,000 Images. Paul Slovic, Mark Layman and James H. Flynn, Decision Research (September 1990) [See chapter19, Fear Sells for rebuttal by Bassett and Hemphil. **[None of the images associated with the Nevada Test Site appear to haunt tourists, contradicts local experience..]**

- * NWPO-SE-029-90 Evaluations of Yucca Mountain: Survey Findings About the Attitudes, Opinions and Evaluations of Nuclear Waste Disposal and Yucca Mountain, Nevada. James H. Flynn, Paul Slovic, C.K. Mertz and James Toma, Decision Research / Coopers and Lybrand (September 1990) **[Polling used as political propaganda vehicle rather than objective science.]**

- * NWPO-SE-30-90 Images of a Place and Vacation Preferences: Implications of the 1989 Surveys for Assessing the Economic Impacts of a Nuclear Waste Repository in Nevada. Paul Slovic, Mark Layman and James H. Flynn, Decision Research (September 1990) **[Overemphasizes negative nuclear imagery despite contrary local evidence.]**

- * NWPO-SE-031-90 The Vulnerability of the Convention Industry to the Siting of a High-Level Nuclear Waste Repository. Douglas Easterling and Howard Kunreuther, Wharton Risk and Decision Processes Center, University of Pennsylvania (September 1990) **[UNLV's school of Hotel Management is world class and could have produced a first rate analysis based on its association with the Nevada gaming and convention industry.]**

- * NWPO-SE-032-90 Risk-Induced Social Impacts: The Effects of the Proposed Nuclear Waste Repository on the Residents of the Las Vegas Metropolitan Area. A. Mushkatel, D. Pijawka and M. Dantico, Management Strategies and Research Inc./ Arizona State University (September 1990) **[Little discussion of positive benefits.]**

- * NWPO-SE-033-90 Major Sociocultural Impacts of the Yucca Mountain High-Level Nuclear Waste Repository on Nearby Rural Communities. Ronald L. Little and Richard S. Krannich, Rocky Mountain Social Science/Utah State University. (September 1990) **[Local expertise in Nevada University system overlooked.]**

- * NWPO-SE-034-90 Nuclear Waste Management: A Comparative Analysis. Jacque Emel, Brian Cook, Roger Kasperson and Ortwin Renn, Center for Technology, Environment and Development, Clark University and Gordon Thompson, Institute for Resource and Security Studies (November 1990) **[Irrelevant to Nevada's socioeconomic needs]**

- * NWPO-SE-035-90 Estimating the Economics of a Repository from Scenario-Based Surveys: Models of the Relation of Stated Intent to Actual Behavior. Douglas Easterling, Vicki Morwitz and Howard Kunreuther, Wharton Center for Risk and Decision Processes, University of Pennsylvania (November 1990) **[Overstates intent to actual behavior.]**

- * NWPO-SE-036-87 Yucca Mountain Socioeconomic Project - Report on the 1987 Risk Perception Telephone Surveys. Wharton School, University of Pennsylvania; Decision Research; and Research Triangle Institute (September 1987) **[Surveys used as political propaganda.]**

- * NWPO-SE-037-91 Yucca Mountain Socioeconomic Project - The 1991 Nevada State Telephone Survey: Key Findings. James H. Flynn, C.K. Mertz, and Paul Slovic, Decision Research (May 1991) **[Surveys used as political propaganda.]**

- *NWPO-SE-038-91 Southern Nevada Residents' Views About the Yucca Mountain High-Level Nuclear Waste Repository and Related Issues: A Comparative Analysis of Urban and Rural Survey Data. Richard S. Krannich and Ronald L. Little, Utah State University and Alvin Muskatel. K. David Pijawka and Patricia Jones, Arizona State University (October 1991) **[Neglects Nevada university expertise.]**

- NWPO-SE-039-91 Native Americans and Yucca Mountain: A Final Summary Report (Volumes I and II). Catherine Fowler with contributions by Maribeth Hamby, Elmer Rusco and Mary Rusco, Cultural Resources Consultants Ltd. / University of Nevada, Reno (1991)

- NWPO-SE-040-91 Socioeconomic Profiles of Native American Communities: Las Vegas Tribe of Paiute Indians. Mary K. Rusco, Cultural Resources Consultants Ltd. / University of Nevada, Reno (October 1991)

- NWPO-SE-041-91 Socioeconomic Profiles of Native American Communities: Yomba Shoshone Reservation. Maribeth Hamby with Appendix on Establishment of the Yomba Reservation by Elmer Rusco, Cultural Resources Consultants Ltd. / University of Nevada, Reno (October 1991)

- NWPO-SE-042-91 Socioeconomic Profiles of Native American Communities: Duckwater Shoshone Reservation. Maribeth Hamby, Cultural Resources Consultants Ltd. / University of Nevada, Reno (October 1991)

- * NWPO-SE-043-91 Native Americans and State and Local Governments. Elmer Rusco, Cultural Resources Consultants Ltd. / University of Nevada Reno (October 1991)

- * NWPO-SE-044-91 A Structural Model Analysis of Public Opposition to a High-Level Radioactive Waste Facility. James Flynn, C.K. Mertz and Paul Slovic, Decision Research and William Burns, University of Iowa (September 1991) **[Overemphasizes unchangeability of public's nuclear perceptions.]**

# 23. Technical Review Committee

The socioeconomic studies conducted for the Nevada Nuclear Waste Project Office were supposedly peer reviewed by the State of Nevada Technical Review Committee. The accompanying table gives the members of the Committee, their affiliations and their not insignificant reimbursement (more than $1 million). Technical review boards on large science projects like Yucca Mountain project are to be encouraged, but it is unclear whether Nevada's Technical Review Committee acted as peer review or as rubber stamps for poor science. The million dollar question is thus whether Technical Review Committee members were chosen to do rigorous oversight, or whether their purpose was to act as cheerleaders for the equity theories of Roger Kasperson and the risk perception theories of Paul Slovic.

The hourly rates of the Committee members may seem high to people outside the world of consulting, but when non-billed hours, office maintenance, etc. are included, those figures are not out of the ordinary. What is out of the ordinary is the amount of peer review work that was done and who was asked to participate in that work:

1. No one connected to Nevada had any hand at all in the peer review. All members of the Technical Review Committee are from out-of-state.
2. The Technical Review Committee appears to have been handpicked to mirror the ideological bias of NWPO.
3. The purpose of the review was supposedly to ensure sound science. Yet, the TRC's criticism of the socioeconomic research was kept secret.

None of the names on the Review Committee have the slightest geographical affiliation with the state of Nevada. If the Universities of Nevada at Reno and at Las Vegas were cow-town colleges with no expertise in local demographic studies this might be acceptable, but both campuses have a deserved national reputation for being up-and-coming as reported in U.S. News and World Report, college review editions.

The exclusion of Nevada residents from academic peer review is in itself mysterious. After all, the socioeconomic studies were meant to explore the social and economic impact of the Yucca Mountain repository on Nevadans, not on people in Pennsylvania, Colorado, Michigan and other distant states from whence the committee members came. NWPO is a Nevada agency entrusted to carry out the best interests of Nevadans, while the federal government and DOE already exist to worry about out-of-state interests.

## TECHNICAL REVIEW COMMITTEE AGENDA

If the academicians on the committee weren't responsible to people in Nevada, then who did they represent? Unfortunately, it appears they mostly represented the cloistered group of scholars encompassing the Mountain West socioeconomic team, serving as rubber-stamps for that groups radical socioeconomic theories. Clues comes from the geographic distribution of some of the participants. Michael Bronzini of Pennsylvania State and Dohrenwend of Columbia University are close to Three Mile Island and indeed Dohrenwend was involved with Roger Kasperson, Paul Slovic and James Chalmers in the Kemeny Commission studies of the Three Mile Island accident in the early eighties. Other Committee members were colleagues of Roger Kasperson or Paul Slovic in the past.

This is not meant to call into question the credentials of the TRC members, but to instead point out that they were formed with an internal bias. Since no Committee members are resident Nevadans, they lacked an internal rational perspective of the socioeconomic situation in Nevada. Being paid large sums for little work, they had little desire to rock the boat if the studies veered away from reality. Since Nuclear Risk Perception is a tiny and specialized field, the group was also subject to academic incest, with no new ideas other than those of the main researchers allowed to enter. The following backgrounds of TRC members thus help to put their biases in an open light.

**GILBERT F. WHITE:** The Technical Review Committee Chairman, White then at of the University of Colorado, but was for a time the president of Resources for the Future, whence Allen V. Kneese derives. Mountain West also took advantage of other members of Resources for the Future, further interlocking the technical review with those doing the research. More importantly, Roger Kasperson got his PhD at the University of Chicago where Gilbert White was then a professor of geography. Kasperson began quoting G.F. White in the sixties and worked with him later. Consequently, the head of the Technical Review Committee was in effect personally chosen by Roger Kasperson.

**ALLEN KNEESE:** A fellow at Resources for the Future, Kneese was one of the first to write a think tank paper condemning nuclear power in a 1972 article titled "The Faustian Bargain" [Resources For the Future, 1972]. This article, later favorably cited by the anti-nuclear zealot Helen Caldicott, was one of the first works to claim the in feasibility of nuclear waste storage. Kneese worried about the creation of what Alvin Weinberg had claimed would be a technological priesthood needed to guard nuclear waste. The Resources for the Future connection to Gilbert White and other opposition to nuclear energy suggest Kneese was not appointed to the Technical Review Committee as unbiased peer review. Indeed, Kneese had provided a paper for Roger Kasperson's 1983 book, Equity Issues in Waste Management, that played fast and furious with the economics and cost-benefit analysis of nuclear waste disposal.

Writing a chapter titled "Economic Issues in the Legacy Problem", Kneese evaluated nuclear energy technology on four ethical criteria: utilitarian, libertarian or Pareto superior, elitist and Rawlsian. Kasperson later requoted the Kneese chapter in a paper for NWPO giving his interpretation:

**The exemplary (Kneese) analyses, using the four criteria, suggest that:**
1. **the utilitarian analysis would reject nuclear power because the costs clearly exceeded the benefits;**
2. **the libertarian or Pareto superior analysis would argue against producing nuclear energy because there is no available mechanism for making compensation which is clearly consistent with the Pareto ethical regime;**
3. **since the production of nuclear energy positively benefits the current generation, the elitist criterion would favor nuclear waste production;**
4. **the Rawlsian criterion leads to any unconstrained maximization of the second generation's expected utility and so the current generation should not produce nuclear power.**
**[Kasperson, Roger; Abdollahzadeh, Sassan; "Distributional Equity Problems at the Proposed Yucca Mountain Facility", NWPO-SE-009-88, 1988, p22]**

Only elitists apparently would approve of nuclear power, although we've seen Kasperson had previously dismissed elitism in favor of Marxist and egalitarian philosophies in his formative monograph "Participation, Decentralization And Advocacy Planning" [Kasperson, Breitbart; Association of American Geographers, 1974]

**EDITH PAGE:** A long time Project Director from the Office of Technology Assessment (OTA), U.S. Congress. Page evidently had no problem pulling in an extra $80,000 over five years doing peer review even while working in Washington for the OTA. This may well have been a conflict of interest if she was a federally salaried employee working against a government sponsored technology that her agency might be called on to offer oversight. At $40 per hour this represents 2000 hours or approximately one year of full time work. One wonders if Page's bosses at OTA encouraged this moonlighting. Since NWPO's avowed goal was to derail Yucca Mountain and nuclear energy in general, we potentially have a federal worker working against the Nuclear Waste Policy Act.

**KAI ERIKSON:** In an opinion printed in the Nevada Nuclear Waste News printed by NWPO [Removing the Solution, . . . Not the Problem Vol 5, No. 1, Jan 1994], Erikson states:

**"To examine the potential social and economic consequences of the repository, the State of Nevada supplemented other federally supported studies by engaging an experienced group of research specialists. . . On the basis of that experience, our committee has become convinced that the federal government has not adequately considered the human element in its thinking about nuclear waste."**

The fact that a social scientist working for the State of Nevada (which spent $15 million on socioeconomic studies and $112,000 on Erikson) finds the human element inadequately considered is somewhat surprising.

**"How can we assume that the environmental envelope in which we live will not be rearranged altogether by advanced technologies? How can we be sure that people will not be attracted to that conveniently packaged waste because they see it as a valuable resource? Perhaps it will be perceived as a weapon buried in enemy territory, needing only to be activated, or as a place of such power that it excites religious awe. We do not know and cannot know the answers to these questions."**

Erikson is right. Indeed, we may not know any sociological fact, for the entire global population might well wake up tomorrow believing that Ronald McDonald is God and Golden Arches have religious significance. A lack of quantifiable criteria for judging the social perception studies conducted by Nevada is one reason these studies are viewed with suspicion. No technology could be constructed under Erikson's restraints for sociological predictability.

**MICHAEL S. BRONZINI** Dept. of Civil Engineering Pennsylvania State University. Why did Nevada's socioeconomic Technical Review Committee need a $75/hour Civil Engineer from Pennsylvania when it could find excellent ones in the state skilled in radiation issues?

**BRUCE DOHRENWEND:** Also present at the Three Mile Island studies, Dohrenwend was already part of the academic bubble that included Mountain West, Decision Research and Clark University.

**E. WILLIAM COLGLAZIER:** From the University of Tennessee, near Oak Ridge National Laboratories. A relatively objective voice, at a February 1, 1989 meeting on the Mountain West Section 175 Report, Colglazier suggested 'a need for an introductory section in the next draft report indicating what is known from the research and what remains unknown. A statement of humility regarding the inability to make projections, given the state of uncertainties, is needed in this introduction.'

**REED HANSEN:**

**RICHARD MOORE:**

**ROY RAPPAPORT:** [As with Colglazier, these three may have been relatively given a lack of evidence of their prior convictions. Nevertheless, as with all the Technical Review Committee, they were not Nevadans. This is more than a small concern, especially since perceived risks came to play such a large part in the socioeconomic analysis provided by the state in the form of so-called 'special effects.' Nevada culture is perceptibly independent, literally a culture of gamblers, so it is difficult to understand how peer reviewers could judge the value of perceived risks among such a population without incorporating distinctly non-Nevada perspectives.]

## Technical Review Committee Opinions

The Technical Review Committee wasn't formed to do the grunt work of basic research, they weren't called on to write voluminous papers, nor were they required to take residency in the state. Some of the Committee members walked away $200,000 richer for doing what might be called an academic sleep walk. The question is whether their main purpose was to do rigorous peer review, or to validate the often radical positions of Mountain West and its contractors.

Sociological theories are accepted not by solid analytical or empirical proof, but through consensus building among professional peers. Unfortunately, one way to build such consensus is by stacking a peer review committee in a theory's favor and awarding lucrative contracts to the reviewers. Whether this is what happened in Nevada will obviously be a subject of debate, though there is a conspicuous lack of record of constructive criticism from the Technical Review Committee that would put this suspicion to rest.

The limited paper trail we do have stops after the February 1, 1989 Technical Review Committee meeting. According to informed sources, observers were not allowed into subsequent meetings, nor were notes taken, because it became obvious there were problems with providing quantifiable results based on the 'special effects' being developed by Mountain West and because the debate had become somewhat rancorous. Instead of allowing these debates to rage in public, the peer review was put behind closed doors. So much for transparency.

Yet, the record from the 1989 meeting suggests even the Technical Review Committee couldn't accept all the liberties taken by the Mountain West in their socioeconomic studies. In preparation for what is termed the Section 175 Report, a large preliminary impact statement required by law to be presented in 1989, the Technical Review Committee met to do peer review over the work of the Mountain West Contingent. An abbreviated set of minutes from this meeting shows that there was substantial dissension in the ranks.

# TECHNICAL REVIEW COMMENT PERIOD 2/1/89

**Gilbert White** (Chair, TRC) introduced the comment period and requested that each member of the TRC indicate the key questions to be resolved.

**B. Dohrenwend** focused on the perception of risk studies and indicated that their strength lies in providing a valuable baseline; their weakness is "everything else". Questions that are likely to be raised by DOE and Congress include the following. Why are the closest residents so favorable to a repository if it is so risky? Why are nuclear images mentioned so infrequently by tourists to Las Vegas, especially given the Nevada Test Site? With responses opposite from those one would intuitively expect, (e.g., the inverse relationship between distance and opposition to a repository), are the results credible? Dohrenwend suggested that the many qualifying statements about the research currently in the report should be taken seriously by the authors, such as "people may not really know how the repository will affect their behavior."

**K. Erikson** stated that the report is very hard to read and is fragmented. He suggested that the social amplification portion of the stigma research is the strongest part, but that a different term should be used. "Amplification" implies that there is a "true" value of risk that has been overexaggerated by ordinary people, which is not correct.

**R. Rappaport** indicated that he agreed that the report is difficult to read and suggested reorganization. The report should indicate how the authors want their research to be used. He also suggested that DOE's effort and the State's research contribute to or create impacts, which can be managed through openness. The critique of DOE should be enlarged to address the organizational ability or inability of DOE to validly conduct site characterization. The State should analyze the organizational culture of DOE. Additionally, Rappaport suggested that the State capitalize on the uncertainty associated with the first-of-a-kind repository and should take a conservative approach: long term uncertainty should not be traded for short term benefits.

**E. Page** suggested that the tone of the report is a fundamental problem. The tone is one of Nevada as victim and the tone is supported by an amalgamation of bits of information rather than a coherent theme. She noted that the report is very weak on transportation analysis, which is Nevadans' primary concern.

**R. Hanson** indicated that the report fails to provide an incisive analysis of the data, although it does provide a summary of work to date. It cannot serve as a companion to the 175 Report. The report's major deficiency is its tone and the lack of a connection between risk perception theory and behavior. Hanson suggested that there is nothing in these studies on which decisions may be made. The research is suggestive, but projections based upon these suggestions are unacceptable. He indicated that the notion of a social welfare function is not addressed. He also noted that a 10 percent long term reduction in tourism, used as the scenario for projections, is simply not credible. The report should contain a discussion of mitigation and compensation.

**R. Moore** noted that DOE bashing, which is prevalent throughout the report, should be deleted. Related to perceptions of risk, Moore does not believe that dampening of future growth is likely. Nevertheless, a short term catastrophe that would close tourism for a brief period would be devastating in the short run. He also indicated that transportation must be emphasized, as well as emergency response needs.

**C. Colglazier** suggested that a separate report may need to be written to address standard impacts without consideration of perceptions of risk effects. He suggested that it is plausible that negative images of Las Vegas or Nevada may not develop as a result of the repository, that such images may be temporary if they do occur, and that the perceptions of risk responses may be altered depending on how DOE responds to them. In order to be credible, all sides of perceptions of risk and stigma arguments need to be presented.

**G. White** concluded that the credibility of the report would be enhanced by clear statements about what is known as a result of the research and what remains unknown. The report also needs a clear statement of the assumptions underlying the research. He noted that alternatives to a repository, such as recycling, are not mentioned in the report, nor is the fact that a repository has no precedence highlighted.

**J. Chalmers** (Mountain West) stated that the research, which is intended for public consumption, needs to be described in clear, fair terms and that there needs to be a delineation between methods of research and the application of results. (Lunch Break)

**P. Slovic** suggested that the potential behavioral responses of unknown probability that may result from a repository, even though uncertainties exist about the nature of these responses, must be taken into consideration in the decision-making process.

**H. Kunreuther** indicated that the research has raised issues that are not considered in traditional benefit-cost analysis. He suggested that the repository is unique and, therefore, the issues and behavioral responses resulting from the repository are unique. Benefit-cost analysis, if employed in a traditional manner, would not capture the uniqueness of the repository or the behavioral responses to it.

**B. Dohrenwend** suggested that many of the researchers talked about an "impressive convergence" of the multiple studies of risk perceptions and intended behavior change. What is not clear in the report is exactly what is converging?

**J. Chalmers** suggested that the convergence of results from multiple studies indicates that something is happening; convergence of findings indicates that one cannot reject the null hypothesis that there is not a perception of risk phenomenon associated with the repository, which may adversely affect behavior.

**R. Kasperson** noted that implicit in the perceptions of risk and stigma studies is the fact that traditional ways of doing social impact analysis are no longer sufficient because the repository is unique. Impact analysis must be innovative in order to include new, emerging issues (e.g., perceptions of risk).

**C. Colglazier** suggested a need for an introductory section in the next draft report indicating what is known from the research and what remains unknown. A statement of humility regarding the inability to make projections, given the extent of uncertainties, is needed in this introduction.

**R. Rappaport** asked the rhetorical question, how does one proceed prudently in the absence of knowledge?

**A. Kneese** noted that the perceptions of risk and stigma paradigm grinds throughout the entire analysis of all communities and impacts, including those communities where studies indicate there would be little or no effect from perceptions of risk. He suggested that standard impacts should be addressed first for the area, then a hypothetical perception of risk scenario should be presented and the risk effects traced.

**E. Page** noted that the notion of the social amplification of risk almost overwhelms those things that can be said with some certainty, such as assessment of standard impacts.

**G. White** noted the difficulty of presenting the research data in ways that will be helpful to locals.

**R. Moore** indicated that the State must be able to show potential or actual impacts in order to continue research funding. He noted that the repository program is continuing, even though it stalls periodically. Therefore, there is a need to identify standard impacts that need to be mitigated.

**D. Bechtel** (Clark County) indicated that he agrees that standard impacts need to be addressed now and need to be separated from special effects stemming from perceptions of risk and stigma. If these components are separated, the assessment of standard impacts may be salvageable. As the report is currently, it is of no value.

**W. Freudenburg** suggested that standard impacts are likely to be twice those currently estimated.

**R. Hanson** agreed that standard impacts are likely to be larger than projected currently. He noted that a comprehensive monitoring program is essential. He also suggested that there is not enough detail in tabular format for readers to discern results.

**J. Chalmers** suggested that the only conclusion to be drawn from the perceptions of risks studies is that it would be unwise to ignore risk-induced behavioral effects. It may be advisable, though, to separate the assessment of standard impacts from the assessment of special effects.

**R. Moore** suggested that the State challenge the Section 175 Report conclusions that sufficient resources exist and sufficient authority exists to mitigate impacts.

**S. Bradhurst** stated that the needs of Nye County should be the basis of future work. Therefore, the assessment of standard impacts needs to be presented in such a way as to identify where additional data and research are needed. The assessment of special effects needs to be separated from the assessment of standard impacts; otherwise, the studies are not usable for impact mitigation. The TRC should assess the adequacy of standard impact analysis. Are there going to be impacts during site characterization?

**D. Bechtel** indicated that he had hoped the report would provide definitive results regarding standard impacts and that he was disappointed with the report. He stated that he was almost ready to simply take the information already collected and continue with analysis of the County's needs. He suggested that he will have difficulty with the State's report even if special and standard impacts are separated.

------------------------------

Considering the fact that the Technical Review Committee was already heavily weighted in favor of accepting the Mountain West contingent's theories and many members had previous professional ties to Roger Kasperson, these minutes can only be viewed as a strong vote of no confidence in the direction of the State's socioeconomic studies. However, even after this blunt rebuke there was very little change in the "DOE bashing", the over-emphasis on dubious risk perception theories and general lack of rigor in NWPO's studies. What did change was that the Technical Review Committee's minutes were no longer generally available for review, possibly because the State was afraid the secret that not everyone viewed their studies as perfect would be discovered.

# 24. Hotels, Gaming & Economic Impact

The good news and the bad news of Yucca Mountain is that its economic impact on Southern Nevada will in the long run be negligible. The good side of this coin is that the impact of the repository on the mercurial gaming and hotel industry will be insignificant. The very few tourists expected to shy away because of their risk aversion to Yucca Mountain repository will be offset by the stabilizing effect of the repository workers on the local economy.

The bad news is that because the socioeconomic impact of Yucca Mountain will be much less than what was originally thought, Nevada's bargaining position with the government for compensation has been severely diminished. If the state's politicians thought they were increasing their bargaining position with the federal government over the years by playing hard-to-get on the benefits question, they obviously miscalculated. This may be because former Senators Bryan and Reid are politician-lawyers, not businessmen, and played their cards based on political rather than economic cues.

The lack of impact is paradoxically highlighted in a report from the Nuclear Waste Project Office titled "Yucca Mountain Employment: A Review of DOE's Data On Jobs And Work force". The report was sent out by NWPO as part of its aggressive ongoing campaign to prove to Nevadans that there would be no positive benefit from the repository. This was a two-edged sword, because it also strengthened the argument that compensating benefits and negotiations with the federal government are the only way Nevada would gain from hosting the repository. Among NWPO's conclusions:

**Yucca Mountain will not be a source of jobs for Nevada Test Site workers displaced as a result of the nuclear weapons testing moratorium. There are too few potential repository jobs and too great a time interval between NTS impacts and peak repository employment." "Employment associated with the Yucca Mountain program is extremely limited and represents only four tenths of one-percent (.004) of the Clark County work force. ["Yucca Mountain Employment: A Review of DOE's Data On Jobs And Work force", NWPO, Aug 1993]**

Obviously, the entire purpose of the NWPO study, as well as its widespread distribution in a sophisticated mailer, was to obtain political mileage from the lack of positive employment impact expected from the repository. This does not in any way relate to the safety of Yucca Mountain or the possibility of mitigating impact, and again points to the use of Mountain West socioeconomic studies for overt political purpose. The effect of this report, and many others produced by NWPO, was to undermine any potential negotiating power Nevada might have had with the federal government. In the event that Yucca Mountain finally proved acceptable, the state will already have demonstrated to its utmost that there is no employment impact on the local economy, either positive or negative!

The fact that gaming and tourism are unlikely to be negatively affected by the Yucca Mountain project is borne out by a study of the effects of the Three Mile Island accident done by Jeffrey Himmelberger, Yelena Obneva-Himmelberger and Mike Baughman. In a draft article titled "Tourist Visitation Impacts of the Accidents at Three Mile Island: Implications for Yucca Mountain spring 1993, the group argues in its conclusions that:

**"Information contained within this report strongly supports the conventional view's assessment that the gasoline crisis rather than TMI or other negative factors was primarily responsible for the 1979 slump in tourism experienced in south-central Pennsylvania. Our findings are compatible with previous qualitative research (discussed previously). Moreover, the state park data provided some evidence linking cool and rainy weather conditions to slowed summer 1979 tourism levels. The polio outbreak in Lancaster County may have benefited tourism at nearby destinations. Finally, none of the regressions associated the nuclear accident at TMI to summer 1979 visitation levels.**

**"Based on these results it is plausible to consider that an accident with similarities to TMI occurring at (or near) the proposed Yucca Mountain repository would generate proportionally large amounts of negative national publicity and would probably translate into reduced tourism immediately following the accident (mainly from select populations that are highly risk adverse). However, assuming that the results of this study can be extrapolated temporally and geographically, it is further possible that such an incident**

would only marginally affect long term visitation levels at nearby tourist destinations. Assuming that little or no radioactive contamination actually resulted from such an accident, recovery of visitation to pre-accident levels would be expected to occur fairly rapidly.

"While our results support the Federal Government's perspective on anticipated tourism impacts of a repository at Yucca Mountain, they do not contradict the State of Nevada's perspective that the perceived repository risks and negative stigma may translate into large secondary socioeconomic effects. However, it may be the case that visitor economies are highly insulated from adverse socioeconomic impacts which may accompany the location of risky facilities and/or accidents at such facilities." [Himmelberger, Jeffrey; Himmelberger, Yelena; Baughman, Mike; "TOURIST VISITATION Impacts of the Accident at Three Mile Island: Implications for Yucca Mountain", Intertech Services Corp., spring 1993]

Thus, while there is little doubt that a significant accident in the transportation or storage of nuclear waste would have a large immediate negative impact on the Las Vegas tourist economy, it is not clear that long term effects would be substantial. In fact, if the engineering projections are correct about the safety margins of waste packages and the repository, damage to a cask might lead to such trivial releases and the cleanup be so moderate that a return to normalcy might be immediate.

It is interesting to note that all three writers of the above report are graduate products of the Clark University geography program, where originated Roger Kasperson's CENTED studies of Yucca Mountain. That their balanced report is not full of the same dire predictions of negative images and a floundering tourist economy as presented by most of the CENTED participants of NWPO's socioeconomic studies is indicative that there exists opposition to the NWPO party line.

Consequently, the socioeconomic studies conducted by NWPO, which concentrated obsessively on the risk perception studies and nuclear stigma studies of Paul Slovic may have caused the gaming industry real harm. Not only were the state's theories not in conformity with past experience in regard to risk perception of tourists to the Nevada Test Site, but they actually have exponentially enhanced the possibility that such nuclear images will grow in the future. NWPO thus hit gamers with a double whammy, their poorly conceived nuclear stigma studies gave the industry false data about possible tourist impacts, and then contributed to making those negative impacts come true.

## REAL THREATS TO NEVADA'S HOTELS, GAMING AND TOURISM

With a little investigation, one finds that an even larger threat to the Southern Nevada gaming and tourist economy may not be Yucca Mountain, but such trends as competition from other states, gambling on Indian reservations and most importantly regulation by the federal government. While the limited impact of Yucca Mountain on the Nevada gaming economy can likely be mitigated by designating transportation routes which completely avoid Las Vegas and other populated areas of the state, solving the other threats to Nevada's gaming industry may require pure political muscle. The Clinton administration provided an example of a potential economic disaster much greater than Yucca Mountain:

### WHITE HOUSE PONDERING DIFFERENT GAMING TAX
The Administration may raise the amount withheld from winnings and extend the rule to new games
WASH. - The Clinton administration has returned to the gaming industry as a potential source of tax dollars to fund welfare reform, officials said Wednesday.

Instead of imposing a 4 percent excise tax on net revenues of casinos and race tracks, as had been considered earlier this year, the White House is mulling a boost in the tax withholding on gambling winnings, Sen. Richard Bryan, D-Nev. said...

Bryan said he asked Panetta about an article in Tuesday's Wall Street Journal that reported the administration is considering a $600 million gambling tax to pay for the General Agreement on Tariffs and Trade...

Under the plan, slot machine players who win more than $7,500 would be required to pay a withholding of 28 percent, Bryan said. Keno and bingo players also would be covered by the rule, which is projected to bring in $250 million over five years...

Gamblers who win more than $50,000 would see an increase in their withholding from 28 percent to 36 percent. This is expected to raise $520 million over five years.

Finally, more stringent reporting requirements are being considered for winnings over $10,000. This would help recover unpaid taxes of about $220 million in five years, officials believe. [Batt, Tony; Las Vegas Review Journal, May 12, 1994, p1A]

After desperate lobbying by Senators Reid and Bryan and Governor Miller with then president Bill Clinton, Congress did set aside its drive to make gambling a revenue source in 1994. Passage of such a bill would have been devastating to Democratic Party politics in Nevada, but it is possible that federal taxation of the gaming industry will be revisited. As the largest gambling oasis in the United States, Nevada would be hurt disproportionately by such taxation.

Thus, the stakes for Nevada at Yucca Mountain are higher than ever if one considers grandfathering of Nevada's tax status as part of the negotiations package. Since taxation might cost Nevada as much as $100 million per year, and the yearly cash compensation from the Nuclear Waste Fund might also be $100 million, the combined "rake" is perhaps lager than $200 million per year. A more inclusive list of negotiated benefits which might impact the Nevada gaming industry's bottom line include:

1.   Grandfathering the tax situation so no new burdens are placed on Nevada's main industry.
2.   Widening of I-15 to Los Vegas to improve the access of Californians to the Strip.
3.   General upgrade of the Las Vegas infrastructure now burdened by a growing population.
4.   Addressing competition from out-of-state casinos and Indian reservations.

Using Yucca Mountain as a bargaining chip for national legislation could therefore have large benefits for the gaming and tourist economy. Furthermore, negotiating for benefits like water rights, infrastructure, educational improvements, etc. might help stabilize the southern Nevada population. These avenues are presently blocked, however, by the political powers of Nevada who refuse to negotiate under any circumstances.

If gambling will no longer be the center of the economic universe in Nevada, the thought that it might be time to diversify the economy is a central subject of state legislative policy. A high-tech venture like Yucca Mountain would then serve to help stabilize the gambling industry in the state as much as the technical community. A line of locals casinos ranging from Arizona Charlie's to the Gold Coast, the Santa Fe, Palace Station and other successful operations argue that resident clientele are able to support sizable gaming operations. While it is true direct Yucca Mountain employment may be small, the wages of those employed will be higher than the state average and worker's impact as a significant part of the non-gaming community can be significantly larger than raw statistics indicate.

The strong suit of Nevada's hotel and gaming moguls has long been knowing when to hold or play their business cards. However they have not been particularly astute on the issue of Yucca Mountain. Rather than forcing Senators Bryan and Reid and Governor Miller to play cards in 1989 or 1990 when the feds were on their knees begging to negotiate with Nevada, gaming executives now have little leverage over their politicians, who in turn have even less leverage in the federal legislature.

# PART IV: GREEN APOCALYPSE

# 25. Technological Priesthood

Unfortunately, the nuclear industry is often its own worst enemy. Caught between a rock and a hard place in its relations with the public, it either puts on an iron mask and bulls its way ahead, or sticks its foot in its mouth when it does try to be sensitive to public needs and perceptions.

An historically important example of this bungled sensitivity were the comments of Alvin Weinberg, former head of Oak Ridge Laboratory, in a series of articles that ran in Science magazine in 1972. Weinberg, often known as the father of the nuclear power industry for his pioneering work, attempted to engage in some philosophy-of-science musings on the subject of nuclear waste that latter were transformed into forty years of nuclear paranoia by the Green movement. According to Weinberg:

**We nuclear people have made a Faustian bargain with society. On the one hand, we offer -- in the catalytic nuclear burner (breeder reactor) -- an inexhaustible source of energy. Even in the short range, when we use ordinary reactors, we offer energy that is cheaper than energy from fossil fuel. Moreover, this source of energy, when properly handled, is almost nonpolluting. . . .**

**But the price that we demand of society for this magical energy source is both a vigilance and a longevity of our social institutions that we are quite unaccustomed to. In a way, all of this was anticipated during the old debates over nuclear weapons. . . . . In a sense, we have established a military priesthood which guards against inadvertent use of nuclear weapons, which maintains what a priori seems to be a precarious balance between readiness to go to war and vigilance against human errors that would precipitate war . . .**

**It seems to me (and in this I repeat some views expressed very well by Atomic Energy Commissioner Wilfred Johnson) that peaceful nuclear energy probably will make demands of the same sort on our society, and possibly of even longer duration. [Weinberg, Alvin; "Social Institutions and Nuclear Energy", Science, 7 July 1972, p33]**

Weinberg's discussions of a "Technological Priesthood" are important because they opened the debate over the disposal of nuclear waste which has evolved into the uproar over Yucca Mountain. Two years later, in 1974, we find Allen V. Kneese (later a member of Nevada's socioeconomic Technical Review Committee) writing a paper titled *The Faustian Bargain* for Resources for the Future, an environmental policy think tank:

**"I am submitting this statement as a long-time student and practitioner of benefit-cost analysis, not as a specialist in nuclear energy. It is my belief that benefit-cost analysis cannot answer the most important policy questions associated with the desirability of developing a large-scale, fission-based economy. To expect it to do so is to ask it to bear a burden it cannot sustain. This is so because these questions are of a deep ethical character. Benefit-cost analyses certainly cannot solve such questions and may well obscure them.**

**These questions have to do with whether society should strike a Faustian bargain with atomic scientists and engineers, described by Alvin M. Weinberg in Science. If so unforgiving technology as large-scale nuclear fission energy production is adopted, it will impose a burden of continuous monitoring and sophisticated management of a dangerous material, essentially forever. The penalty of not bearing this burden may be unparalleled disaster. This irreversible burden would be imposed even if nuclear fission were to be used for a few decades, a mere instant in the pertinent time scales. [ Kneese, Allen V.;"The Faustian Bargain", Resources, Resources For the Future, 1974, p1]**

By concluding nuclear waste is an ethical question rather than one of science, Kneese apparently delinked his analysis from objective benefit-cost analysis eleven years before his work for NWPO. Seven years after Kneese's article in 1979, Roger Kasperson, the philosophical core of NWPO's socioeconomic studies, also sympathetically referenced the priesthood symbolism. Kasperson notes the following in an article titled Institutional Responses to Different Perceptions of Risk in the book "Accident at Three Mile Island":

"To this list [of issues the ACLU believes make the nuclear energy a threat to democratic processes] can be added other issues that have arisen in the past: the centralization of decision making involved with a complex technology that few understand and the priesthood role that could develop for specialized managers and guardians of safety.["Institutional Responses to Different Perceptions of Risk", Accident at Three Mile Island: The Human Dimension, ed. David Sills, Westview, 1981]

A full eighteen years after the Weinberg articles, we see quotes like those of Sierra Club activist Jerry Mander still referencing the nuclear priesthood threat:

"The existence of nuclear energy, and nuclear weaponry, in turn requires the existence of what Ralph Nader has called a new "priesthood" -- a technical and military elite capable of guarding nuclear waste products for the approximately 250,000 years that they remain dangerous. So if some future society, tiring of the present path, should determine to move away from a centralized technological society and toward, say, an agrarian society, it would be impossible. The technical elite would need to remain, if only to deal with the various wastes left behind. So it is fair to say that nuclear technology inherently steers society toward greater political and financial centralization, and greater militarization."[Interview in Whole Earth Catalog, Fall 1991]

Interestingly, Mander quoted Ralph Nader as the source of the "technological priesthood" theory, although it was the nuclear scientist Alvin Weinberg who proposed this scenario long before most environmentalists were involved in nuclear issues. We'll find that Ralph Nader's paranoia of nuclear technology is a germinal organizing force in the anti-nuclear movement, expressed through his many spinoff organizations including Public Citizens Critical Mass Energy Project, the Safe Energy Communication Council, etc.

Mander's comments are instructive in that they link anti-nuclear environmental paranoia to an underlying ideological core. Mander is apparently conjecturing that:

1.   A technological and military priesthood are inevitable given the existence of nuclear energy, and this priesthood is sinister.

2.   A decentralized agrarian society is to be preferred to America's industrialism and centralized farming. Presumably Marxist North Korean agrarian technology would be an apt substitute because it is decentralized and ubiquitous (most of the population is forced to farm to avoid starvation).

3.   Nuclear technology, and technology in general, are inherently flawed because they create a technical elite. Evidently the world should reject high-tech endeavors and reeducate technologists to be non-elitist agrarians.

4.   Nuclear technology inherently steers us towards greater political and financial centralization (why this is necessarily bad we are not told), and towards greater militarization (which doesn't explain the very decentralized non-nuclear feudal wars of Serbia, Bosnia, Iraq and Afghanistan).

Decentralization as a political philosophy is entrenched not only in Mander's opposition to nuclear energy, but in that of much of the environmental movement. Yet, the unanswered question is why our centralized technological society would care to move back towards a decentralized agrarian society. Decentralized agrarianism would only make sense if our sources of energy were disrupted, making our present lifestyles impossible (perhaps by Green revolutionaries attempting to forcefully restructure society along the lines of a decentralized environmental utopia?). Moreover, if nuclear technology inherently steers society toward greater political and financial centralization, why are the greatest centralized dictatorships in the world agrarian states like China and Cuba?

Ironically, a decentralized agrarian society imposed on America by destruction of our nuclear energy capacity would likely require a "Green priesthood" to prevent taboo technological advances from being made. The anti-nuclear worldview is thus potentially coercive in nature; nuclear technology must not only be opposed, but religiously rooted out to prevent future temptation (such as at Yucca Mountain). The Green movement may have an emotional if not moral advantage in pushing this position, although historically attempts to stop technological advance have proved futile. For example, emotional moralistic opposition to Galileo's attempts to forward a scientific worldview were soon overcome and are now a mere historical curiosity.

Thus, Mander's association with the environmental moralists of the Sierra Club does not automatically make his analysis of the evils of centralized nuclear technology synonymous with the common man's best interests. Nor do environmental opponents of Yucca Mountain necessarily own the moral high ground merely by opposing nuclear technology. Both may simply want to replace a technological priesthood with an equally intrusive and coercive Green priesthood.

It takes a well honed sense of paranoia to go from Alvin Weinberg's statements many decades ago to a mystic fear of the creation of a 250,000 year nuclear priesthood carrying out rites on top of Yucca Mountain. The modern Druids of Stonehenge, worshiping an ancient astrological instrument, have turned out to be much less threatening than once supposed. There is reason to suspect that a nuclear priesthood would also prove to be a paper tiger.

Nevertheless, the extent of technological fear created by Weinberg's statements has been substantial. The biologist John Edsall suggested forty years ago that slowing economic growth in America was the only way to avoid Weinberg's Faustian Bargain. In an open letter to Weinberg in the journal Science in 1972, Edsall questioned:

**What then should we do? I would make several suggestions, none original, but several still unheeded.**

**1.    Stick to fossil fuels for the present, with drastically improved antipollution equipment, as our major energy source for some years to come.**

**2.    Slow down the rate of growth of electric power; emphasize economy in the use of power, and work to the utmost to improve efficiency. Set a sliding scale of costs, with rates charged for use of power increasing as use increases; this would promote both efficiency and economy.**

**3.    Discourage manufacture of products that require large amounts of electrical energy (for example, aluminum cans) if satisfactory alternatives that require less electricity exist.**

**4.    Greatly intensify research to develop sources of energy that are nearly pollution-free, notably nuclear fusion and solar energy.**

**5.    Create a national energy commission to formulate and promote national policy on the total energy problem and, as Alfven has suggested (Bull. At. Sci. 28 (No 5), 5 (1972)), create also an international energy commission, to deal with such problems in global terms. [Edsall, John; Letters, Science, vol. 178, Dec. 1, 1972, p933]**

Of course, Edsall's list of suggestions would require a Green priesthood of regulators and commissioners and centralized governmental apparatus to enforce them, the very nightmare people like Jerry Mander twenty years later claim nuclear technology threatens. One might argue such a Green priesthood may in fact be in place at the Environmental Protection Agency. Anti-technological solutions have an inherent threat of coercion; society has to give up things like fossil-fuel cars, electric heat, aluminum cans and myriad other conveniences to make such a system work. This in turn requires very centralized planning to make sure no one shirks their environmental responsibility by, for example, running a gas powered generator rather than a more expensive solar photovoltaic system.

Dismantling the nuclear option thus carries its own political and environmental risks. Weinberg produced a remarkable rebuttal to Edsall in 1972 which is still a timely reply to many of the present critics of Yucca Mountain:

**In focusing so sharply on the negative side of the Faustian bargain implied in nuclear energy, Edsall all but ignores the primary and positive aspects of the bargain. The simple fact is that mankind can avoid the catastrophe predicted by the Club of Rome only if an essentially inexhaustible energy source is developed. Of the possibilities that are visualized, only one, the nuclear breeder, now appears to be technologically and economically realistic. That this route to the inexhaustible energy source carries with it certain risks is unfortunate; but the Club of Rome catastrophe that will befall man if he cannot find such an energy source is a risk of much greater magnitude.**

**Edsall proposes, among other things, that we intensify research on fusion and solar energy. I agree; yet suppose, as is quite possible, that neither of these sources is found to be feasible, either for technical or**

economic reasons. **The two generations we would thereby lose could make the energy-environment crisis much graver than it now is.**

**On the other hand, if fusion or solar energy becomes practical, the problem of safely dismantling the fission technology is nowhere near as serious as Edsall suggests. The only surveillance that would then be required would be the rather minimal guarding of a few burial grounds for radioactive wastes. As I explained in my article, I do not consider this degree of surveillance to be at all unreasonable, especially since, even without surveillance, the likelihood of widespread contamination from wastes buried in salt (geologically) is extremely small. [Weinberg, Alvin; Letters, Science, vol. 178, Dec. 1, 1972, p933]**

The debate from 1972 suggests that members of the nuclear establishment had already thought long and hard about the costs and benefits of nuclear energy projected into the future before any environmentalists showed up on the scene. It appears the deepest thought given to the possible dangers of long-term storage of nuclear wastes originated with Alvin Weinberg and only later did lawyers like Nader and environmentalists like Jerry Mander mimic Weinberg's thinking.

Unfortunately, Weinberg miscalculated the amount of disruption dedicated advocacy groups like those formed by Nader could create in the lines of communication between the scientific community and the community at large. Nader and the environmental movement are not composed of lay people, but are dedicated professional activists with an agenda far beyond community representation. Weinberg was perhaps naive in advocating public discussion of technological issues, not realizing how easily special interests could hijack the process:

**We scientists value our republic of science with its rigorous peer group review. The uninformed public is excluded from participation in the affairs of the republic of science rather as a matter of course. But when what we do transcends science and impinges on the public, we have no choice but to welcome public participation. Such participation by the uninitiated in matters that have both scientific and trans-scientific elements may pose some threat to the integrity of the republic of science. To my mind, however, this is a lesser threat than is the threat to our democratic processes that would be posed by excluding the public from participation in trans-scientific debate. [Weinberg, Alvin; "Science and Trans-Science", Science, July 21, 1972, p211]**

Unfortunately, democratic participation processes carried to their extremes in Nevada have led to domination of the debate by aggressive special interest groups. Weinberg's qualification that "Such participation by the uninitiated in matters that have both scientific and trans-scientific elements may pose some threat to the integrity of the republic of science." appears to have been an understatement. In fact, such participation can lead to the paralysis of scientific, political and technological institutions.

At Yucca Mountain, participation has been manipulated by sophisticated advocates claiming to be the voice of the uninitiated public. These groups, variously composed of leftist academics, Naderites, radical Greens and the peace movement, have challenged not only scientific integrity, but also the welfare of the public they claim to represent. Blind obedience to "participation by the uninitiated" thus may not solve the problems at Yucca Mountain, but instead cater to the whims of social scientists, lawyers and special interest environmentalists whose objective is to subvert the participation process for political ends quite distinct from the goals of objective science.

# 26. Ralph Nader: The Green Hive

In a hive, the seemingly random actions of many individuals take on an order and synergy that no individual within the hive could possibly understand fully nor orchestrate. In nuclear issues, Ralph Nader has been the premier hive builder for nearing 50 years.

If there was a coordinated effort to destroy nuclear energy in this country by subverting the Nevada Nuclear Waste Project Office and by influencing certain politicians in the State of Nevada, the national leader of that conspiracy would certainly be Ralph Nader. Nader's web of interconnecting environmental groups and activists, culminated in the Safe Energy Communication Council which while closing its doors in 2003, had served as the prime anti-nuclear guerrilla organization providing commitment, institutional savvy and long range strategic planning to carry out such a plan.

Whether or not Ralph Nader is the grand anti-nuclear conspirator who has personally wreaked havoc on the state of Nevada or whether the protest movement is driven by an independent hive mentality towards a collective anti-nuclear goal is difficult to determine. It appears that Ralph Nader plays the part of Queen Bee within the anti-nuclear community, so that while it may not be true that he directed the war, he exerted an enormous organizing influence. As explained in *The Antinuclear Movement*:

**The prime mover for the reassessment of nuclear power by environmentalists is Ralph Nader, also a consumer advocate. Nader sponsored a convention of nuclear power critics in Washington D. C. ("Critical Mass '74) in order to coordinate antinuclear activities throughout the nation. Initially, Nader portrayed the emergence of a garrison state to protect plutonium. Only by not using fission power would this possibility be negated. The energy program advocated by Congresswatch, the Nader organization, in fact became federal energy policy during the administration of President Carter. Conservation and the use of coal, with environmental safeguards, were interim solutions, while in the later part of the century solar, geothermal and fusion power would be available. [Price, Jerome; The Antinuclear Movement, 1982, p52]**

Nader has been involved in the anti-nuclear movement since at least 1972 when he helped organize the Critical Mass nuclear protests, which repeated in 1974 and 1978. It was from these efforts that Nader's central organization, Public Citizen's Critical Mass Energy Project, grew. Public Citizen is now no longer focused solely on the nuclear issue and was headed for 26 years by Joan Claybrook (Jimmy Carter's head of transportation), but is now headed by Robert Weissman. The organization now takes on everything from pesticide use to the North American Free Trade Agreement, with enviable success.

In 1979, Nader wrote a book titled *The Menace of Atomic Energy* in which he presented most of his case against the dangers of nuclear power, bringing up many of the same anti-nuclear arguments that would be tried in Nevada over the following decade. Of special note are some of the people Nader chose to quote in The Menace of Atomic Energy, especially four people who a decade later come to play a role in Nevada's anti-nuclear circus:

Amory Lovins (Friends of the Earth, Rocky Mountain Institute), Marvin Resnikoff (NYPIRG, Sierra Club Radioactive Waste Campaign and later the Nevada Nuclear Waste Project Office), Helen Caldicott (founder of Physicians for Social Responsibility) and Robert D. Pollard (Union of Concerned Scientists) all tied the Naderite past to the Nevada present. These Nader proteges all reemerged on the Nevada stage, showing the Yucca Mountain protest has at its core a small but dedicated kernel of true believers, who professionally opposed nuclear energy for as many as forty years.

Marvin Resnikoff's work for Nevada Nuclear Waste Project Office, Helen Caldicott's speech in Nevada October 1991 sponsored by Judy Treichel's Nevada Nuclear Waste Task Force, Amory Lovins' support and advocacy for Senator Bryan's CAFE bill and Pollard's testimony in Casper, Wyoming against an MRS facility at which he played advocate for Marvin Resnikoff, subtly link all these actors to the Yucca Mountain debate. This is the core of the professional anti-nuclear elite whose entire careers and purposes in life has been to fight nuclear energy to the death. The interesting question is how such a small number of dedicated protesters came to have such an inordinately large impact on our nation's energy future.

The answer to that question lies in part in Ralph Nader's seemingly unstoppable formula for creating interlocking organizations both in the environmental and consumer spheres. Most of these groups alone are seemingly innocuous consumer-oriented dwarfs, but their ability to fuse and gang tackle their opposition makes them a dangerous amoeboid killer of technology. The span of Nader organizations is truly a modern wonder of the Washington D.C. based environmental movement.

- Public Citizens Critical Mass Energy Project
- Congresswatch
- Environmental Action
- Public Interest Research Group (affiliates in New York, New Jersey, Maine , etc. PIRG )
- Safe Energy Communication Council

By establishing these and other organizations in an ever expanding maze, Nader was able to camouflage his activities while creating a protest organism that grew with each court injunction and with each unsuspecting donor contribution.

A political sore point in Nevada since 1991 was the existence of the *Nevada Initiative*, an admittedly secret plan created by the nuclear industry through the American Nuclear Energy Council, to conduct a political lobbying campaign in Nevada. Much was made of this plan by NWPO and the Sawyer Commission in an attempt to portray the industry as evil plotters and schemers. What was missing from this analysis is the fact that the anti-nuclear forces had equally secretive political campaigns in place since the late seventies. We will later quote from the Nevada Initiative, but it is instructive to cite some of the secret plans of Ralph Nader's groups.

A manuscript called "Shutdown Strategies: Citizen Efforts To Close Nuclear Power Plants" written by Joseph Kriesberg in May 1987 for Nader's Public Citizen Critical Mass Energy Project is not itself secret but is in fact a 68 page compendium of other secret plans. Shutdown Strategies both covers the strategies of a national network of anti-nuclear organizations and lists key contacts among those groups. What this document shows is that the anguish shown by the anti-nuclear establishment over ANEC's Nevada Initiative is really crocodile tears shed because the environmentalists no longer had a monopoly on secret plans in the area of nuclear energy policy. In fact, ANEC's campaign was at least a decade late in coming compared to those of Nader's allied environmental factions.

Historically interesting about "Shutdown Strategies" is that it connected a number of later players in the Yucca Mountain war to Nader's organizations. Scott Denman (Safe Energy Communications Council), Ken Bossong (previously with Citizen's Energy Project with Denman and late with Public Citizen), Caroline Petti (organizer of the National Nuclear Waste Task Force, parent to Judy Treichel's NNWTF and later with the EPA), Joan Claybrook (eventual president of Public Citizen) are all listed in the acknowledgments. Marvin Resnikoff (late head of Waste Management and consultant to NWPO), Bob Halstead (transportation consultant to NWPO) are both listed in Shutdown Strategies as anti-nuclear contacts. Thus, it could be argued that NWPO employed secret planners from the environmental movement on its staff (Halstead, Resnikoff) and is the pot calling the kettle black when it condemned the *Nevada Initiative*.

To illustrate the goals of Nader's organizations, we quote directly from "Shutdown Strategy":

**GOALS OF THIS REPORT**

**In September 1986, representatives from over 40 local, state and national safe energy groups convened in Washington, DC to develop strategies for phasing out nuclear power. There was a clear consensus that these efforts should focus primarily on the local and state levels. . . .**

**This report is therefore designed for citizen groups, individual activists, and state and local government officials who are trying to either cancel plants under construction or to close existing plants.**

**This report has three specific objectives:**

**1. Inform readers of different strategies and tactics being used at the state and local levels.**

2.   Provide case studies so readers can determine which strategies may be appropriate to their communities; and

3.   Direct readers to sources of further information about possible strategies and particular case studies. Readers are strongly encouraged to contact the people listed for particular case studies in order to learn why a strategy was or was not successful and to learn what nuclear proponents did to block citizen efforts. [Kriesberg, Joseph; "Shutdown Strategies: Citizen Efforts To Close Nuclear Power Plants, Public Citizen Critical Mass Energy Project, May 1987]

It should be noted that this shutdown campaign was absolute, there was no discussion of the possibility that even some minor levels of nuclear technology might be beneficial to humanity. The decision was already made in many activist's mind that all nuclear technology must be shutdown no matter what technical arguments are brought to bear. The worry at Yucca Mountain is that oversight was completely compromised by NWPO's ties to a Luddite environmental lobby.

The close links of others on NWPO's staff to Naderite groups is evidenced by another report from Public Citizen called "Nuclear Legacy: An Overview of the Places, Problems and Politics of Radioactive Waste in the U.S." published in 1989 and written by Scott Saleska (later affiliated with the Institute for Environmental and Energy Research, another anti-nuclear think tank). Again we find acknowledged Bob Halstead (NWPO transportation), Bob Fulkerson (Citizen Alert), Arjun Makhijani (IEER and NWPO consultant), Caroline Petti (National Nuclear Waste Task Force and now a WIPP regulator with the EPA), Marvin Resnikoff (Radioactive Waste Management, Sierra Club, NYPIRG, and NWPO consultant), Bob Loux (executive director NWPO). While a disclaimer is made that the report is not necessarily endorsed by those acknowledged, the associations are clear.

All these circumstantial bits of evidence are not in themselves sinister, but they do point to the fact that the Nader organizations held as much political leverage over popular politics in Nevada concerning the Yucca Mountain repository as anyone from the nuclear power industry. But we are left with the nagging question of why Ralph Nader and his spinoff environmental organizations feel so compelled to oppose nuclear energy.

We need to return to Nader's philosophical roots. His father worked in a textile mill and later owned a bakery and restaurant where he talked politics with his customers. According to some biographical accounts, Nader's father was a staunch Marxist. Ralph Nader's concern for the oppressed consumer may well have been shaped by this upbringing, but we don't care to make too much of this, other than to comment that this fits the pattern of most of the environmental activists we have traced so far. Obviously, Nader's motives regarding nuclear energy are more complex.

Nader's forty year overreaction to Alvin Weinberg's suggestion that a technological priesthood would be necessary to watch nuclear waste for many generations gives us more clues. Nader's importance in the battle over nuclear energy stems not only from his ability to organize opposition, but also from what he believes about our future. Comments made by Nader in a debate with James A. McClure, Morris Udall and Carl Walske in 1980 held by the American Enterprise Institute seem consistent with his present views:

**MR. EDWARDS: My question is, Mr. Nader, how do you reconcile a total stoppage of nuclear power in light of the facts, in light of what is available today, and also in light of the fact that you said several years ago that we should take care of nuclear first, and then take care of coal?**

**MR. NADER: Eventually, in the next fifty years, our economy will be solar-powered. That will certainly be the case if we extend the wisdom of present knowledge. Now we have co generation and many other sources to avoid nuclear.**

**As to shutting down nuclear plants, they contribute 12 percent of our electricity. We waste at least half of our electricity. Can we replace the 12 percent of our electricity with the 50% we waste? We have 35% excess generating capacity above peak load. . . .**

**It can be done. We do have an electric pool, not the best interconnected system that we would like, but we do have one. It is quite important that we lay out the facts on this issue and show that it is much better to**

shut nuclear plants down now, giving us the initiative to express our efficiency capabilities in electricity production and use. Better that than to wake up some morning and read news about several hundred thousand people being contaminated and several hundred square miles being uninhabitable. . . .

PETER VEREKION, Rhode Island Public Interest Research Group: My question is directed to the panel. Is nuclear power as job intensive as solar?

[ Nuclear Energy, A Reassessment, American Enterprise Inst., 1980]

Again we find the presence of "decentralist" thinking and also an almost naive optimism for solar energy. Almost forty years after Nader's comments, the likelihood of a mass conversion to solar are just as distant, for sound engineering reasons. The setup question from Peter Verekion of Rhode Island PIRG, a Nader affiliate, is included to show the pervasive misconception of Nader groups that creating make-work jobs in the solar industry (there would be many in high-tech endeavors like washing mirrors) is more productive than efficient centralized nuclear energy. Despite much effort and subsidies since Nader's comments in 1980, those jobs have not materialized to any great extent.

What we seem to be observing is a forty year vendetta against the nuclear industry orchestrated by Ralph Nader using his many spinoff environmental and consumer organizations to bear his torch. Consequently, the view that the nuclear industry is an evil industrial monolith intent on destroying human lives with radiation must be tempered by the knowledge that there exists an opposing anti-nuclear monolith whose own motivations and rules of engagement are quite murky. For example, according to the physicist Bernard Cohen, Nader may at times purposely overestimate the risks of radiation:

When my paper on plutonium toxicity was first published, including its estimate of 2 million cancer deaths per pound of plutonium inhaled, Ralph Nader asked the nuclear Regulatory Commission to evaluate it. Judging from the number of telephone calls I received asking about calculational details, they did a rather thorough job, and in the end they gave it a "clean bill of health." Nevertheless, Nader continued to state, in his speeches and writings, that a pound of plutonium could kill 8 billion people, 4,000 times my estimate. In fact, he accused me of "trying to detoxify plutonium with a pen" [Cohen, Bernard; The Nuclear Energy Option, Plenum, 1992, p250]

# 27. SECC Shell Game

SECC was the acronym for the Safe Energy Communication Council, a shell organization for a rainbow of environmental and Naderite groups. The SECC coordinated the activities of the Green movement on energy issues, for solar and conservation but especially against nuclear energy during the 90s and into the new millennia, closing its doors in 2003 as funds became less available. It has since been superseded by other environmental networks, but serves as a template for any future inevitable opposition to nuclear projects.

Coordinating national opposition to the Yucca Mountain repository is obviously a much larger agenda than could be handled by local grass roots efforts. The leadership of the groups that exist in Nevada (Citizen Alert, Nevada Desert Experience, the Nuclear Waste Task Force, etc.) are comparatively lightweight operatives and the SECC filled the central leadership void. Headquartered in Washington D.C. the SECC pulled the strings that turned a relatively straightforward nuclear disposal problem in Nevada into a major test of national political will.

If your organization is politically correct, it belonged to the SECC:
- Environmental Action
- Environmental Action Foundation Environmental Policy Institute
- Friends of the Earth
- Fund for Renewable Energy and the Environment
- Greenpeace Action
- Institute for a Sustainable Future
- Media Access Project
- Nuclear Information and Resource Service
- Public Citizens Critical Mass Energy Project
- Public Media Center
- Renew America
- Sierra Club
- Telecommunications Research and Action Center
- Union of Concerned Scientists
- U.S. Public Interest Research Group

Notable among the SECC affiliated groups are a number of organizations backed by Ralph Nader and his comrades in arms from previous anti-nuclear protests (e.g., Friends of the Earth). Nader was long at or near the center of nuclear protest as shown in the previous chapter.

It is therefore unsurprising to find that the SECC was an outgrowth of the Citizen's Energy Project, a subset of Nader's Public Citizen's Critical Mass Energy Project, which operated in the late 70s. Two members of the early Citizen's Energy Project became a part of the debate over Yucca Mountain: Ken Bossong of Public Citizen and Scott Denman, director of the Safe Energy Communication Council

The SECC's political savvy and institutional muscle allowed it to orchestrate much of the Yucca Mountain protest through its environmental and Naderite coalition. Clearly the SECC energy policy had a strong impact on Yucca Mountain politics and on the actions of Nevada's two Senators. The SECC influenced national energy policy under the Clinton presidency in much the same way Nader's groups negatively influenced earlier Carter administration energy policy. The result was a scorched earth effort to derail nuclear energy in favor of a solar utopianism that carried through the Obama administration.

# SECC Publication List
## MYTHBUSTERS Series

**Demand Forecasting**
**Nuclear Waste Disposal**
**Foreign Oil Dependence**
**The Greenhouse Effect**
**Renewable Energy**
**Energy Efficiency**
**Nuclear Reactor Safety Supplement: Reactor by reactor chart of specific chronic safety problems.**
**Other SECC Publications**

- **Turning Down the Heat: Solutions to Global Warming (Public Citizen's Critical Mass Energy Project/SECC).**
- **Widespread Confusion: Why Congress Must Codify the Fairness Doctrine (U.S. Public Interest Research Group/SECC)**
- **Media Skills Manual (comprehensive guide to press relations and media strategies). (By SECC.)**
- **Americans Speak Out on Energy Policy (SECC and Frederick/Schneiders Inc.)**

**Viewpoint**
**Energy commentaries by various experts.**

- **N1 Scott Denman: A Lesson Lost? America's Response to Chernobyl**
- **N2 Jill Lancelot: Nuclear Accidents: Should Taxpayers Pick Up the Tab?**
- **N3 Jim Hightower [newspaper editor, Texas Agriculture Commissioner]: Restoring Confidence in Our Nuclear Waste Program**
- **N4 Scott Denman: Nuclear Power No Solution to Foreign Oil Dependence**
- **N5 Scott Denman: Three Mile Island: The Accident Continues**
- **N6 David Kraft: Safety Last at Nuclear Power Plants**
- **N6 Robert Pollard: [nuclear safety engineer] Just Say No to Stello**
- **N8 Maria Holt [politician] and Joan Bozer [politician]: Radioactive Waste: Coming Soon To Landfill Near You**
- **N9 Michael Mariotte: NRC Finds Few Takers for Deregulated Radwaste**
- **N10 Kay Drey: Nuclear Power's Dirty Secret**
- **N11 Ralph Nader [lawyer] and Eric Glitzenstein [lawyer]: One Stop Nuclear Licensing - One Step Backwards for Democracy**
- **N12 Scott Denman: U.S. Energy Strategy: Danger or Opportunity?**
- **N13 Gov. Bob Miller [lawyer, politician]: Surrendering Democracy to Nuclear Waste**

**Energy Efficiency**

- **E1 The Honorable Claudine Schneider [politician]: Energy Efficiency: Giving the U.S. a Competitive Edge**
- **E2 Christopher Flavin: Memo to President Bush: Rx for Energy Policy**
- **E3 Armond Cohen [lawyer]: Energy Efficiency: Good News from New England**
- **E4 Bill Magavern: Higher Gas Mileage - And a Better Environment**
- **E5 Scott Denman: America's Oil Addiction: The Road to Recovery**
- **E6 James S. Adams: Energy Efficiency: A Blueprint From California**
- **E7 David Zucker [comedy film writer]: U.S. Energy Policy: Theater of the Absurd**
- **E8 Ken Bassong: Celebrating Sun Day 1992 and Beyond**

**Global Warming**

- **G1 Michael Oppenheimer, PhD[atmospheric physicist]: Safe Energy Options: Best Hopes For Global Warming**
- **G2 Michael P Walsh: Cars, Trucks and Global Warming**
- **G3 Jeremy Rifkin [lawyer]: Global Warming: What You Can Do**

**Utility Issues**

- **U1 Alan Nogee: Utilities Go Overboard: Unneeded Plants and Rising Rates**
- **U2 Scott Ridley: Dimming the Lights on Public Power**
- **U3 Leon Lowery [lobbyist]: Utilities High-Flying Deals Recall Hollywood Thrills and Chills**

**Communications**

- **C1 Senator Timothy Wirth [politician]: Congress Should Reinstate Fairness Doctrine**

**Alternative Fuels**

- **M1 Jay Harris: Move Over Gasoline, Solar Cars Are Coming**

At first glance, the SECC coalition appeared to form a formidable opposition to nuclear energy and in the sense that the coalition groups were able to concentrate political energy and sway the media, the SECC was truly an 800 pound gorilla. However, while the coalition claimed to be the prophets of a new energy future, their energy philosophy saw limited popular support as it would lead to significant reorganization of our national energy policy. The public shows little enthusiasm for the extreme energy conservation measures the SECC is willing to take in order to force an energy utopia on America. Changing to energy efficient lightbulbs (at a not so modest $10.00 a bulb) and increasing home insulation is one thing, but downsizing cars, eliminating nuclear energy and plowing massive amounts of money into costly solar energy projects requires another magnitude of sacrifice.

We've duplicated the SECC publication list from the mid 1990s on the previous page because it shows clearly who was involved in this advocacy group. Those writing for the SECC were mostly lawyers, politicians, professional environmentalists or some combination of the above. Other than Michael Oppenheimer (who is an atmospheric scientist, not a nuclear energy expert) and Robert Pollard (who does have nuclear industry credentials, but has spent the last forty years fighting nuclear energy for the Union of Concerned Scientists), the experts of the Safe Energy Communications Council in general appear to be policy wonks. Lacking professional engineers on their staff, the SECC lacked practical expertise in energy matters.

Research into the backgrounds of the energy policy advocates who write for the Safe Energy Communication Council suggests energy policy had become a secondary issue in relation to social reordering for these writers. Quotes from SECC authors in their unguarded moments display a political inclination that seems more revolutionary than environmentally concerned.

**Ken Bossong:**
"We're losing. Or at least those of us who once saw solar technologies as tools for stimulating social change have gotten somehow sidetracked.

Early supporters of solar energy were attracted to the technology because it seemed capable of promoting individual and community self-reliance as well as decentralization of energy production and control. Solar technologies appeared to be vehicles for redistributing income and thereby benefiting low-income citizens as well as other Americans." [Bossong, Ken; A Solar Critique, Citizen's Energy Project, 1980, p1)


**Jeremy Rifkin:**

. . . Today we are being forced to make a transition from the Industrial Age of nonrenewable resources to a new and still undefined age based once again on renewable sources of energy, and we will have to do so in little more than one generation. The radical change in the worlds view required to make this transition will have to be accomplished virtually overnight. There will be no time for polite debate, subtle compromise, or momentary equivocation. To succeed will require a zealous determination -- a militancy, if you will -- of herculean proportions. [Rifkin, Jeremy; Entropy - A New World View, Viking Press, 1984, p186]


Interestingly, Governor Miller of Nevada shows up on the SECC list as author of an article titled Surrendering Democracy to Nuclear Waste:

"High-level nuclear waste presents a future risk for a longer time than all past recorded history. It is imperative that democratic process, cool heads and objective science prevail in seeking to solve the problem." [Miller, Bob; "Surrendering Democracy To Nuclear Waste", Viewpoint, SECC, Bob Miller, 1992]

In fact, democratic process has been followed at Yucca Mountain, as witnessed in the fall of 1992 when Senator Bryan's filibuster to prevent language favorable to Yucca Mountain in the 1992 Energy Bill was defeated 86-6. Neither have "cool heads and objective science" necessarily been evident in the Nevada Nuclear Waste Project Office which Miller has directly overseen.

Miller's advocacy for the Safe Energy Communication Council ignored the fact that the people and organizations under the SECC umbrella also vehemently oppose the Nevada Test Site, a source of jobs and patriotic pride for many Nevadans. Greenpeace, Friends of the Earth, and others from the SECC coalition have regularly illegally invaded the Nevada Test Site in opposition to NTS, creating conditions for a shutdown which would lead to the loss of approximately 30,000 jobs in Southern Nevada. The SECC affiliated environmental groups also have positions that are antithetical to mining interests, another crucial industry in Nevada. Former Secretary of the Interior Bruce Babbitt's crusade to increase grazing fees, a sore point in Nevada, also incubated among these groups.

Was the Safe Energy Communications Council an environmental group, or a movement bent on massively reshaping society? It is hard to tell from their literature which encompasses not only nuclear and solar issues, but also the fairness doctrine. It seems that energy was almost an after-thought in the SECC world view. Only if the world's energy supply is non-renewable and finite (i.e., only if nuclear power is canceled) can the Green Revolution be forced to occur. One gets the feeling that the members of the SECC prayed each night that a nuclear reactor would fail, simply so they could justify forcing America to become an energy and environmental banana-republic dependent on solar collectors.

It is not clear Nevadans or the nation as a whole would knowingly agree to being returned to an agrarian solar-age society by the SECC. In fact, the SECC itself may not believe it has popular support for its positions because it rarely discussed its Utopian Green world view openly, instead focusing on Washington lobbying to obtain its goals.

## SECC GUERRILLA WAR AGAINST YUCCA MOUNTAIN

Primary among the SECC's objectives was convincing Americans of the untrustworthiness of the Department of Energy, the evil intentions of the nuclear industry and the coming age of solar power. It did this by actively courting support from politicians like former Senator Wirth, Congresswoman Claudine Schneider, Senator Bryan and Reid and Governor Miller in a traditional lobbying effort. More interesting in regard to its affect on the Yucca Mountain battle has been its non-traditional political insurgency techniques.

An example of this political gamesmanship is the coup the SECC pulled in exposing what is known as the Nevada Initiative. In 1991, the American Nuclear Energy Council (ANEC), the Edison Electric Institute, the American Committee on Radwaste Disposal (ACORD) and other trade organizations interested in the outcome of the Yucca Mountain repository began planning a media and political campaign, the Nevada Initiative, to combat the negative position of Nevada's politicians regarding the repository, and to educate the public about Yucca Mountain. Since the hysteria over Yucca Mountain was then at a fever pitch, revelations of an industry campaign to promote the repository were politically explosive.

Kent Oram of Oram, Ingram and Zurawski Advertising of Las Vegas along with the lobbyist Ed Allison prepared a proposal which outlined the strategy and necessary budget for the Nevada Initiative. One of the members of the Edison Electric Institute whose company's energy was derived from non-nuclear sources, leaked this document to the Safe Energy Communication Council. The New York Times broke the story first:

## NUCLEAR INDUSTRY PLANS ADS TO COUNTER CRITICS

**. . . The blueprint, prepared by the American Nuclear Energy Council, an industry trade association, was obtained by the Safe Energy Communication Council, an anti-nuclear group in Washington that said it received it from an executive in the nuclear industry who is critical of the campaign. [Schneider, Keith; New York Times, Nov 12, 1991]**

In a follow-up press release dated November 13, 1991, Scott Denman of the SECC blasted the Nevada Initiative:

**NUCLEAR INDUSTRY MEDIA BLITZ OF NEVADA REVEALED Washington, D.C. - A massive three year, $8.7 million nuclear industry campaign is underway to manipulate Nevadans into accepting a high-level nuclear waste repository. The plan is detailed in confidential documents provided to the Safe Energy Communication Council (SECC), and released today. In addition to a saturation advertising program to bludgeon Nevada with pro-nuclear ads, the industry sponsored campaign also employs a covert strategy of lobbying and public relations "attack/response" teams. Ratepayers from across the United States may foot the bill for the industry's campaign.**

"The nuclear industry treats Nevadans like pawns in its self-serving and desperate chess game for revival," said Scott Denman, Executive Director of SECC. "Their plan hinges on a grossly cynical belief that with enough advertising and money, Nevadans will inevitably accept anything - including a nuclear dump in their state."

The three-year industry plan budgets nearly $8 per Nevada citizen: more than $4.4 for advertising; $3.36 million for consultants, political operatives and lobbying; $480,000 for a "media response team"; and $400,000 for tracking polls.

"The nuclear industry needs a quick fix to the radioactive waste problem before it can sell the American public on more nuclear power," said Denman. "It's counting on selling out Nevadans first."

The industry documents reveal that the industry has prepared "attack/response teams" of scientists to speak in television, radio and print advertising in an effort to convince Nevadans that a high-level nuclear waste dump would be "safe." A second "team" of media professionals would "be the vehicle for generating positive free media coverage," and convert "the press away from its opposition to the repository."

Studded with militaristic jargon, the major strategy document refers to the establishment of a "political beachhead" and programs to "neutralize the political resistance." The plan is to create a "sustained advertising program aimed at Nevadans" and "by softening the public opposition, the campaign will provide 'air cover' for elected officials who wish to discuss benefits." "A professional media attack/response team will be deployed" to "counter those who counterattack with misinformation." According to the document, "the primary target audience will be women, aged 25 to 49 - the group with the highest statistical potential for favorably affecting the polls if they can be informed, reassured and moved."

The industry campaign would be funded by assessments against each utility that owns nuclear power facilities, including the bankrupt Public Service of New Hampshire, and utilities that own reactors that have been shut down. The Edison Electric Institute, the utility industry trade organization, would collect the assessment with its regular dues and pass the campaign fees through to the American Nuclear Energy Council, the nuclear industry's lobbying arm, which would direct the "Nevada Initiative" media campaign. The U.S. Council for Energy Awareness, the industry's public relations group, is also involved in this effort.

"Nuclear utilities have evaded a reasoned debate about the real dangers of radioactive waste," said Denman. "But the industry is either incredibly naive or intolerably ignorant to assume that by simply unleashing an avalanche of propaganda, it can buy the hearts and minds of Nevadans." [Attachment I, Report of the Nevada Commission on Nuclear Projects (Sawyer Commission Report), 1992]

Interestingly, this press release was published as an attachment to the 1992 Report of the Nevada Commission on Nuclear Projects, showing just how dependent the state was on the SECC to generate negative press releases. However, this points out again that the opposition to Yucca Mountain was hardly a grass roots affair, for the SECC turned up at the center of many such press releases.

The SECC's hyper-reaction to the Nevada Initiative was curious. Until the Nevada Initiative, the pro-nuclear side of the story was distinctly absent from the debate because of lockstep anti-nuclear unanimity by the political and media culture of the State. This animosity was in part the result of a decade of efforts by the SECC to shut down all nuclear energy.

The SECC's real complaint about the Nevada Initiative seems to be that it was losing its monopoly status in the propaganda wars in Nevada. As a Washington Beltway headquartered organization devoted entirely to being an anti-nuclear "Communications Council", the SECC deserved little pity for suddenly being confronted by a similar, though opposite, Nevada based pro-nuclear political and educational campaign. After all, the SECC's own vigorous and well-funded anti-nuclear propaganda campaign had gone unchallenged in Nevada for many years and was in many ways responsible for the nuclear hysteria found in the Silver State.

There are a number of interesting things to note about the SECC response to the Nevada Initiative. First, Scott Denman's focused on linking the Nevada Initiative to an attempt to "sell the American people on more nuclear power." Nevadans themselves aren't necessarily against nuclear power, though polls show them to be opposed to the Yucca Mountain radioactive waste repository. In essence, Denman's comment was a Freudian slip in that it shows the true intent of the SECC was to cripple the nuclear power industry by opposing Yucca Mountain, not to save the people of Nevada from the dangers of the nuclear waste repository.

Also interesting is that the SECC was chosen as a recipient of the leak in the first place, showing the central position of this organization in the anti-nuclear movement. The question this poses is why the Nevada Initiative documents weren't leaked first to local Nevada news sources, if there is indeed a "grass roots" opposition to Yucca Mountain unmanipulated by outside special interests. Actually, Scott Denman, the principle source of the press release against the Nevada Initiative was a paid, full-time anti-nuclear activist with a staff of three and a number of unpaid assistants. Combined with the rest of the Yucca Mountain opposition coalition coordinated by the SECC, this creates a powerful organizational structure in most ways equal to the pro-nuclear campaign run by the American Nuclear Energy Council.

If the Nevada Initiative was studded with military jargon, it is little wonder. Greenpeace Action, affiliated with the SECC, have long run sophisticated infiltration and disruption activities at the Nevada Test Site and elsewhere and would need to be countered, if at all, with similar militaristic tactics. In fact, the title Greenpeace Action is meant to indicate this is the militaristic tactical wing of Greenpeace. Moreover, the SECC had engaged in counterespionage of its own to obtain the Nevada Initiative documents and it would be naive to believe that the battle for Yucca Mountain is anything but a full blown political war in which both sides have fought for advantage.

One advantage the SECC held over the public relations firms working for ANEC was that the environmentalists were given an inordinate amount of free press. For example, in a September 1993 article, Mary Manning of the Las Vegas Sun promoted the views of Martin Gelfand, research director of the the the Safe Energy Communication Council:

"It could be the end of Las Vegas as we know it," he said. "It certainly isn't going to help the Las Vegas economy. . .

"Everybody comes up to me and says, 'What are we going to do with the waste?" I don't know.

"I'm not a scientist," he said. "I'm not an engineer. The industry just wants to dump it quick. The real answer is, we don't know what to do with it. That's why we need time."

A Cleveland resident, Gelfand said he can sympathize with Nevadans in their fight against the nuclear dump. He lives near a nuclear power plant built on an earthquake fault near Lake Erie. [Manning, Mary; Las Vegas Sun, Sept 1993]

Gelfand was a non-scientist, non-engineer, non-resident quoted authoritatively by Manning on nuclear issues in Nevada. Why this was not viewed as external manipulation of Nevada's residents is unclear. In another authoritative article by Manning, the SECC's position is again forwarded:

YUCCA FUNDS SLASH SOUGHT
"A national coalition of energy, environmental and public-interest groups has called for the Clinton administration to cut $305.7 million from its high-level nuclear waste project until an independent panel reviews the program.

The Safe Energy Communication Council recommended in its "Sustainable Energy Budget" Wednesday that the Department of Energy stop work and spending for underground tunneling at Yucca Mountain, temporary storage facilities, transportation, quality assurance and other programs." [ Manning Mary;, Las Vegas Sun, Nov. 18, 1993 p5A]

The question this poses is whether Nevada and national nuclear energy policy has been formulated through normal political and technical institutions, or whether a small energetic staff of solar energy activists in Washington has dictated policy. Through proper positioning in the media, connections to a strategically placed state agency (NWPO), and coalition building among powerful environmental lobbies, it appears the SECC and a few allies have come quite close to strangling the nuclear industry by crippling the nuclear waste repository program.

# 28. Resnikoff & Audin: The Camel's Nose

Marvin Resnikoff and Lindsay Audin were two technical consultants who demonstrated the subtle links between the Washington based anti-nuclear groups and Nevada's Nuclear Waste Project Office. Resnikoff and Audin first joined forces in 1983 in writing *The Next Nuclear Gamble,* an analysis of the dangers of nuclear waste transportation produced by the progressive New York Council on Economic Priorities. Later, the two were brought into NWPO's studies by the socioeconomic researchers under Mountain West.

It is a curious anomaly of the way NWPO conducted its business that transportation impact studies were originally under the umbrella of the Mountain West consortium and its political geographers and psychologists rather than attached to an engineering department skilled at analyzing transportation issues. It was only after considerable effort that a Nuclear Waste Transportation section was set up at the University of Nevada Las Vegas, although its funding was always under pressure. The reluctance of the state to fund these studies apparently stemmed from UNLV's academic independence from the State's political posturing.

Transportation studies are generally a subset of the Civil Engineering discipline; those who make roads tend to study what travels over them and how to design the appropriate traffic flow patterns. Yet Marvin Resnikoff was brought in to critique a modeling software package called RADTRANS and Audin was hired to give an analysis of transportation cask design. Unsurprisingly, their conclusions were a rehash of their 1983 *The Next Nuclear Gamble* findings. The bibliographic sketches of Marvin Resnikoff from *The Next Nuclear Gamble* give us some insights:

**MARVIN RESNIKOFF: the project director of the Nuclear Waste Transport and Storage Project, received a PhD in high energy theoretical physics from the University of Michigan in 1965. He has been a technical consultant on nuclear waste matters to the New York and Illinois Attorneys General, the State of Lower Saxony, West Germany, and numerous environmental organizations. Since 1974 he has testified on numerous occasions before the U.S. Congress and State Legislatures on nuclear fuel reprocessing, waste management and transportation. Previous to his work at the Council, he taught at Rachel Carson College and the Department of Physics at the State University of New York at Buffalo. [Resnikoff, Marvin; Audin, Lindsay; The Next Nuclear Gamble, New York Council On Economic Priorities, 1983]**

Despite his credentials as a theoretical physicist, Resnikoff is not known to publish academic papers outside those for environmental groups. Instead, Resnikoff appears to be a one-man anti-nuclear consulting firm whose credentials are used to bolster anti-nuclear groups around the nation who need expert opinion. Resnikoff has appeared in this anti-nuclear consulting capacity since at least 1972 and appears to be the anti-nuclear expert of last resort.

For example, in the Casper Wyoming Star Tribune of April 17, 1992 in an article titled "MRS Conference Criticized As One-Sided By Nuclear Scientist" we see a confluence of anti-nuclear activists:

**"A May 12 conference sponsored by the Wyoming Heritage Foundation will provide "full debate and discussion" on the proposal to store radioactive nuclear waste in Fremont County....**
**Both a panelist scheduled to participate in the conference and a nuclear scientist formerly with the Nuclear Regulatory Commission - now with the Union of Concerned Scientists - say the conference's schedule is weighted in favor of the nuclear industry.**
**Schilling (Heritage Foundation executive director) said he searched for scientific experts who might have views differing from those within the industry and DOE, but could not find them.**
**"There clearly are scientifically trained people who could provide a different point of view" from those perspectives likely to predominate at the Heritage conference, said Dan Reicher, an attorney for the Natural Resources Defense Council who has opposed nuclear waste disposal elsewhere in the country.**
**"But they (the Heritage Foundation) simply haven't made that attempt" to find such credentialed scientists, Reicher said. Although interviewed separately, both Reicher and Pollard (Union of Concerned Scientists) cited the same scientific expert who is known to differ with the industry -- a Dr. Marvin**

**Resnikoff, who operates a waste management consulting firm in New York." ["MRS Conference Criticized As One-Sided By Nuclear Scientist", Casper Wyoming Star Tribune, April 7, 1992]**

Reicher, Pollard and Resnikoff are professional anti-nuclear activists all well known to each other. Pollard especially, as a former scientist for the Nuclear Regulatory Council until 1977, should have been able to cite a dozen scientists with opposition views of nuclear waste management if such opposition was wide-spread. Marvin Resnikoff was instead cited, the trusted expert who could be expected to toe the anti-nuclear party line.

This points to the limited state of technical opposition to Yucca Mountain since most scientists are not excessively concerned about the nuclear waste safety issue. While there is heated scientific debate about how to optimize nuclear waste safety and technology, there is little debate that eventual solutions to problems are feasible. This allowed Resnikoff a near monopoly on anti-nuclear consulting. It was therefore unsurprising that NWPO found a way to include Resnikoff in its transportation and risk studies at Yucca Mountain.

Joining Resnikoff in writing The Next Nuclear Gamble was Lindsay Audin, whose bio stated:

**LINDSAY AUDIN: consultant to the Nuclear Waste Transport and Storage Project, graduated from Rensselaer Polytechnic Institute, in the aeronautical engineering program in 1970. He is an engineer at Goldman, Sokolow and Copeland, specializing in energy studies. Mr. Audin served as a consultant for the New York Attorney General on the transportation of irradiated fuel through New York City and has researched nuclear transportation issues since 1975. [The Next Nuclear Gamble, 1983]**

A theoretical physicist and an aeronautical engineer thus became the main opposition witnesses for the state of Nevada on the transportation of nuclear waste. Brought out at news conferences and Sawyer Commission meetings by Bob Loux and NWPO, Audin and Resnikoff were used to contest the safety of waste transportation casks, based on a nearly decade old analysis. The media accepted Resnikoff and Audin as bonafide experts and independent consultants, although neither of the two had ever been involved in the actual testing or transport of nuclear waste canisters.

Marvin Resnikoff got his start in the anti-nuclear movement by providing the scientific testimony that helped shut down the West Valley nuclear reprocessing plant in New York. Resnikoff continued his involvement in the anti-nuclear movement, later becoming involved with Ralph Nader's efforts to shut down the industry. In the book *The Atomic Menace*, Nader quotes Resnikoff's studies:

**A recent study by the New York Public Interest Research Group (NYPIRG) indicates that even these costs may be seriously underestimated. The study was conducted by Marvin Resnikoff, professor of physics at the State University of New York at Buffalo, with the aid of four engineering students. In contrast to industry statements that mothballing would take 120 to 200 years, Resnikoff's group found that at least 1.5 million years would elapse before the radiation in reactors had decayed to safe levels.**

**The difference between the estimates of the atomic industry and of Resnikoff's group comes from the industry's apparent oversight of the isotope nickel-59. [Nader, Ralph; The Menace of Atomic Energy, Norton, 1979, p141]**

Public Interest Research Groups are Nader offshoots that at one point in the 1970s were being started at universities across the country. The PIRG satellites essentially confiscated student funds because once the student body voted in a PIRG, individuals could not opt out of the system. This became a forced contribution to Nader's advocacy war machine.

Resnikoff's early connection with the Public Interest Research Group implies his opposition to nuclear technology is thus long developed. As a consultant to the State of Nevada, his findings were consequently preordained. His anti-nuclear work for the Public Research Interest Group in the 70s led to the starting of the Sierra Club Nuclear Waste Campaign, the authoring of his book, *The Next Nuclear Gamble*, in 1983, and later the creation of Waste Management which is still active, his own consulting firm.

In a 1988 publication, Deadly Defense, Military Radioactive Landfills, we find more about Resnikoff's evolution as a career antagonist to nuclear technology:

**Marvin Resnikoff is research director for the (Sierra Club Radioactive Waste) Campaign. He received a Ph.D. in high energy theoretical physics from the University of Michigan in 1965. He was a founder of the**

**Campaign in 1978. . . . . (Resnikoff, Marvin; et. al.; Deadly Defense, Military Radioactive Landfills, Radioactive Waste Campaign, 1989]**

The list of those co-writing Deadly Defense with Resnikoff, as presented in the preface, includes:

**Minard Hamilton . . . Formerly, she sat on the board of Greenpeace, USA and was the Northeast organizer for "Sun Day." . .**
**Ed Hedemann . . . . He is author and editor of the Guide To War Tax Resistance, and editor of the War Resisters League Organizer's Manual.**
**William McDonnell . . He was an activist in opposition to a proposed radioactive dump in Warwick, New York, and a coordinator of the First Global Radiation Victims Conference.**
**Jennifer Scarlott is assistant director of the Lawyers Committee on Nuclear Policy. Previously, she was public affairs coordinator for the Union of Concerned Scientists.**
**Jennifer Tichenor . . She is a registered nurse and is director of the Nurses Alliance for the Prevention of Nuclear War.**

It would appear that the Sierra Club's Radioactive Waste Campaign was more of an anti-war group than a scientific entity interested in the safety of radioactive waste disposal. This once again points to the confusion of Yucca Mountain, a nuclear waste repository, with protest against the nuclear weapons complex. In their zeal to make a political statement against war, environmental activists have often set aside science at Yucca Mountain. This larger anti-war political goal may overshadow scientific truth for investigators like Resnikoff.

Resnikoff's authorship of research papers for the Sierra Club's Radioactive Waste Campaign in the mid 80s can thus not be viewed as part of normal academic debate and peer review, but as advocacy for a political agenda. Resnikoff started the Radioactive Waste Campaign, so it is not unusual that he would be also published by them. This doesn't mean his work isn't sincere, but that Resnikoff and many others in the Green movement who claim scientific respectability have not subjected their ideas to rigorous peer review by other qualified scientists. Resnikoff and Audin are no friends of the nuclear industry. Consequently, their appearance in 1991 as consultants to the Transportation studies being conducted by NWPO and paid for by the Nuclear Waste Fund, was a significant coup by the Greens. The anti-nuclear movement was able to place two of their most rabidly opposed members at the very core of the studies being conducted and get them paid for being in this position.

Further clouding the story, it would seem from his online bio that Resnikoff derives significant funding from his expert testimony in radiation exposure cases:

**Marvin Resnikoff, Ph.D. is Senior Associate at Radioactive Waste Management Associates and is an international consultant on radioactive waste management issues. He is Principal Manager at Associates and is Project Director for dose reconstruction and risk assessment studies of radioactive waste facilities and transportation of radioactive materials. Dr. Resnikoff has concentrated exclusively on radioactive waste issues since 1974. He has conducted studies on the remediation and closure of the leaking Maxey Flats, Kentucky radioactive landfill for Maxey Flats Concerned Citizens, Inc. and of the leaking uranium basin on the NMI/Starmet site in Concord, Massachusetts under grants from the Environmental Protection Agency. He also conducted studies of the Wayne and Maywood, New Jersey thorium Superfund sites and proposed low-level radioactive waste facilities at Martinsville (Illinois), Boyd County (Nebraska), Wake County (North Carolina), Ward Valley (California) and Hudspeth County (Texas). He has conducted several studies of transportation accident risks and probabilities for the State of Nevada and several Nevada counties and dose reconstruction studies of oil pipe cleaners in Mississippi and Louisiana, residents of Canon City, Colorado near a former uranium mill, residents of West Chicago, Illinois near a former thorium processing plant, and residents and former workers at a thorium processing facility in Maywood, New Jersey. In West Chicago he calculated exposures and risks due to thorium contamination and served as an expert witness for plaintiffs A Muzzey, S Bryan, D Schroeder and assisted counsel for plaintiffs KL West and KA West. He is presently serving as an expert witness for plaintiffs in Karnes County, Texas, who were exposed to radioactivity from uranium mining and milling activities. He also evaluated radiation exposures and risks in worker compensation cases involving G Boeni and M Talitsch, former workers at Maywood Chemical Works thorium processing plant. He recently completed work on several major personal injury cases involving former**

uranium mines and mills in South Texas.  In June 2000, he was appointed to a Blue Ribbon Panel on Alternatives to Incineration by DOE Secretary Bill Richardson. Under a contract with the State of Utah, Dr. Resnikoff is a technical consultant to DEQ on the proposed dry cask storage facility for high-level waste at Skull Valley, Utah and proposed storage/transportation casks.  He is assisting the State on licensing proceedings before the Nuclear Regulatory Commission.  In addition, at hearings before state commissions and in federal court, he has investigated proposed dry storage facilities at the Point Beach (WI), Prairie Island (MN) and Palisades (MI) reactors.  He is also presently preparing studies on transportation risks and consequences for the State of Nevada and Clark and White Pine Counties.

## BATTLING ANEC

When the American Nuclear Energy Council (the nuclear trade organization in the early 90s) began to air commercials produced by OIZ Advertising of Las Vegas showing films of transportation casks being subjected to damage testing, Resnikoff and Audin were ready. The media were given information by NWPO based on an Audin paper titled "Nuclear Cask Testing Films Misleading and Misused" Lindsay Audin October, 1991. Audin's analysis created an uproar in the state by claiming the films used in the ANEC commercials were rigged.

The Nuclear Waste Project Office funded Audin under contract NWPO-TN-012-91 with money from DOE Grant number DE-FG08-85-NV10461. To quote from the article's title page: "As part of its oversight role, NWPO has contracted for studies designed to assess the socioeconomic implications of a repository and of repository related activities. Since Audin's article was a response to an advertisement run by the nuclear industry done outside the normal nuclear waste transportation studies at UNLV, this response was actually part of NWPO's media relations campaign. Audin never worked in the nuclear industry, was never present at container tests, and was never employed as a design engineer. In fact, in a statement to the Nevada press, Audin stated:

**"I don't claim to be a PhD. What I do claim is that I have credentials of analysis of the documents and finding errors and as you can see from most of my reports, virtually all the references in there were straight from the DOE and NRC. I'm basically a librarian collecting information they don't want you to have. [Press conference, 10/4/91]**

What was unknown to the local media was that Audin's associate, Marvin Resnikoff, had taken part in a similar scenario in 1984 in England. British environmental activists claimed tests of nuclear transportation casks were rigged and inadequate, implicating not only British designs but U.S. casks as well as well. Interestingly, Dr. Marvin Resnikoff had just attended a conference on nuclear transportation issues in Britain representing the Sierra Club Radioactive Waste Campaign.

British Greenpeace activist George Pritchard charged that a crash pitting a spent fuel cask against a train was stage managed to make the effects less severe than they might have been. It was asserted that engine mounting bolts had been removed from the locomotive to reduce the force of impact. It was also claimed that weights were put in the train's carriage to minimize secondary impact. CEGB, the utility which underwrote the demonstration demanded proof of the claims or an apology. They got the apology.

Pritchard conceded that he had acted "somewhat naively" in making the allegations. The apology went as well to British Rail and to the engineering firm which set up the demonstration. In closing, Pritchard added that he looked forward to "a more understanding relationship between Greenpeace and the CEGB."

This attack by Greenpeace likely was advised by Resnikoff and the Sierra Club Radioactive Waste Campaign. After all, Resnikoff was the only environmentalist on either side of the Atlantic claiming expertise in nuclear transportation through his work on the book, *The Next Nuclear Gamble*. Moreover, the British representative of Friends of The Earth, Amory Lovins, had worked with Resnikoff to halt plutonium shipments in 1979 and the anti-nuclear movement had long been internationalized. In Nevada, anti-nuclear hysteria was so rampant by 1991 that the campaign against the ANEC commercials run by Audin in NWPO's behalf was greeted with media acclaim. Later, Dan Burns, director of Channel 3 in Las Vegas also did an expose claiming the ANEC films were rigged. Little did the media understand how they were being manipulated by anti-nuclear professionals who had honed their tactics worldwide.

Resnikoff often consulted for the Safe Energy Communication Council. In the *SECC Myth Busters #8*, Resnikoff discusses low-level waste disposal and claims all present sites are contaminated or inadequate. This is consistent with other reports issued by his firm, Radioactive Waste Management, and comes as little surprise. Among the recommendations of the Myth Busters report authored by Resnikoff:

**6) The production of "low-level" nuclear waste must be phased out in an orderly and economic manner by ending U.S. dependence on nuclear power as a source of energy.**

**7) The federal government, individual states, communities and utilities should significantly increase research, development and commercialization of safe, clean and affordable energy sources such as solar, wind, biomass, geothermal and solar hydrogen. The government's highest funding priority for meeting our nation's energy needs should be these renewable energy sources, energy efficiency and conservation. Many of these resources are cost-competitive and available today. [Myth Busters #8, Safe Energy Communication Council, summer 1992]**

Resnikoff appears to be an absolutist in regard to utilizing solar energy in place of nuclear energy technology. However, Marvin Resnikoff has been remarkably short in suggesting practical solutions to the nuclear waste problem beyond leaving the waste aboveground for future generations. Similarly, Lindsay Audin is in essence a technical librarian and has no hands-on experience with nuclear shipping casks, though claiming to be an expert in cask design deficiencies.

In testimony to the US House of Representatives in 2002, Resnikoff pushed two nuclear transportation disaster scenarios: 1) tunnel fires and 2) terrorist attack.

**Our report for the State of Nevada traced the progressive degradation of a hypothetical rail cask in the tunnel fire. We estimated the release of radionuclides, primarily cesium, from the cask. We determined that a single rail cask in such an accident could have contaminated an area of 32 square miles. Failure to cleanup the resulting contamination, at a cost of $13.7 billion, would cause 4,000 to 28,000 cancer deaths over the next 50 years. Between 200 and 1,400 latent cancer fatalities would be expected from exposures during the first year. The Baltimore Tunnel fire report is attached to this testimony.**
. . . .

**An attack on a GA-4 truck cask using a common military demolition device could cause 300 to 1,800 latent cancer fatalities, assuming 90% penetration by a single blast. Full perforation of the cask, likely to occur in an attack involving a state-of-the art antitank weapon, such as the TOW missile, could cause 3,000 to 18,000 latent cancer fatalities. Cleanup and recovery costs would exceed $17 billion. It would be easier for terrorists to attack these shipments than to attack storage facilities at power plants, and these DOE shipments may be symbolically more attractive targets than civilian facilities.**
**Testimony of Dr. Marvin Resnikoff on behalf of Radioactive Waste Management Associates, United States House of Representatives Committee on Transportation & Infrastructure, April 25, 2002**

The problem with the tunnel disaster scenario is that it can be avoided completely by restricting passage through tunnels to not coincide with shipments of volatiles. The problem with the terrorist attack scenario is that on-site storage sites are similarly vulnerable and numerous, arguing for burial rather than against.

During the Fukushima incident, Resnikoff went on the offensive with a widely dispersed article provocatively titled "Doomsday Scenario At Fukushima", Resnikoff was short on predicting the final death toll, which was in the end dwarfed by the earthquake and tsunami deaths, but was almost gleeful in his conclusion that nuclear reactors would be shut down:

**Take this out of the nuclear realm. Imagine another harmful poison, botulism. Imagine a botulism reactor, reproducing botuli fast enough to produce heat and steam to turn turbines. Then imagine having to contain these billions of botuli so the public is not harmed. This is essentially the friendly atom that has now come full circle in Japan and that the Nuclear Regulatory Commission will relicense for an additional 20 years at Vermont Yankee and at 30 other Fukushima-type reactors in the United States. Fortunately, the State of Vermont has taken matters into its own hands and has decided not to allow Vermont Yankee to run past 2012..**

http://www.huffingtonpost.com/marvin-resnikoff/fukushima-nuclear-meltdown-japan_b_835932.html,
March 15, 2011

The glaring problem for Resnikoff and the environmental movement is that they have not been part of real world demonstration projects that give evidence that their theories produce *economical* solutions to disposal problems. While Resnikoff can point to reasons he believes nuclear technology cannot be utilized safely, absent from his literature are alternatives. Should we allow nuclear waste to remain on-site indefinitely in dry casks and in storage pools? How do we replace the 20% of our energy now provided by nuclear reactors? How do we compete economically with China and India once they are powered by advanced nuclear technology?

On these subjects, Resnikoff seems oddly mute. In 2013 Resnikoff gave an extensive lecture "Decommissioning Nuclear Power 101". Ironically he called the decommissioning plans for reactors the "Stonehenge Concept" where unsafe canisters are stored onsite in standing containers. In sort, after working for 40 years since 1974 on the issues, he still was unable to provide a positive scenario for dealing with the existing problem of long term nuclear waste storage.

# 29. Prophets of Doom

A number of institutes and think tanks are dedicated to the idea that environmental collapse is eminent due to unrestrained technology, epitomized by nuclear energy. Based mostly in Washington, these groups work in synergy and many of their members floated in and out of association with the Safe Energy Communication Council and the anti-nuclear Naderite core. Among the most interesting groups who determine the national flavor of environmental activism and opposition to nuclear technology are Amory Lovins of the *Rocky Mountain Institute* (formerly a representative of Friends of the Earth), Dianne D'Arrigo of *Nuclear Information Research Service* and Jeremy Rifkin of the *Foundation on Economic Trends*. All have contributed to the efforts to foreclose the nuclear option by opposing Yucca Mountain.

## ROCKY MOUNTAIN INSTITUTE

**"Eschatology is the religious doctrine or study of the last or final state of affairs, such as the end of the earth or resurrection . . . . Eschatological thinking in the environmental movement characterizes the Friends of the Earth." [Price, Jerome; The Antinuclear Movement, 1982]**

One of the major actors during the 1970's and 1980's opposing the nuclear industry was Amory Lovins, then of the Friends of the Earth and now of the Rocky Mountain Institute in Snowmass Colorado. Lovins and Marvin Resnikoff worked together in New York in the late 1970's on anti-nuclear action within New York, protesting the shipment of plutonium within the state. Lovin's close ties with Marvin Resnikoff provide historical clues to why and how Yucca Mountain is now opposed.

The origins of Friends of the Earth are deeply rooted in anti-nuclear activism:

**"The emphasis on conservation reflects the origins of Friends of the Earth. It was established by David Brower in New York City in 1969 as an alternative to the Sierra Club. Brower believed that traditional conservationist societies were not concerned with nuclear proliferation, the most serious environmental threat in both a military and ecological sense. Neither would the traditional organizations concern themselves with inflation, unemployment, or similar inequities that are often produced by environmental abuse." [Price, Jerome; The Antinuclear Movement, 1982]**

The Friends of the Earth have proven to be a highly effective thorn in the side of not only the nuclear industry, but fought the Boeing SST, the Concorde and the trans-Alaska pipeline as well. They also helped the late Senator Edward Kennedy's attempt to block passage of the Price-Andersen Act in 1977 which provides insurance for nuclear operators. In other words, they're old hands at trying to choke nuclear power and have a hand in devising strategy against Yucca Mountain through their position as a coalition member of the Safe Energy Communication Council.

Lovins' biography from his 1984 book "Energy Wars" points out some of the limitations of resumes from the anti-nuclear lobby:

**"Amory Lovins, born in Washington DC in 1947, is a consultant physicist who has lived in England since 1967. After two years each at Harvard College and at Magdalen College, Oxford he became Junior Research Fellow of Merton College, Oxford in 1969, but resigned in 1971 to become full-time British Representative of Friends of the Earth, Inc. (and, in 1979, Vice President of Friends of the Earth Foundation). He received an Oxford MA degree by Special Resolution in 1971 and a DSc degree honoris causa from Bates College in 1979. Twice appointed Regents' Lecturer in the University of California (Berkeley, energy policy, spring 1978, and Riverside, economics, spring 1980), he was 1979 Grauer Lecturer in the University of British Columbia. In 1980 he was appointed to the Energy Research Advisory Board of the US Department of Energy."**

**A consultant experimental physicist since 1965, Mr. Lovins has concentrated on energy and resource strategy since about 1970. His current or recent clients, none of whom is responsible for his views, include**

several United Nations agencies, the Organization for Economic Cooperation and Development (OECD), the MIT Workshop on Alternative Energy Strategies (WAES), the International Federation of Institutes for Advanced Study (IFIAS), the Science Council of Canada, Petro-Canada, the US Energy Research and Development Administration, the US Congresses' Office of Technology Assessment, the US Solar Energy Research Institute, Resources for the Future, the governments of Colorado, Montana, Alaska and Lower Saxony, and other organizations in the US and abroad. He is active in energy affairs at a technical and political level in about fifteen countries, and has published several books, several monographs, and many technical papers, articles, and reviews. [Lovins, Amory; Lovins, Hunter; Energy Wars, Friends of The Earth, 1980]

If Lovins was truly a consultant physicist in 1965, he would have been eighteen years old at the time. By our count, Lovins would have had twenty-three or more affiliations when the above bio was written in the mid eighties, which added to the books (which take approximately a half year at least in preparation) might bring the consultations and major accomplishments to thirty. At the then ripe age of approximately thirty-eight, over a twenty year working period if we include college, this brings Lovins devotion to any major aspect of his career to approximately nine months.   No one can accuse Lovins of not keeping busy.

Despite the length of this resume, as with many of the other resumes of environmental activists, it appears Lovins has little practical experience. That is, his Aspen Institute has at times promoted all sorts of alternative energy schemes from solar cells to all-plastic cars to electric vehicles.  While this is all a lot of fun, there seems to be a lack of evidence of projects that actually make it all the way through to commercial viability.  The Aspen Institute is, however, a great way to raise funds.

The National Geographic gave Lovins coverage in a February 1981 Special Report:


**"How much energy do we need? Just enough to do each task, balancing the cost of getting more energy against the cost of wringing more work from what we already have. Investing this way over the next 20 years could reduce energy use. in the U.S. by a quarter and nonrenewable fuel use by nearly half - with two-thirds increase in gross national product, unchanged lifestyles and more jobs. . . . .**

**Available renewable sources are not cheap, easy or instant, but they are cheaper, easier and faster than synfuels plants or still costlier power stations" [National Geographic, Energy Special Report, 1991]**

If Amory Lovins had been right in 1981 that alternative energy technologies which stress conservation were capable of boosting the economy by two thirds, certainly he should be the alternative energy tycoon of the 1990's. A sizable amount of energy conservation did take place since the 1980s, principally through market mechanisms that put a premium on efficiency, but it is not clear that we had no change in lifestyle. Energy utopia and the real world are two different places and we have yet to see any of our professional alternative energy gurus actually turn a profit making the various devices they promote as energy cure-alls.

The significance of Amory Lovins in relation to Yucca Mountain is not only his advocacy of a non-nuclear "soft energy path", but equally important was his activity at the origins of the anti-nuclear protest movement as an organizer. Some of Lovins' strategies, as well as those of his friend Marvin Resnikoff, were later reincarnated at Yucca Mountain. One of Lovins' most significant victories, and a harbinger of the tactics implemented in Nevada, was his intercession in Germany at the proposed Gorleben geologic waste repository. According to Luther Carter in *Nuclear Imperatives*:


## THE GORLEBEN INTERNATIONAL REVIEW

**. . . . in the fall of 1977 (Count Hermann) Hatzfeldt, attending a conference in the United States, came in contact with Amory Lovins of Friends of the Earth, prominent proponent of solar energy and other soft technologies as alternatives to nuclear power. From Hatzfeldt's discussions with Lovins there emerged the idea for the Gorleben international review at which plans for the integrated fuel cycle center would be subjected to analysis and criticism by foreign and German experts knowledgeable about reprocessing and waste management. . . . . .**

.... the state announced the list of panelists who supported the Gorleben proposal. There were thirty-seven of them, twenty from outside Germany. . . . The foreign panelists came from a half dozen countries and included nuclear physicists, fuel cycle specialists, and experts on health physics. . . . . The distinguished physicist and philosopher Carl Friedrich von Weizacker, then director of the Max Planck Society's Institute for Research on Life in the Scientific Technical World, was asked to preside over the meetings. . . . .

Before the review was to open in Hannover, Hatzfeldt invited Lovins and a number of other Gorleben critics to meet von Weizacker during a weekend retreat at Schloss Crottorf, Hatzfeldt's ancestral estate some fifty miles east of Cologne. Hatzfeldt respected von Weizacker even though not sharing his acceptance of nuclear technology. . . . . Hatzfeldt believed that if the critics came to know Weizacker at least part of the suspicion would be relieved.

Over the weekend the critics gave von Weizacher, in broad outline, a preview of their critique of the Gorleben project. Lovins, leading the discussion on the crucial question of whether Germany needed reprocessing and an expanded program of nuclear energy argued as follows: The German economy did not necessarily require more energy; if more were needed, it probably could be obtained by further gains in energy efficiency; if a convincing need for new energy supplies should arise, it would probably not be for electrical energy, and if it were the best way of meeting it would not be from nuclear reactors but from more benign sources. Lovins and von Weizacker are said to have established a rapport. Von Weizacker seemed much engaged, for example, when in an informal evening recital that Hatzfeldt arranged, Lovins played Chopin nocturnes and Beethoven sonatas on one of Crottorf's Bechstein grand pianos. Jonathan F. Callander, a Gorleben critic and geologist from the University of New Mexico, told me that through this recital Lovins established a new level of communication with von Weizacker, who "was very taken by this other dimension of Lovins."

The Gorleben International Review opened in Hannover on March 28, 1979 -- the same day that the Three Mile Island accident began grabbing headlines around the world. To say the least, this coincidence affected the atmosphere of the hearings, and no doubt had a bearing on its outcome. As Lovins has put it, "simultaneously you had on page one stories about the Gorleben hearings and stories about whether the Three Mile Island reactor was going to do itself in. That certainly helped create some atmosphere of skepticism about [the project proponents'] safety claims. " But the impact of the events at Hannover and Harrisburg went beyond news coverage. The Gorleben Review had opened in the middle of the week; by the week's end up to 140,000 anti-nuclear protesters from all over Germany had converged on Hannover (peacefully, as it turned out), according to police estimates . . . .

The proceedings took the form of a debate in which the critics' statements were always followed by rebuttals by the pro-Gorleben panel of experts. Inasmuch as Prime Minister Albrecht did not want the debate focused on the details of the Gorleben project, the critics found themselves frustrated in presenting their case, especially as it related to the merits of the Gorleben salt dome as a repository site.

Yet Albrecht was to find reason not to allow licensing to proceed for the integrated fuel cycle center DWK had proposed. The tens of thousands of protesters who had gathered in the streets of Hannover could hardly be overlooked . . . . [Carter, Luther, Nuclear Imperatives And Public Trust, Resources For The Future, 1987, p270-275]

Clearly, Yucca Mountain was only the latest stage in a movement dedicated to stopping nuclear technology and nuclear waste disposal worldwide. The coalition of anti-nuclear activists that Lovins represents are professionals, with an international record of success at disrupting the advance of nuclear technology. There are two issues to note about the Gorleben incident:

- Technical proceedings were used as a publicity and delay tactic.

- Organized mass protest, sometimes verging on mob rule, can be effective and intimidating. 140,00 protesters converged on Hannover; Yucca Mountain may at some point also experience such "spontaneous" protest.

## FOUNDATION ON ECONOMIC TRENDS

Jeremy Rifkin is one of the most prominent anti-technologists of our time, a vortex of radical environmentalism. Rifkin's occasional ties to the Safe Energy Communication Council link his philosophy to the opposition against Yucca Mountain. Rifkin is currently president of the Foundation of Economic Trends (FOET), a Washington based environmental group, which according to his website:

**The Office of Jeremy Rifkin is operated by Jeremy R. Rifkin Enterprises, a sole proprietorship with the purpose of advancing Mr. Rifkin's written work and lecturing. The Foundation on Economic Trends is a non-profit 501(c)3 organization whose mission is to examine emerging trends in science and technology, and their impacts on the environment, the economy, culture, and society. [http://www.foet.org/]**

As well as being a center for advocating a Third Industrial Revolution, FOET appears to be a money making venture funded by Rifkin's many books (18 so far). This long string of books however gives us the ability to see just how accurate Rifkin's prescience has been (not very). Among FOET's current concerns are stopping the use of bovine growth hormones and halting genetic engineering as well as creating a worldwide "empathic civilization". Rifkin touches Yucca Mountain because of his writing for the Safe Energy Communication Council:

**The year: 2035. Massive dikes around New Orleans, Miami and New York are holding back rising sea water. Phoenix is baking in its third straight week of temperatures above 115 degrees. Decades of drought have laid waste to a once-fertile Midwestern farm belt. Hurricanes batter the Gulf Coast, and forest fires continue to blacken thousands of acres across the country.**

**Science fiction? Hardly. These are the sobering global warming or "greenhouse effect" scenarios that many scientists believe may happen if we continue to pollute our environment [Rifkin, Jeremy; Global Warming: What You Can Do, Viewpoint, Safe Energy Communication Council, 1989]**

Rifkin's book, The Emerging Order: God in the Age of Scarcity, is a Rosetta Stone that helps connect some anti-nuclear activism and Christian evangelical/environmental groups. Vice-president Al Gore has many of the same philosophical foundations as Rifkin through the environmental organizations of the Southern Baptist church. In fact, in the early 80's, then Senator Gore was the first to bring Rifkin before Congress to testify on environmental matters. Rifkin and Gore seem to hold in common what might be called an apocalyptic Southern Baptist thermodynamic environmentalism! In a chapter titled "Limits To Growth", Rifkin lists our energy options:

**"It's hard for most of us to imagine that the usable oil on this planet will be gone in the next twenty years or so. . . ."**

**". . the external costs associated with extracting and harnessing coal energy make it impossible to entertain even the prospect of mining even a fraction of what remains. . . . the environmental dangers associated with burning massive amounts of coal make it prohibitive. . ."**

**"The fact is, the energy required (in terms of costs) to retrieve shale oil in a usable form is so high that the net energy return is minimal."**

**"It is estimated, even at current rates, that worldwide expansion of nuclear plants will within twenty years generate enough fissionable material in international transit to make 20,000 atomic bombs. . . . . Even if United States nuclear power continues to level off, it will be necessary to find new burial sites as often as every two or three years after the turn of the century to accommodate all the waste. This in turn will necessitate strict monitoring and around-the-clock armed guards on each site for thousands of years to insure against leakage into the biosphere."**

**"Solar energy is diffuse, unevenly distributed and unavailable at night, and it varies with climate, the seasons and the geography of the planet. For all these reasons solar power does not lend itself to massive centralized systems of collection and dispersal - the kinds of energy grids that our highly industrialized economies call for." [Rifkin, Jeremy; The Emerging Order: God in the Age of Scarcity, Putnam, 1977, p47]**

Written in 1977, Rifkin mistakenly predicted the end of the oil economy by 1997 and assumed a Yucca Mountain would need to be built every two or three years. Although the Emerging Order was written in 1977 at the height of an energy crisis, Rifkin's thinking seems not to have changed much since then. We are left with the uneasy feeling that there are no solutions to impending energy doom. Rifkin based his pessimistic world view on a distortion of a law of physics called the Second Law of Thermodynamics.

**Industrial technology, then, creates temporary order, but at the expense of speeding up the overall process of moving from low entropy to high entropy. In other words, the more we exploit and expend the low-entropy matter and energy around us in the natural world in order to create a more efficient order in a concentrated time span and place, the greater the overall chaos we ultimately create in a larger world. [Rifkin, Jeremy; The Emerging Order: God in the Age of Scarcity, p67]**

The larger world Rifkin should be talking about is the entire universe, a volume so grand that we cannot even calculate its full extent. Throwing chaos off into this larger volume only threatens humanity on the time frame of many billions of years, not within the foreseeable future.

Unfortunately, Rifkin's views drift through the environmental lobby in Washington. Rifkin's impact on Yucca Mountain comes because his philosophy of impending environmental collapse is part of the milieu which drives anti-nuclear lobbying within the Beltway. In fact, the only way to avoid thermodynamic collapse, at least as Rifkin envisions it, is by developing new energy resources. Thus, nuclear energy would be a prime candidate for alleviating environmental chaos, not a source of environmental problems. However, Rifkin's utopia is all solar with no hydrocarbons or nuclear energy. From his website in 2010:

**The Third Industrial Revolution**
*"We have the science and technology to do it, but it will mean nothing unless there is a change in will."* — Jeremy Rifkin
**We are on the cusp of a Third Industrial Revolution that could give us a door open to a new post-fossil fuel era. It was the first Industrial Revolution that brought together print and literacy with coal steam and rail. The second combined the telegraph and telephone with the internal combustion engine and oil. What we now have now is the possibility of a distributed energy revolution. We can all create our own energy, store it, and then distribute it to each other. Twenty five years from now millions of buildings will become power plants that will load renewable energy. We will load solar power from the sun, wind from turbines and even ocean waves on each coast. We can also make the power grid of the world smart and intelligent; we call it inter-grid. Not far from now, millions and millions of people will load power to buildings, store it in the form of hydrogen and distribute energy peer-to-peer; just like digital media and the internet. The first inter-grids are going up in the United States this year in Houston, Boulder Colorado, and Southern California. The "Third Industrial Revolution" is an economic game plan. We have the science and technology to do it, but it will mean nothing unless there is a change in will. [http://www.foet.org/tir.html]**

The fact that all of this has been predicted since the early 70s, yet the engineering to make it practical seems nowhere near emerging because of cost and physical limitations, seems to be of no concern to Rifkin, whose books provide a steady income despite a record of mistaken predictions.

In short, both Amory Lovins and Jeremy Rifkin have had forty year careers built on forecasting the rise of a solar utopia. Along with the Nader groups, their philosophy inspired much of the energy policy of the Obama administration. The only thing standing in the way of this utopia seems to be physics and the fact that solar energy densities make alternative energy sources prohibitively expensive. We will cover the problems of solar energy versus nuclear energy in a later chapter, however it has been important to expose the ideological roots of the anti-Yucca Mountain opposition for what they are, an extreme political movement which will not be content with anything less than a total reordering of civilization to meet their Utopian dreams.

# 30. Nevada Test Site Protest

The connections between those protesting nuclear weapons at the Nevada Test Site and those protesting the construction of the nuclear waste repository at Yucca Mountain are stronger than one supposes. Among the most prominent Test Site protest organizations are the following:

- Nevada Desert Experience / Lenten Desert Experience
- American Peace Test
- Citizen Alert
- Atomic Veterans
- Greenpeace
- Hundredth Monkey
- Sagebrush Alliance (defunct)
- Western Shoshone

The merging of pacifist nuclear-weapons protest with protest against nuclear waste disposal was apparent from the beginning as activists in the Southwest banded together to form the Cactus Alliance.

**CACTUS ALLIANCE Over the weekend of 1-2 October, 1977, people from the states of Colorado, New Mexico, and Nevada came together with the intention of forming a regional alliance of groups working against all aspects of nuclear energy and weaponry. We were successful and are happy to share with you the results. Naming ourselves the Cactus Alliance, we identify with our western mountain states region, as well as with the struggle throughout the United States and the world.**

**The Cactus Alliance is a coalition of citizens dedicated to the betterment of life through the advancement of ideals and values of respect for life and health, and of sensitivity to the earth and its systems. These bring us into opposition with:**
**\* the high cost and risks, especially health risks, of nuclear energy.**
**\* the introduction or radioactive wastes into the environment, and**
**\* the production, proliferation, and use of nuclear weapons.**

**We actively support the alternatives of strict conservation practices, the redirection of technology to meet human needs, and the full development of alternative energy sources along with decentralization of energy systems. To this end, we pledge to further our goals by means of education, communication, direct action, and community organizing. . . . [Anna Gyorgy and friends; No Nukes: Everyone's Guide To Nuclear Power, South End Press, 1979]**

Among other books published by South End Press at that time were *Science, Technology and Marxism* (Stanley Aronowitz), *Ecology as Politics* (Andre Gorz), and *The Sun Betrayed: The Corporate Seizure of Solar Energy* (Ray Reece). Clearly the mood at the time was one of intertwined Marxist and decentralist ideas being merged into environmental and technological issues.

Among the members of the Cactus Alliance at its formation was the Sagebrush Alliance, a Nevada group that later proved effective in opposing the MX missile system which was proposed for Nevada. Members of the Sagebrush Alliance eventually merged with Citizen Alert as the MX missile issue was resolved. Protests at the Nevada Test Site over weapons testing only began in earnest after the formation of the Lenten Experience, later known as the Nevada Desert Experience. Franciscan Louis Vitale and Sister Rosemary Lynch were critical in starting the Lenten Desert Experience. Friar Vitale's 1978 tour of the test site evidently sparked the idea of a Franciscan vigil on the road to Mercury, Nevada. A history of some of these events was recorded in 2009 when Sister Lynch passed away:

In California Lynch had witnessed the difficulties faced by her low-income African-American and Mexican-American students, and in Montana she had seen the struggles that the Native American students underwent. These experiences had sensitized her to social inequity. Her exposure to unspeakable misery around the world incalculably deepened her awareness of systemic injustice and violence. This growing conscientization was reinforced by the work of a fellow Franciscan, Sister Klarita, who worked for the Pontifical Commission of Justice and Peace during those years. Through the work of the commission, Lynch became more aware of specific examples of social-structural injustice and an emerging theology that emphasized the sacredness and dignity of all human persons coupled with systematic social analysis and strategies for social change. After finishing her assignment for her congregation in 1975, she spent an additional year in Rome working with an international education association. All of these experiences led to a profound deepening of Lynch's personal commitment to nonviolent social change.

After returning to the United States in 1977, Lynch settled in Las Vegas where she joined the staff of the Franciscan Center. That summer, President Jimmy Carter announced that he was seeking funding from Congress to develop the "enhanced radiation" or "neutron" bomb. Soon afterward, news was leaked that the neutron bomb had already been developed and tested at the Nevada Test Site. Lynch decided to do some research on this program and the test site in general. In the course of her exploration, she discovered that a group of Quakers, including Larry Scott and Albert Bigelow, had held the last demonstration at the test site on August 6, 1957.

Spurred by this, she and a group of friends in Las Vegas organized an event at the gates of NTS to mark the 20th anniversary of this activity, to protest the impending production of the enhanced radiation weapon developed there, and to remember the bombing of Hiroshima thirty-two years earlier. They dubbed themselves "Citizens Concerned about the Neutron Bomb." As it was later reported:

*Nineteen people met at the main gate of the NTS before dawn to hold a prayer vigil and conduct a teach in about Hiroshima. The vigilers held signs along the road that led into the Test Site and they were very careful to make signs that supported the workers but objected to testing. One sign read: "NTS Workers Yes, Nuclear Bombs No." The vigil was highlighted by the visit of Japanese Hibakusha [survivors of the atomic bombings] who wanted to present a book of drawing of the bombs dropped on Japan to the Test Site officials. The vigilers went directly to the guard house at Mercury Station. The Japanese approached the gate house but the guards refused to accept their book. An older Japanese lady, a Hibakusha, extended her hand to the guard and he refused to shake her hand. The small group began a chant, "Take her hand. Take her hand." Finally the guard gave in and shook her hand.* (Michael Affleck, *The History and Strategy of the Campaign to End Nuclear Weapons Testing at the Nevada Test Site, 1977-1990* (Las Vegas, NV: Pace e Bene, 1991.)

One tell tale sign of what Sister Lynch and the other Franciscans were involved in is the use of the word *conscientization* which comes from Portuguese and derives from Liberation Theology developed in Brazil by the Franciscans. In short, it is Marxism rough a catholic lens.

Questions of Soviet involvement in the Test Site protests were raised in a 1983 article written by Mary Manning who interviewed an outgoing DOE official:

### LV ANTI-NUKE ACTIVISTS DENY LINKS WITH SOVIETS
The U.S. Department of Energy has evidence the Soviet Union is supporting some of the growing anti-nuclear protests at the Nevada Test Site, retiring Nevada Operations Manager Thomas Clark said Wednesday, but local activists say they have no Russian ties.

Spokesmen for such local peace groups as American Peace Test, Clergy and Laity Concerned and Nevada Desert Experience vehemently denied communist ties, either funding or moral support. . . .

"The communists are supporting some of these protesters here, and doing so very aggressively," Clark said. "That's bad news."

Asked who collected proof for DOE, Clark replied, "All you have to do is go out there and buy a copy of "The Daily Worker.'" The Daily Worker is a Communist Party of the U.S. newspaper.

Clark also refused to name government agencies that have linked test site protesters to communist sympathizers, but offered a letter written on State Department stationery by Kathleen C. Baily, the wife of Dr. Robert Barker, the assistant secretary of defense for nuclear energy and a liaison between the DOE and the Defense Department.

The letter contained information linking 15 of 60 Danish students who visited Las Vegas April 23 to 25 with communist youth groups.

However, Judy Treichel, coordinator of Clergy and Laity Concerned, said the Danish Communist Party is not tied to the Soviet Union.

"Everybody thinks it's an automatic Soviet tie, but some new communist groups are working in the communal sense of that word and don't want any super-power attachments . . . " [Manning, Mary; Las Vegas Sun, July 2, 1987, p 1A]

Of course, in 1983, the Cold War was still very much a reality and attempts by citizen's groups to affect weapons testing, no matter how sincere, were still threats to national security. It is not apparent whether there was any Russian support for the disruption campaigns engaged in by the members of the Cactus Alliance and the Sagebrush Alliance, however it is unlikely the Soviets would have been ignorant of such an opportunity. Attempts to pursue a Freedom of Information Act request on this subject met with limited success but obviously deserve to be pursued.

Later, the national FREEZE Campaign, which also proposed moratoriums on nuclear testing, made its presence known in Nevada, eventually spawning the American Peace Test (APT).

"The August 1985 Nevada Desert Experience (NDE) action was the event which inspired the APT. Although the desire for direct action had been growing within the FREEZE Campaign, it was this action which provided the inspiration for Nancy Hale and Jessie Cocks to convince others of what they saw as an opportunity at the NTS. . . .

Hale commented on the August NDE action where she explained that daily diversions do not exist in the desert. She indicated it provided the experience to combine intellectuality with spirituality, and if that experience could be recreated for others, then a successful campaign would be possible." [Mann, David; The Historical Origins of the American Peace Test, Masters thesis Dept. of Pol. Sci., University of Nevada Las Vegas, 1991]

While the original protesters from Nevada Desert Experience had been strictly pacifist, more activist groups like Greenpeace also began joining the protest marches and added to the ranks that American Peace Test had brought into the fray. An important history of these events has been provided by Michael Affleck of the Nevada Desert Experience, *The History And Strategy of the Campaign to End Nuclear Weapons Testing at the Nevada Test Site, 1977 - 1990*. Near the center of this protest had been Judy Treichel. According to Affleck:

Judy Treichel was a member of a union associated with NTS and a Mormon, who was working locally for a test ban. Judy had worked on the first LDE and had taken considerable risk within her religious community for her outspoken views on social issues. She was influential in organizing the second LDE. [Affleck, Michael; The History And Strategy of the Campaign to End Nuclear Weapons Testing at the Nevada Test Site, 1977 - 1990, Nevada Desert Experience, 1991]

It is little wonder that Treichel's views as the information officer for the Nevada Nuclear Waste Project Office were biased by her previous anti-test site activist history. It is also unlikely Bob Loux, executive director of NWPO, was blind to this activity when he chose the Nuclear Waste Task Force to provide information on the Yucca Mountain repository to Nevadans. Treichel thus provides the first link between the NTS protest movement and protest of the repository. More generally, Citizen Alert as parent to the Nuclear Waste Task Force has long been heavily involved in supporting anti-NTS protest

The religious, environmental, indigenous people and pacifist motivations of Yucca Mountain and Nevada Test Site protesters were summed up in what was called the *Healing Global Wounds* campaign of 1992. Sponsored by Citizen Alert, American Peace Test, Nevada Desert Experience and Walk Across America for Mother Earth, this protest at the Nevada Test Site neatly combined all the philosophical threads evident in the non-scientific anti-repository movement as shown in an article from the event newsletter:

HEALING GLOBAL WOUNDS *culminates our international Walk Across America For Mother Earth*
Ever since Columbus' arrival 500 years ago, there has been a terrible disparity between the "discoverer's" and Native People's attitudes towards care of the Earth and respect for Life. Traditional Indigenous Peoples see the Creation as a home, a sacred "Thou" to take care of, to respect and love;

contemporary "Industrial Peoples," on the contrary, regard Nature as a mere "It", a warehouse or resources to be used and exploited to suit their own ends, regardless of the effects upon the environment and other living beings. Nowhere is this difference in world views more obvious than in today's struggle over nuclear weapons. Although humankind has inhabited this planet for millions of years, in just the past fifty years has so-called "civilized man" poisoned vast areas of the Earth, making them uninhabitable - such as through manufacturing and testing nuclear weapons.

No "Industrial Governments" test nuclear weapons on their own lands. All test bombs on Indigenous People's lands.

[Walk Across America for Mother Earth Newsletter, Issue Five, August 1992, p1]

Similar arguments were applied at Yucca Mountain, that "Industrial Peoples" are harming both Mother Earth and "Indigenous Peoples" by poisoning the land. Yet, combining protest of weapons testing and nuclear waste makes little sense. As Iran's accumulations of calutron separators for uranium processing shows, separating bomb grade plutonium from spent fuel or natural sources is not something an arm-chair chemist can do in his garage. Nuclear waste is not bomb-making material and halting the building of the nuclear waste repository is not inherently linked to weapons control. Nevertheless, the quasi-religious belief of anti-nuclear activists that all radioactive substances are environmentally evil and a social injustice directed against indigenous peoples has motivated not only their opposition to nuclear weapons of war, but also against peaceful nuclear technology as well.

# 31. Citizen Alert

**"Citizen Alert organized 18 years ago (1974) to fight the U. S. government's plans to site a high-level nuclear waste dump in Nevada. We're still fighting. In that time, we have joined forces with citizen groups around the country in a common effort to bring accountability to this country's nuclear waste disposal policies."[frontpiece of Nuclear Waste Special, Citizen Alert, 1992]**

Citizen Alert was the thorn-in-the-side activist group whose activities have been most responsible in Nevada for throwing roadblocks into the Yucca Mountain machinery. That is not to say it is the only local activist group; CAN-WIN, Rural Alliance, Nevada Desert Experience, American Peace Test, Southwest Information etc. also come to mind. However, Citizen Alert has the most bodies to throw into the breach at DOE meetings and is the most frequently quoted in press releases regarding Yucca Mountain and other nuclear issues.

Citizen Alert had modest beginnings in 1975. Two of its members, Susan Orr and Kathryn Hale, traveled the state holding forums to get Nevadans to stop the low-level nuclear waste dump that U.S. Ecology was already operating in Beatty, Nevada. Consequently, nuclear waste protest in Nevada is the child of Citizen Alert. The turning point for the group occurred in 1981 when the Great Basin MX Alliance, under the leadership of Citizen Alert, was part of the successful effort to convince Congress and President Reagan to scrap plans to deploy the MX missile system in Nevada. In reality, practical and political questions stopped the MX rather than just the opposition of Citizen Alert. Nevertheless, the MX protest effort was critical in defining the organization's character and gave it confidence in its ability to influence political events in Nevada.

## CITIZEN ALERT AGENDA

Citizen Alert has a broad set of issues that it presses, most of which can be viewed as anti-military and anti-government. Beyond Yucca Mountain, their agenda connects issues from nuclear testing, to desert tortoise habitat protection. However, this agenda should not be viewed as being composed of fragmented pieces. From a 1992 Citizen Alert flier:

**Some of the issues Citizen Alert will be monitoring include: Radioactive waste disposal; nuclear weapons and the military budget; comprehensive test ban and the Nevada Test Site; radiation victims; lands issues, such as privatization and militarization of public lands; return of Western Shoshone lands to the Western Shoshone; Air Force and Navy proposals to expand supersonic airspace over rural Nevada; and the needs of the low-income citizens, such as Indian Health Services and public health services in general; hunger and housing, particularly in rural areas - issues resulting from use of our national resources.**

This broad agenda contains some consistent threads:
1) Radiation paranoia
2) Land reform issues.
3) Opposition to use of the Nevada desert by the military.
4) Advocacy for the poor in relation to natural resource utilization.

Taken separately, these are environmental issues. Taken as a whole, this is a social movement which merely finds environmental issues convenient levers. A 1990 flier from Citizen Alert expresses their position on nuclear waste:

## FUNDAMENTALS OF NUCLEAR WASTE MANAGEMENT
**1) No acceptable disposal method for nuclear waste exists. Proposed methods will not dispose of the waste, only postpone contamination of the environment.**

**2) No level of radiation exposure is safe. Every dose carries a risk of damage. Epidemiological studies into cause/effect relation of radiation and health are premature and inconclusive. We need baseline health profiles for any community which has an existing or planned nuclear facility so we can start gathering information.**

**3) Radioactive waste is radioactive waste. The distinction between types and levels, such as high level, low level, transuranic, etc., are artificial and arbitrary and should not determine handling procedures. All**

**radioactive material should be treated with the same care.**

**4) Our present inventory of nuclear waste should be isolated in retrievable, aboveground facilities which are as durable as we know how to make them. All necessary political, social and financial arrangements should be made to ensure, to the best of our ability, its careful guardianship.**

**5) Existing nuclear waste should not be moved; it should be isolated on-site in retrievable facilities. Exceptions may have to be made in rare cases where not even a carefully designed, retrievable facility is able to secure public health and safety; for instance when waste is located on an active earthquake fault or flood plain.**

**6) No more waste should be created . . . from weapons, from electricity generation or from industry. Wastes from nuclear medicine area minuscule portion of the radioactivity of our waste stream. Their disposal poses a manageable problem because of their short half lives and low concentrations; they can be tolerated until reliable substitutes can be found.**

Bob Fulkerson, executive director of Citizen Alert until 1994, denied Citizen Alert held some of these positions, realizing that they were perhaps too radical. Still, Citizen Alert has a history of scientific naiveté which makes this position paper more representative of the views of the organization's member than not. The nuclear waste position paper, as with many Citizen Alert positions, at first seems to be admirable environmental concern over radiation. Closer point-by-point analysis shows their position would call for revolutionary readjustments in our relationship to nuclear technology:

1) "No acceptable disposal method" --While nuclear waste is not benign and does require millennia to degrade, contamination of the environment is no more inevitable nor dangerous from radioactive waste than it is from other artifacts of civilization. Coal plants spew by far more uranium, thorium, carbon 14 and other radioactive substances into the atmosphere than the nuclear industry. Radioactive waste from nuclear energy production is compact, solid, traceable (even a cheap Geiger counter can detect a leak) and can be isolated permanently from the environment in geologic repositories. In contrast, pollution from other energy sources is spread around the globe.

2) "No level of radiation exposure safe" --If Citizen Alert truly believes no radiation level is safe, then they should protest human exposure to the sun, television sets, airline travel, outdoor cosmic ray exposure, x-rays, cigarettes, and their own skeletons, which are all sources of radiation (more on this in later chapters).

3) "All radioactive material should be treated with the same care" – All radioactive waste does not have the same intensity and biological effect, that's why there is the special measurement called the *rem* used to distinguish different intensities of biological exposure. All radioactive waste should NOT be handled with the same care. It is lethal to handle spent fuel direct from a reactor, yet the minute amounts of Carbon-14 caused by cosmic rays striking nitrogen in the air is not something we even consider a waste and would never consider isolating.

4) "Nuclear waste ... should be isolated in retrievable, aboveground facilities". Above ground storage is NOT safer than geologic storage. Institutional instability and nature argue that aboveground facilities could remain a hazard to future generations for millennia. As for the statement "All necessary political, social and financial arrangements should be made to ensure, to the best of our abilities, its careful guardianship", in reality calls for a totalitarian Green police state, though this may not have occurred to Citizen Alert in quite those words.

5) "Existing nuclear waste should not be moved" – If Citizen Alert considers radioactive waste so deadly, why would they endorse on-site storage when this statistically increases the risk of exposure by the population? On-site storage makes little sense unless political decentralization is the true goal since almost every major nuclear power plant is near some combination of rivers, faults, oceans or other critical areas.

6) "No more waste should be created..." This is a fine sentiment, but the problem is hardly limited to nuclear waste. Every industry, even the protest industry, creates waste (everything from spent fuel to discarded protest posters). However, nuclear waste as contained at Yucca Mountain will be environmentally isolated compared to other energy and chemical industries. As for radioactive nuclides being replaced in medicine, this is highly unlikely unless we wish to return to invasive procedures with sharp scalpels in everything from cancer treatment to medical imaging.

## CITIZEN ALERT FOOTSOLDIERS

During the early 80's, Bill Vincent, an activist journalist with an interest in environmental issues, served as director of Citizen Alert. By 1984., Bob Fulkerson Citizen Alert, an English major from the University of Nevada at Reno with an emphasis in medieval studies, had become the executive director. Intensely political, Fulkerson left Citizen Alert in early 1994 to start his own populist movement which appears to be in the same mold as the progressive (i.e., anarchist, decentralize) populism of Jim Hightower and elements of Ralph Nader's Public Citizens' Critical Mass Energy Project.

Chris Brown, a founding member of the peace group American Peace Test and former Southern Nevada Coordinator for Citizen Alert, became executive director in early 1994 when Bob Fulkerson stepped down. Brown was originally a theology major at Oberlin College, sharing with Fulkerson a liberal arts background lacking in practical expertise in energy matters. While naive technically, Brown did bring an encyclopedic knowledge of the rules and regulations of both DOE and the EPA to Citizen Alert, talents useful in the protracted war of legal delay carried on by environmental activists. Brown's testimony at hearings is well prepared, but being half-educated in engineering is often worse than knowing nothing at all.

September 15, 1992, Chris Brown was the featured speaker on NWPO's radio show on KDWN radio, moderated by Dennis Baughman, a representative from NWPO. One of the most amazing things Brown said was in response to a listener request for his credentials:

**CALLER: Could Chris Brown tell the audience his background in engineering and his degree.**

**BROWN: One of the rights and obligations of an American citizen is to get involved in issues that may have an impact on health and well being. My degree happens to be in religion. But the issue should not be to have so-called experts tell us what is safe or not safe. That has been the history of DOE. Parading experts in front of the public, telling them that things that eventually cause death or disease to many people were safe and you didn't have to worry about them.**

**BAUGHMAN : Yeah, I don't think you need a degree in engineering or any kind to read congressional reports and realize we ought not be putting our eggs in a basket with someone who has a miserable track record. ["Yucca Mountain: Fact Not Fiction", sponsored by NWPO, KDWN Radio, Sept. 15, 1992]**

Brown's close ties to the Nuclear Waste Project Office made his expertise very much an issue because of his impact on NWPO policy. Some of the data on which Yucca Mountain is to be judged by Congress is being generated by NWPO. Consequently, the expertise of NWPO staff and their environmental advisers has an obvious bearing on how much weight this data is given. The question is whether NWPO was using Nuclear Waste Fund money to conduct a rigorous engineering assessment of Yucca Mountain or whether a theology major from a vehemently anti-nuclear group was generating NWPO's position:

**BAUGHMAN: What do you think of the Yucca Mountain project?**

**BROWN: It is clearly an engineering project. They are clearly intending on building this project regardless of how safe it is. There are serious problems with Yucca Mountain: earthquakes, corrosive water, hot springs and volcanoes. They should put it in a place with less likelihood for disaster. DOE isn't interested in science or study for truths, only engineering the project. [KDWN Radio, Sept. 15, 1992]**

Ironically, the claim made by Chris Brown and Dennis Baughman for NWPO was that DOE cannot be trusted, and that the only reliable people to make that judgment are non-engineers with no background in nuclear technology of any kind. Why this criticism didn't also apply to NWPO is unclear. The blind leading the blind is perhaps too kind a label to place on this, and shows once again the dominant influence of uncredentialed energy analysts over the nuclear energy policy of the State of Nevada.

At a Sawyer Commission meeting in March of 1992, Brown made the claim in a casual conversation that the Nevada Test Site could be turned into a giant solar farm capable of supplying the nation with energy. This later turned up as Senator Harry Reid's plan to reconfigure the Nevada Test Site as a giant solar-hydrogen energy complex. The question of course is who supplied Brown and Reid with this duplicate image of solar utopia. Senator Reid's later statements that he had consulted with vice president Al Gore (a former journalist) on the matter of solar-hydrogen technology and was assured that solar-hydrogen technical solutions are already available (they are not) is a matter of concern.

Of course, NTS does receive enough solar energy to light the nation, if it could be bottled and transported economically. The problem is, as we show later, solar energy is resource intensive and using it on a large scale is anything but low-tech. The anti-repository forces tried to modify this impression by claiming hydrogen technology (derived from solar energy) would save the day. Yet, a state-of-the-art hydrogen-solar plant is unproven and likely more costly than run-of-the-mill coal or nuclear plants.

## INCESTUOUS CONNECTIONS TO NWPO

The power that Citizen Alert wields is further evidenced by a faux pas that occurred in October of 1992 that disturbed even Governor Miller. Miller's position on a nuclear rocket project the U.S. Air Force was proposing was that if environmental concerns could be addressed, then the test site would be the best place to put it. Of course, Miller was not in charge of the State's position on nuclear issues, Senator Bryan's political appointee Bob Loux was. A report was prepared by NWPO that effectively placed Nevada in opposition to the nuclear rocket. The fact that the governor was being pushed around by an agency he supposedly controlled was bad enough, but even more embarrassing was Citizen Alert's involvement in producing the NWPO opposition paper. An article in the Review journal describes what happened:

**COMMENTS CONCERN GOVERNOR**
**Nearly identical written comments from a state agency (NWPO) and an environmental group concerning the Air Force's plan to test nuclear rocket engines at the Nevada Test Site prompted concern Wednesday from Gov. Bob Miller.**
**"I don't think that's good public policy to share a document with an interest group, in advance of submission, even if there might be a commonality of opinions," Miller said.**
**Miller said he doesn't want state agencies to preclude discussions with citizens or citizens groups "but we don't want them taking our products."**
**His statement came after copies of the two documents show that about 90 percent of the written comments from the environmental group, Citizen Alert, were taken word for word from a document similar to one submitted to the Air Force by Bob Loux, executive director of the State's Nuclear Projects Agency.**
**Loux, who was returning from Washington, D.C. Wednesday, said "I'm surprised and disappointed that Citizen Alert would have taken a draft document and used it on their own."**
**"We're going to be much more careful. We're not going to share paper" with Citizen Alert again, Loux said....**
**"I'm kind of embarrassed about it. I didn't know they were going to submit their (draft remarks)," Brown said Tuesday, noting, "It wasn't Loux's doing." [Rogers, Keith; Las Vegas Review journal, Nov. 29, 1992]**

But it was someone's doing, and if Loux was not responsible for allowing Citizen Alert to collude and perhaps even dictate his agency's policy, then who was? Certainly not Governor Miller, who had long ago given up trying to direct NWPO. Even more telling is the stance that Citizen Alert and NWPO took in lockstep, also described in the Rogers article:

**Both documents recommended that only the "no action alternative" is acceptable. They both say the project should be shelved until there is a clearly defined mission or purpose for the project, a point both say is lacking.**
**Altogether, 177 others signed signature cards supporting a similar, three point stance, Brown said. The three points are that the project does not have a mission, a preliminary environmental impact statement needs to be completed, and the project would generate high-level radioactive waste, which is unacceptable to Citizen Alert. [Las Vegas Review Journal, Nov. 29, 1992]**

While the question of an environmental impact statement has some validity, the other two objections are indicative of greater issues:

1) Unsurprisingly, NWPO and Citizen Alert saw no mission for the nuclear rocket because they see no mission for anything nuclear. The past opposition of Citizen Alert and NWPO staff to the Test Site, to the MX missile system, to Yucca Mountain and to the military in general suggests this is a peace movement that has confused itself with the environmental movement.

2) The fact that both NWPO and Citizen Alert objected to the nuclear rocket simply because it generates high-level nuclear waste suggests there are no technical criteria for nuclear waste safety that relation to radioactive substances either NWPO of Citizen Alert will accept as reasonable. It may therefore be a wasted effort to expect rational guidelines for the disposal of nuclear waste to emerge from NWPO.

NWPO's entire objective, as for Citizen Alert, seemed to be to ratchet their objections to the point where all nuclear technology is banned, regardless of any counterbalancing benefits. It follows that these organizations have less interest in assuring the safety of Yucca Mountain than in shutting down the nuclear industry and weapons production of all sorts. This may be a commendable goal in moral terms, but it does suggest that the nuclear hysteria these two organizations create in synergy has little to do with reasonable engineering standards at Yucca Mountain.

## CONNECTIONS TO NATIONAL ANTI-NUCLEAR NETWORK

Citizen Alert's involvement with NWPO should not be thought of as simply local grassroots politics. In fact, there is a coordinated national to shut down nuclear energy which naturally focuses on Yucca Mountain. At a June 29, 1993 CAN-WIN meeting highlighted by a speech by Steve Frishman of NWPO, Chris Brown added the following comments:

CHRIS BROWN: "I think it's important to realize that we're not alone in this fight. The tag that's been given by a national network of organizations is a 'Blue Ribbon Commission'. The idea that there should be a review above the level of DOE of nuclear waste programs. And it's not just about high-level nuclear waste.

People around the country have seen that the low-level waste program has come to a grinding halt ... The state compact is in disarray. Beatty is the death knell of the compact on low-level waste."

At WIPP, the people of New Mexico have been able to keep the site closed because it's not safe. Slabs have fallen from the ceiling ...

All of the programs are in disarray. We have allies around the country. What we see and hear are the ads from the nuclear industry. But you should know we have strong allies around the country. Already in the House of Representatives there is legislation in support of the Blue Ribbon Commission is being formulated. And it's being formulated not by Nevada's legislators, but by legislators from other states who see that this is a national issue and that the review has to be higher than DOE level. John Gibbons, the National Science Advisor, has expressed support of a review like this.

Our biggest obstacle is once again the guy who's got his cross-hairs targeted on Yucca Mountain, J. Bennett Johnston. But even here, Paul Wellstone (Sen. Minn.) has said he would be interested in introducing the review concept into the senate committee.

So things are happening that will help us out a great deal. We aren't alone, we aren't just the twenty or so people in this room trying to stop the dump. There are people - in fact this effort on a national level is largely being coordinated out of New Mexico. The people there see it is in their best interest to get Yucca Mountain reviewed because along with that review will come a review of WIPP. The people that are working on low level sites see that if Yucca Mountain is reviewed, there sites will be reviewed too. I think it's in all our interests to get the review. All of these programs are in serious trouble. [CAN-WIN meeting, June 29, 1993]

Consequently, Citizen Alert is not a lone operator in its opposition, but part of a well established anti-nuclear network that includes national environmental organizations, key Congressional politicians and the Nevada Nuclear Waste Project Office. The evolved tactic for obstructing nuclear technology was to call for a "Blue Ribbon" Commission:

WHAT WE WANT

To rectify this morass and set the country on a scientifically sound, politically untainted course, we are calling on President Clinton to convene a Blue Ribbon Commission to review the current state of nuclear waste management and disposal and to make recommendations on the best means to move forward

The Commission should include representatives from the DOE, the Nuclear Regulatory Commission, the Environmental Protection Agency, tribal governments, Governors, community organizations, local and national environmental organizations and independent technical experts unaffiliated with DOE or NRC

[Citizen Alert flyer, May 1994]

This seemed innocuous enough, but would actually have opened up every aspect of nuclear technology to obstructionists. First, such a Blue Ribbon Commission would eviscerate the power of the Secretary of Energy whose expertise would be sidestepped. Second, it is interesting to note that the nuclear industry was completely left out of the potential actors in the Blue Ribbon Commission, a sure way to divorce the proceedings from reality. Finally, leadership of the Blue Ribbon Commission is also a problem. As far back as 1988, Vice-President Al Gore used this issue in his aborted presidential campaign:

### GORE WOOS NEVADA VOTES WITH TALK ABOUT NUKE DUMP, TEST SITE, 474TH

**Tennessee Sen. Albert Gore Jr., emphasizing that Nevada could provide him with a bridge to the West on "Super Tuesday," promised Saturday that as president he would re-evaluate the nuclear waste dump siting process "from start to finish." [Ralston, Jon; Las Vegas Review Journal, Feb. 21, 1988)**

Al Gore, a natural chairman of a Blue Ribbon Commission, but also a somewhat Utopian environmentalist, would likely lead such a nuclear review into completely uncharted territory.

### LONG TERM STRATEGY

Citizen Alert has prospered in its forty year battle against things nuclear by conducting an intelligent and Machiavellian war against its opponents. Its creation of the Nevada Nuclear Waste Task Force was a coup worthy of textbooks, for it allowed key Citizen Alert allies (Judy Treichel, Abby Johnson, Chris Brown, Marla Painter, etc.) to claim respectability as supposedly neutral citizen advocates under the control of the Nevada Nuclear Waste Project Office. In reality, the Nuclear Waste Fund was used as a $2 million slush fund by Citizen Alert, through its subsidiary Nevada Nuclear Waste Task Force and the offshoot Rural Alliance for Military Awareness. The fact that the Nevada Nuclear Waste Task Force was a front for Citizen Alert is evident from Citizen Alert's continuing activity with the National Nuclear Waste Task Force:

### MOCK RADWASTE CASK HITS THE ROAD
**The recent tour of the National Nuclear Waste Task Force through California was a success. After a two year hiatus, we took our 20X7 foot life-size mock nuclear waste cask out of mothballs and towed it through Sacramento and the Bay Area. We spoke to the media and local grassroots organizations about the hazards of transporting nuclear waste across our interstates to a nuclear repository.**

**Speakers for the tour ... included Don Hancock, the nuclear safety coordinator for the Southwest Research and Information Center. [Citizen Alert, Spring 1993, p15]**

Southwest Research and Don Hancock represent the Citizen Alert of New Mexico and were co-creators of the National Nuclear Waste Task Force, helping fund Caroline Petti (later of EPA) as a lobbyist from 1985 to 1988. In August of 1994, Chris Brown was fired as executive director of Citizen Alert, reportedly over a contract dispute but likely for other reasons not discernible from this somewhat secretive organization. The Review Journal announced that Brown was leaving a $33,000 per year job and that:

**Citizen Alert has 1,100 members and an annual budget of about $300,000. [Las Vegas Review Journal, Aug. 31,1994, p4B]**

These inner workings of Citizen Alert show that it was a well funded and savvy activist organization with twenty years of experience in finding the jugular of nuclear energy in general and the nuclear waste repository program in particular. Knowing this adds perspective to the later attempts of the nuclear industry to run a pro-repository public relations campaign in Nevada. The industry did not face a starved kitten when it began its advertising campaign in 1991, but a full grown tiger whose claws were sharpened by years of battle with the establishment.

# 32. Rural Alliance, CAN-WIN, SRIC & IEER

Four organizations which show how tightly the web of anti-nuclear protest against Yucca Mountain was woven are exemplified by Rural Alliance, Citizens Against Nuclear Waste In Nevada (CAN-WIN), Southwest Research Center (SRIC) and Institute for Environmental and Energy Research (IEER).

## RURAL ALLIANCE FOR MILITARY ACCOUNTABILITY (RAMA)

This organization was a spinoff of the Nuclear Waste Task Force created by Judy Treichel, her friends Marla Painter and Abby Johnson, with the support of Citizen Alert. Marla Painter, who along with Treichel was a board member of Citizen Alert in the mid 80's, became RAMA's executive director. Abby Johnson, also a Citizen Alert alumnus, became RAMA's newsletter editor. According to Treichel's Nuclear Waste Task Force log:

**7/28/1988 - I met with Abby Johnson . We took care of some loose ends for the task force and discussed her possibilities for working with Marla in rural areas of the state. . . At 3:00 on 7/28, I met with Marla Painter at Foresta Institute. She submitted a proposal for organizing in rural Nevada. I felt that it would greatly enhance the success of the task force.**

But was Rural Alliance created to provide objective informational outreach on Yucca Mountain, or did it have a different purpose? According to RAMA's newsletter, RAMA Resources:

**"RAMA is an informal working alliance of rural organizations and individuals in the United States who are adversely affected by the U.S. military, including the Department of Energy. RAMA's goal is to change the way the U.S. military operates by making military installations responsive to environmental, public health, economic, and human rights concerns." [RAMA Resources (Winter 1993)]**

RAMA listed itself as a "project of the Tides Foundation", a politically correct philanthropy which in 1992 "made 845 grants totaling $5,464,700 to some 500 organizations in 42 states and 8 countries." Besides Rural Alliance, among the Tides Foundation's anti-nuclear grantees are the Southwest Research and Information Center and the Rocky Mountain Institute, both of which have been active in the opposition to Yucca Mountain.

Despite the Tides sponsorship, RAMA owed its existence to Judy Treichel and Marla Painter who in 1989 were organizing a front in Northern Nevada for protest against Yucca Mountain. It was Nuclear Waste Task Forces resources which supported RAMA's startup. Consequently, claims that Rural Alliance was an independent entity uninvolved with Treichel's NWTF and the state were diversionary. In fact, a flow chart drawn by Judy Treichel in 1988 and presented in her bid proposal to NWPO shows just how tightly incorporated Rural Alliance has been in the state's anti-nuclear hierarchy:

Hidden in this chart are clues which show how the Nuclear Waste Task Force promoted an activist agenda hidden behind Yucca Mountain. Rural Alliance was created not to provide informational outreach, but to actively oppose not only the repository, but military activities nationwide. Tod Bedrosian was brought on by the Task Force to organize further political opposition. Listing the League of Women Voters as a neutral organization in the flow chart hides the fact that Abby Johnson was then the president of the League of Women Voters and a co-founder of the Nevada Nuclear Waste Task Force with Judy Treichel. From Treichel's 1998 log, we see Johnson was instrumental in founding RAMA and was listed as RAMA's newsletter editor. Citizen Alert's motives are already well known.

Whether RAMA was a part of NWTF, a part of Citizen Alert, an independent entity or a part of the Nuclear Waste Project Office is difficult to tell. For example, in 1993 RAMA submitted a letter to the Secretary of Energy Advisory Board Task Force on Radioactive Waste Management (TRWM) investigating levels of trust within the DOE, but failed to indicate its connections to the Nuclear Waste Task Force:

It is remarkable that this task force (TFRWM) was formed to study the public trust and confidence issue, one of the many problems facing the Department of Energy (DOE) in the management of nuclear waste. The report's most striking finding is that no one - not even nuclear industry representatives - trusts the DOE. We certainly agree with that conclusion. [Letter to Task Force Radioactive Waste Management, March 10, 1993]

RAMA's lack of trust in the DOE is evidently absolute; the organization has operatives opposing Los Alamos, WIPP, Savannah River, Hanford and other military reservations. In fact, RAMA appeared to be primarily anti-military but used toxic waste issues to pursue its pacifist agenda.

Abby Johnson, RAMA's newsletter editor, also wrote a letter to the TFRWM, but did so as a consultant to Nye County Nevada (which surrounds Yucca Mountain) without admitting her ties to RAMA. Clearly, a small cell of activists centered in the Nuclear Waste Task Force, the Rural Alliance for Military Accountability and Citizen Alert leveraged their input into nuclear waste issues far beyond their limited numbers.

RAMA also pursued the DOE's and military's use of depleted uranium. Grace Bukowski, who worked for both Citizen Alert and Rural Alliance, was the principle author of "Uranium Battlefields Home and Abroad", issued by RAMA in 1993.

"Amid Kuwait's war litter lies the unseen danger of radiation from uranium-tipped shells used to knock out hundreds of Iraqi tanks. It was the first time DU shells had been used in combat by U.S. and British forces. Radioactive Depleted Uranium debris left behind by the allied coalition in the Persian Gulf War have almost certainly contaminated the air, water, land and food chain of the region." [Bukowski, Grace; Uranium Battlefields Home and Abroad, Rural Alliance for Military Accountability, 1993, p2]

Whether the vast desert regions of Iraq and Kuwait were irreparably contaminated by the low-level radioactivity of depleted uranium shells is dubious. Depleted uranium environmental damage must also be contrasted with the huge amount of pollution caused by the Kuwait oil fires. While Bukowski pointed to health concerns about depleted uranium exposure by U.S. soldiers, she did not address the question of how many U.S. (and even Iraqi) soldiers would have died if depleted uranium shells had not decimated the Iraqi tank battalions and led to an early conclusion of the war.

A discussion of depleted uranium links back to Yucca Mountain because Marvin Resnikoff of Radioactive Waste Management and Arjun Makhijani of the Institute for Energy and Environmental Research were two of the technical advisers for the Uranium Battlefields book. Both these gentlemen also worked for the Nevada Nuclear Waste Project Office and both made careers out of opposing nuclear technology on all levels. The mixing of heartfelt anti-nuclear pacifism with scientific objectivity on the waste disposal issue has proven difficult.

The question all this raises is whether the state of Nevada in effect sponsored the creation of an anti-military, anti-nuclear activist organization with tentacles stretching across the nation. RAMA listed as its field office a phone number and mail box in Questa, New Mexico. This turned out to be near the nuclear laboratories at Los Alamos, New Mexico, and also served as an office for opposition to the Waste Isolation Pilot Project in Carlsbad. Since Rural Alliance was originally part of the NWTF's anti-Yucca Mountain campaign, and the NWTF represented the state of Nevada, it appears Nevada supported a covert operation in another state to affect that state's dealings with the federal government. Whether legal or not, it is unlikely New Mexico would have been pleased to find they were infiltrated in such a manner. Other states and governmental entities may find they have been similarly influenced.


## CITIZENS AGAINST NUCLEAR WASTE IN NEVADA (CAN-WIN)

Led by Tom Polikalas, this group also had a strange pedigree. Derived in part from a former group, Nevadans Against the Dump, the name CAN-WIN was a derivative from a similar group which existed in New Mexico (New Mexicans Against Nuclear Waste In New Mexico). CAN-WIN for a short while seemed to be the pet of a number of Nevada's influential politicians, taking special attention in the 90s from Las Vegas mayor Jan Laverty-Jones and former Governor Grant Sawyer. Both those luminaries were present to give kickoff speeches at a special press release for CAN-WIN.

"Some people ask me how I can get involved. Most thinking citizens in Southern Nevada feel about this (Yucca Mountain) as deeply as I do. They're deeply concerned. And every day they come to me and say to me, what can I do? And I have nothing to tell them. Now we have a group. [Grant Sawyer, Channel 3 interview, 3/3/92]

The curious thing about Grant Sawyer's presence was that as head of the Nevada Commission on Nuclear Projects, he oversaw $50 million in federal grant monies to NWPO, but was there engaged in helping inaugurate a political action committee to derail the federal government's site characterization in Nevada. Mayor Jones was also a member of the Commission, so apparently any legal issues were ignored.

## INSTITUTE FOR ENERGY ENVIRONMENT RESEARCH (IEER)

Run by Arjun Makhijani, a transplant from Bombay, India, this institute is a curious cross between a scientific consulting firm and an antiwar protest group. Makhijani's technical credentials are respectable, doing a PhD on controlled nuclear fusion at the University of California at Berkeley in 1972. Although the anti-nuclear movement consults with Makhijani regularly, some of his views are inconsistent with the goals of the environmental movement. At a Citizen Alert "Town Hall" meeting, March 16, 1993, Makhijani called for seabed disposal of nuclear waste and burning excess weapons-grade bomb material in existing reactors. He called the nuclear waste issue a "Messy problem with no good solution", which would not seem to rule out the possibility that Yucca Mountain might still be the best of many poor solutions. Indeed, in a paper prepared for NWPO under Judy Treichel's guidance, Makhijani stated:

"We should emphasize that on-site storage is not a long-term solution, and cannot take the place of a long term method of managing nuclear waste, such as that envisioned for an appropriately structured repository program.
There are a number of difficulties with on-site storage that make it less than ideal:
* The need for continuous maintenance, monitoring and surveillance, which cannot be guaranteed for the hazardous life of the waste.
* The need for security to prevent access to wastes by intention or accident; and
* The lack of assurance that the storage facilities used today will themselves be safe for the very long term. In addition, interim storage facilities are not without potential safety or environmental problems. For example, spent fuel casks may be subject to the release of gaseous radionuclides due to rough handling, and at some nuclear sites there may also be concerns from seismicity hazards." [DOE contract DE-FG-08-85-NV10461, High Level Dollars, Low Level Sense, Arjun Makhijani, Scott Saleska]

Among those listed in the credits of High-Level Dollars, Low-Level Sense are Steve Frishman of the Nevada Nuclear Waste Project Office and Don Hancock of Southwest Research and Information Center (see above). Co-author Scott Saleska appears to also have ties to the Safe Energy Communication Council and Ralph Nader's Public Citizen. This again demonstrates the existence of a large, long term professional anti-nuclear network which mixes the nuclear waste issue with protest against the nuclear weapons complex. It also shows how NWPO, through funding of Makhijani and others, became involved in nuclear weapons complex issues, beyond its mandate.

In a report called "Facing Reality, The Future of the U.S. Nuclear Weapons Complex", in part funded by the Tides Foundation (also benefactor of Rural Alliance), Makhijani (IEER), Hancock (SRIC) and others from the Military Production Network, called for a major downsizing of America's nuclear weapons production complex. "The Complex should be designed to support a future arsenal in the range of zero to 3,000 at most, subject to further revision downward." [p4]

While members of this group had every right to advocate a zero nuclear weapons position, the difficulty for Yucca Mountain came when representatives from this network, such as Makhijani, imposed an anti-weapons agenda on the practical science of creating a geologic nuclear waste repository to hold spent-fuel from peaceful commercial endeavors. Linking energy policy to weapons policy thus threatened to derail energy production, without reference to the merits of the underlying science. Given the proliferation of nuclear powers creating their own weapon grade materials (North Korea, Iran, etc.), these protests no seem obsolete.

A recent listing of subjects from the IEER website in 2010 showed they continued to be anti-nuclear everything and solar Utopian:

**WHAT'S NEW / WHAT'S HOT (But not radioactive!)**
**North Carolina Study Shows that Wind and Solar Power Can Supply Three-Quarters of the State's Electricity. Read full report by John Blackburn (March 4, 2010)**
**Testimony in favor of Minnesota's moratorium on new nuclear reactors, by Arjun Makhijani, delivered by Lisa Ledwidge before a committee of the State Senate (March 2, 2010)**
**DOE: Take Back Depleted Uranium Sent to Utah and Send No More. IEER's Memo to HEAL Utah (February 16, 2010) | Radioactive Waste Profile Record for DU shipped to Utah from DOE's Savannah River Site | Press Release & HEAL Utah's Letter to DOE Secretary Chu (March 3, 2010) | NRC's Low-Level Waste Classification Rule**
**Can Obama Re-energize Climate?, Arjun Makhijani joins others at the National Journal Expert Blogs: Energy & Environment (January 28, 2010)**
**Do Hacked E-mails Change Climate Debate?, Arjun Makhijani joins others at the National Journal Expert Blogs: Energy & Environment (December 7, 2009)**
**Depleted Uranium Waste: NRC is on the Wrong Track, Read IEER Comments by Arjun Makhijani (October 30, 2009)**
**Depleted Uranium Waste Rulemaking, Arjun Makhijani's Notes on NRC's Meeting, held September 2-3, 2009.Carbon-Free and Nuclear-Free: A Roadmap for U.S. Energy Policy**
**[http://www.ieer.org/]**

## SOUTHWEST RESEARCH & INFORMATION (SRIC)

Southwest Information & Resource Center, headquartered in Albuquerque, New Mexico, was another of the prime coordinator's of the anti-Yucca Mountain campaign. Its lead spokesman was Don Hancock, who had been active also in the *Don't Waste U.S.* movement Close proximity to Los Alamos and the Waste Isolation Pilot Project in Carlsbad New Mexico has spawned in SRIC an activist anti-nuclear environmental group whose roots stretch back to the mid-1970's, in many ways paralleling the history of Citizen Alert in Nevada. In fact, the National Nuclear Waste Task Force and the Nevada Nuclear Waste Task Force seem to be a joint SRIC and Citizen Alert effort, hiring Caroline Petti (now at EPA) as their lobbyist.

The main publication of Southwest Research was the Workbook, a journal of alternative publications centered around anti-nuclear and Gaia type articles. The most active writer on the WIPP and Yucca Mountain situation appears to be Don Hancock. Among his Workbook articles are:

- **Getting rid of the Nuclear Waste Problem: the WIPP stalemate**
- **The Wasting of America: Target / Nevada - Target / New Mexico**
- **Nuclear Waste: Another Washington Scandal**
- **How Not To Find A Nuclear Waste Site**
- **The Nuclear Waste Legacy: How Safe Is It?**

From the SRIC website in 2010 we find that SRIC's sole purpose seems to be defeating nuclear energy in all its forms:

**David Meets Goliath**
**Since the late 1970s, SRIC has been a party to five major court cases (three with New Mexico Attorney General Tom Udall) to ensure that the WIPP facility met federal and state health and safety standards. It has provided technical assistance and training to affected community groups along the route to the site, and worked hard to stimulate and support informed citizen involvement. Virtually every citizen group in the state working on this technical issue has ties with Don Hancock at SRIC, and benefits from research collected by the Center.**

"There's no one I'd rather talk to about nuclear issues than Don," says Mary Lou Cook, a minister in the Eternal Life Church in Santa Fe. "He's our mentor. He's a hidden treasure. We call on him and SRIC for all kinds of help," says Cook, who has been heavily involved in Concerned Citizens for Nuclear Safety, a Santa Fe-based group that has been fighting WIPP since the late 1980s. [http://www.sric.org/index.html]

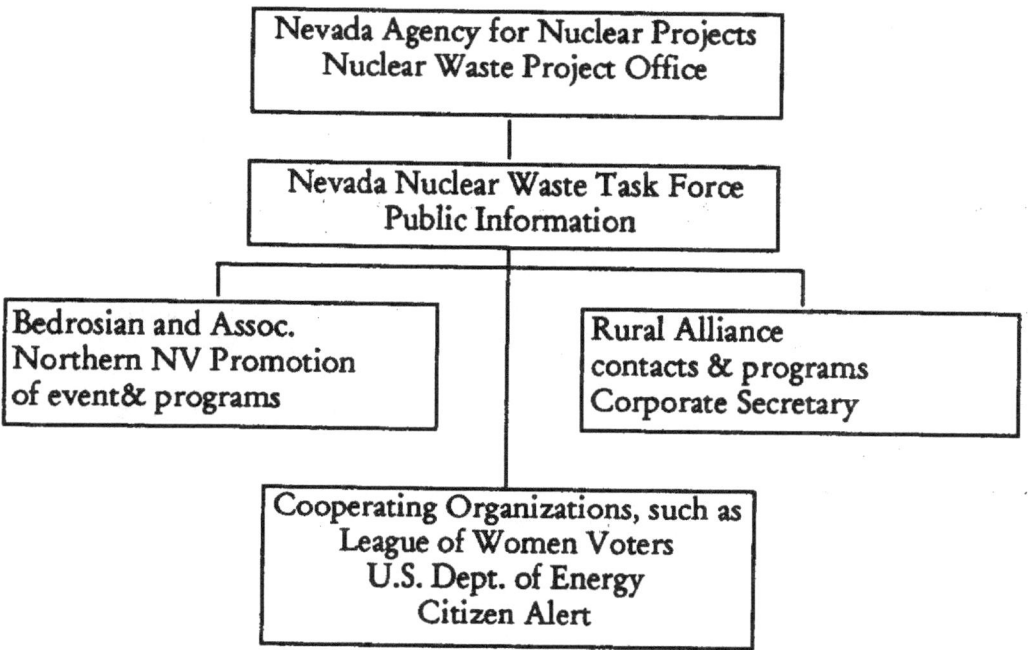

New Mexico is where the first shot in the transportation "perceived risk" battle was fought with the award of $300,000 to the Komises of Albuquerque for claimed loss in property value due to proximity to a WIPP route. Don Hancock felt the transportation "perceived risk" problem would stop any possible implementation of a MRS facility on the Mescalero Indian reservation in New Mexico as well [McNeil Lehrer, PBS, Sept. 1, 1994]. As in Nevada, it appears a relatively small group of activists have found ways to strangle the national nuclear waste policy, perhaps as much as a hobby as from an in depth analysis of costs and benefits.

# 33. Indian Territory

*"We have only one planet: one water, one air, one land.*
*If we don't take care of it, it won't take care of us.*
*I pray for forgiveness from the water, the air, Mother Earth.*
*I pray for healing for the people."*

**[Corbin Harney, Western Shoshone Spiritual Elder, Nevada Desert Experience flier announcing 1992 Lenten protest at the Nevada Test Site]**

For better or worse, the Western Shoshone and Southern Paiute Indian tribes became inextricably bound into the Yucca Mountain debate, both through their geographical and cultural heritage and through an uneasy symbiotic relationship with fringe environmentalists. Earth Worshipers, of both the environmental movement and the anti-nuclear religious sects, adopted Nevada's Indians as lovable pets, whose naturalistic culture is a convenient manifestation of their own sociological theories.

Sadly, this means Indian input into the Yucca Mountain debate has often been limited to their doing sacred rain dances for the benefit of environmentalists intent on showing that using the geological formations beneath Yucca Mountain for radioactive waste storage is spiritual sacrilege. As Mary Manning of the Las Vegas Sun reported in May of 1984:

**"The Shoshone Indian tribe may have a claim to Yucca Mountain, one of nine sites proposed as a high-level nuclear waste site.**

**The Indian Land Commission awarded $23 million to the Shoshone under the Ruby Treaty in the mid 1970s for land taken by white settlers, but the tribe refused the money, said Bill Vincent, Citizens Alert Southern Nevada coordinator, during a public hearing in Las Vegas.**

**They claimed that the land was still theirs. The Indian lands issue still is under appeal.**

**"Their land claim may still be open," Vincent said. Included in that claim is the remote volcanic tuff mountain targeted by the U.S. Department of Energy as a choice location for a high-level nuclear waste repository.**

**Indian tribes have a stake in DOE plans, Judy Treichel of Clergy and Laity Concerned said. Under the Nuclear Waste Policy Act, Indians are entitled to federal funds when their land is used for nuclear waste disposal.**

**"There are large amounts of money available to the tribe," she said.**
[Manning, Mary; Las Vegas Sun, 5/23/84]

What is disturbing about this article is that most of it doesn't deal with Indian issues or Indian spokespeople, but with Citizen Alert and its white environmental activists. The question is whether Citizen Alert and Clergy and Laity Concerned (our notorious Judy Treichel) were speaking for the best interests of Nevada's Indians, or whether these peace protesters were pushing an anti-nuclear agenda by any means possible, and found the Indians convenient puppets.

Actually, it is a misconception that all Indian tribes are alike and oppose nuclear technology unequivocally. While the Western Shoshone from northern Nevada opposed Yucca Mountain without qualification, their neighbors to the south, the Southern Paiute, took a much more flexible position, taking active part in the cultural resources studies that were conducted at the site. Jim Arnold, a Southern Paiute, was a key cultural resources consultant to the repository study.

Other tribes have had mixed reactions to the nuclear issue. According to an Indian periodical called Indian Voices:

Other Tribes who have declared their lands "Nuclear Free Zones"

The Salish and Kootenai tribes of the Flathead Nation in Montana declared their land free in 1984. The Inuits of Alaska appealed to the United Nations to establish a Nuclear Free Zone in the Arctic. In 1979 and 1981 the Cheyenne River Sioux of Eagle Butte, South Dakota barred the location of nuclear wastes within the boundaries of the reservation. The White Mountain Apache of Arizona passed a resolution in opposition to the construction of a nuclear power plant in their area. [NIC, Indian Voices, p4, Woozhch'IID 1994]

On the other hand:

$Tribes invite Nuclear Waste$

Two Native Nations, the Mescalero Apache (NM) and the Skull Valley Goshute Tribe (UT), are moving forward in the process to select tribal lands for nuclear waste storage despite widespread criticism from, environmentalists who are advocating their reservations be declared "Nuclear Free Zones." Both the Goshute and Mescalero Apache of New Mexico have so far received $300,000 in DOE funding and are each applying for an additional $2.8 million to continue studies for potential waste storage sites. [NIC, Indian Voices, p4, Woozhch'IID 1994]

## WESTERN SHOSHONE / PAIUTE RIVALRIES

The two tribes involved in the Yucca Mountain dispute are the Western Shoshone and Southern Paiute. Historically the two tribes were often enemies, who through a quirk of their boundaries came to divide claim over Yucca Mountain (see Figure following page). Paiute traditions view Yucca Mountain as near the sacred center of their universe, giving them an especially good reason to be wary of disruption of the area. However, the Paiutes have not pressed their claims as vigorously as the Western Shoshone, whose position seems to be part of a larger battle with the federal government based on disagreement over legal settlement of the *Ruby Valley Treaty of 1873*. Because Shoshone opposition to Yucca Mountain has been most vocal, they have become the favorites of national and international environmental groups. To show just how indispensable the Western Shoshone have become in the anti-nuclear and peace and justice mythology, we quote from a Las Vegas Review Journal World Brief from July 14, 1993:

## ANTI-NUKE PROTESTERS CLIMB PALACE WALLS

London - Fourteen women who breached the walls of Buckingham Palace in an anti-nuclear protest Tuesday while Queen Elizabeth was inside were charged with conspiracy and disorderly conduct, police said.

The women, from a group calling itself the Women's Nuclear Test Ban Network, scrambled up ladders to scale a perimeter wall topped with barbed wire.

The Network group said it was demanding royal recognition of the Western Shoshone Indian reservation near the U.S. nuclear weapons testing site in Nevada. The group said 23 British nuclear explosions since 1957 at the site had brought poverty and sickness to people in the area. [Las Vegas Review Journal, July 14, 1993]

The international portrayal of the Western Shoshone as helpless victims has larger political implications than merely concern for the plight of Nevada Indians. Various social liberation movements require an oppressed people as a catalyst for political action. Among the environmental movement are those who believe toxic dumps are purposely placed in disadvantaged communities as a mechanism for subjugating those populations, rather than simply as the practical result of economic and technical considerations. Liberation theology, which has philosophical roots in Brazilian land reform among dispossessed indigenous populations, has had a profound influence on the Test Site protests of the Franciscans of the Nevada Desert Experience. The Franciscan movement in many ways requires an oppressed class displaced from the land for its validity. Claims that the Shoshone have been disenfranchised from the land are therefore a prerequisite catalyst for political protest.

In reality, the Western Shoshone's economic problems more likely stem from their preoccupation with protesting the Nevada Test Site and Yucca Mountain to the exclusion of more lucrative claims for benefits from the federal government. While the Southern Paiute settled their claims against the *Ruby Valley Treaty of 1873* in the mid-80's, the Shoshone resisted.

Chiefs Frank Temoke and Frank Brady adamantly refused the government payoff at Battle Mountain, Nevada on December 11, 1992. Temoke was sure that the Shoshone would lose their claim to the lands if they accepted the funds. He said, "I did not sign any agreement for money. The actions of the federal government are unconstitutional, immoral, genocide and against international law." Brady urged his people to refuse the settlement also, saying, "The people need land, not money." They both faced immense pressure from their own people to sell out because many of the Shoshone wanted the money. Brady said, "Some say we've lost the land already and that may be so, but we still have a fighting chance if we don't take the government payment."

The United States Federal Government passed the Western Shoshone Claims Distribution Act of 2004, which authorized payment of $145 million for the transfer of 25 million acres (101,000 km²) to the United States. Seven of the nine tribal councils within the Western Shoshone Nation passed resolutions opposing the legislation.

On March 10, 2006 the United Nations Committee on the Elimination of Racial Discrimination stated "credible information alleging that the Western Shoshone indigenous people are being denied their traditional rights to land". On January 17, 2006, the U.S. District Court for the District of Nevada dismissed a lawsuit filed by the Western Shoshone National Council against the United States of America that sought quiet title to lands whose boundaries were defined by the Treaty of Ruby Valley.

WESTERN SHOSHONE

NEVADA

UTAH

YUCCA MOUNTAIN CULTURAL RESOURCES STUDY AREA

OWENS VALLEY PAIUTE

SOUTHERN PAIUTE

ARIZONA

CALIFORNIA

LEGEND
1  BENTON PAIUTE RESERVATION
2  TIMBISHA SHOSHONE RESERVATION
3  BISHOP PAIUTE SHOSHONE RESERVATION
4  BIG PINE PAIUTE SHOSHONE RESERVATION
5  FORT INDEPENDENCE RESERVATION
6  LONE PINE PAIUTE RESERVATION
7  YOMBA RESERVATION
8  DUCKWATER
9  PAHRUMP PAIUTE TRIBE
10 LAS VEGAS PAIUTE INDIAN COLONY
11 LAS VEGAS INDIAN CENTER
12 CHEMEHUEVI RESERVATION
13 COLORADO RIVER RESERVATION
14 MOAPA RIVER
15 SHIVWITS
16 CEDAR CITY
17 INDIAN PEAKS        } PAIUTE INDIAN TRIBE
18 KANOSH                  OF UTAH
19 KOOSHAREM
20 KAIBAB PAIUTE RESERVATION

This settlement for lands that were stripped from the Shoshone by the federal government, lands which include parts of the Nevada Test Site and Yucca Mountain, has been adjudicated through the courts to the Supreme Court level and is not likely to be much changed in the future.

In retaliation, the Shoshone Tribal Council has over the years made allies with American Peace Test, Citizen Alert, Nevada Desert Experience, Greenpeace and a number of anti-nuclear protest groups, going so far as to issue permits to Test Site protesters for access to Indian lands!

Of course, the Shoshone and Paiute may very well deserve a much better settlement for Nevada Test Site and Yucca Mountain lands than was offered; it would not be the first time Indian tribes had been offered glass beads for commercially valuable property. However, this points to the possibility that just compensation for land is more central to the Western Shoshone position in regard to Yucca Mountain than its radiological hazards:

Ian Zabarte, Manager of the Western Shoshone Nuclear Waste Program, has strong opinions regarding Native American Sovereignty and Nuclear Waste issues.

**Zabarte says, "I want to challenge the Department of Energy attempt to load and control a process of involving Stakeholders. I also hope to convince interested parties to take a stand in support of Native American Stakeholders to make decisions themselves without interference. I believe that if you understand the legal and historical cases of Native American Sovereignty, particularly Western**

Shoshone land rights, you will chose to take action to work with Native Americans to change the laws which keep them disenfranchised and compel the U.S. Government to follow the laws . . . not just those that provide some benefit to the United States Government.

[Western Shoshone Offer Strong Medicine to the DOE, Indian Voices, p3, Wooshch 'IID 1994]

Both the pacifist religious environmentalists and the Western Shoshone know a shotgun wedding when they see it and their alliance is shaky at best. Indian land claims are supported by the environmentalists who in return are allowed to carry on NTS protests with tribal approval and occasional logistical aid. The problem for the Indians is that by using this tactic they haven't married just one protest group, but were forced to bring a whole range of protest groups under the tent flap.

In the Western United States, the original Indian way of life fell as much by its confrontation with the Iron Horse and technological acceleration as by confrontation with white men. While by 2010 the Indian tribes of Nevada were celebrating having stopped the iron wheels of progress at Yucca Mountain and the Nevada Test Site, the victory appeared Pyrrhic since no benefits have become available to a still poor tribes. The question then becomes, if Yucca Mountain is revived, will the tribes at some point break with the environmentalists and the state and negotiate separately? In that case, Citizen Alert, Nevada Desert Experience and Nevada may not prove to be the best and strongest voice for the Indians, and the Western Shoshone may find more forceful voices among a new generation of elders who can speak for the economic and political interests of the Indian reservations.

Section 118 of the Nuclear Waste Policy Act does give affected Indian Tribes some say in the repository siting process and access to compensation. Among a number of important statements regarding Indian rights:

**Sec. 118 (a) Participation of Indian tribes in repository siting decisions. Upon the submission by the President to the Congress of a recommendation of a site for a repository located on the reservation of an affected Indian tribe, the governing body of such Indian tribe may disapprove the site designation and submit to Congress a notice of disapproval. . . . .**

**(b) (3) (A) The Secretary shall provide financial and technical assistance to any affected Indian tribe requesting such assistance and where there is a site with respect to which the Commission has authorized construction of a repository. Such assistance shall be designed to mitigate the impact on such Indian tribe of the development of such repository. . . .**

**(6) Financial assistance authorized in this subsection shall be made out of amounts held in the Nuclear Waste Fund established in section 302 [42 U.S.C. 10222]**

While Yucca Mountain is not legally on tribal reservation land as per settlement of the Ruby Valley Treaty by the Supreme Court, their cultural links to the land are still quite strong. Unfortunately, access for the Western Shoshone and Southern Paiute to the financial and technical assistance designated by the Nuclear Waste Policy Act has been lost in the shuffle as Nevada's Senators, Governor, ex-governors, mayor and assorted anti-nuclear environmentalists pursued their own political self interests. Tribes not shackled by the Nevada political situation will receive substantial benefits.

For example, the Mescalero and Goshute Indian tribes received substantial funds for studying possible acceptance of a monitorable retrievable storage complex on their reservation land. Interestingly, Nevada tribes (specifically the Fort McDermitt Paiutes on the Nevada Oregon border) who might have accepted an MRS facility on their land were opposed by the Nevada Nuclear Waste Project Office on the grounds that Nevada is not legally obligated to accept both a repository and a MRS facility. The stakes are substantial:

COUNTY COULD BENEFIT FROM MCDERMITT MRS SITE
**Nevada or a neighboring state may be forced by the federal government to host a facility for spent reactor fuel if negotiations with Indian tribes fall through, U.S. Nuclear Waste Negotiator Richard Stallings warned Monday. . .**

Fort McDermitt is one of four Indian tribes across the nation still actively negotiating with Stallings' office. The tribe, which has already received $300,000 in study grants, must now decide whether to welcome a host site. . .

These benefits, according to Stallings, include the creation of approximately 500 long term jobs and between $50 and $60 million in research money, which would come from the nation's nuclear waste fund. [Albert, Corey; Humboldt County Sun, May 18, 1994]

A false impression among at least the Shoshone was that the Nevada Nuclear Waste Project Office acted as their benefactor. Because the state's position is to oppose Yucca Mountain at all costs, rejecting all suggestions of benefits, the Shoshone may in fact have been hobbled by their association with the state. NWPO did produce a respectable series of socioeconomic studies on Nevada's Indians, the result of work done at the University of Nevada Reno by Maribeth Hamby, Elmer Rusco and Mary Rusco, some of the few socioeconomic researchers who were Nevada residents [see Native Americans and Yucca Mountain: A Final Summary Report (Volumes I and II), NWPO-SE-041-91]. However, the tribes certainly did not gain monetarily from their partnership with the NWPO in opposing the Yucca Mountain repository.

## CULTURAL HERITAGE

To understand why Yucca Mountain should be so important to Nevada's Indian tribes and carry cultural meaning beyond that of a desert rock outcrop. Some history is necessary:

### TRADITIONAL HOLY LAND

The Yucca Mountain area is located on the northern boundary of the Mojave Desert and southern boundary of the Great Basin Desert, and is an important area to many Native American ethnic groups. These groups resided there for thousands of years, using the land and its resources and building these into a cultural definition of themselves as a people. Most of these groups perceive that they were created in these two deserts and that, in so doing, the Creator gave them a special supernatural responsibility to protect and manage the land and its resources. In western cultural terminology, these deserts are their Holy Land.

The Southern Paiutes, for example, believe that they were created by the supernatural near Charleston Peak - called Nuvagantu - located in the Spring Mountains, twenty-five miles southeast of the proposed Yucca Mountain high-level radioactive waste repository. According to Laird:

In pre-human times Nuvagantu was the home of Wolf and his brother, Mythic Coyote. It was the very heart of Tuwiinyaruvipu, the Storied Land.

There was and is no place in the Southern Paiute traditional territory more sacred than the Spring Mountains and the areas around them. Concerns for this sacred area have been expressed repeatedly in the cultural resources studies involving the Southern Paiute people because of its relationship to the Spring Mountains. [Literature Review and Ethnohistory of Native American Occupancy and Use of the Yucca Mountain Area, Science Applications International Corp. January 1990, p14]

Another subliminal current to the debate is the mistreatment of Southern Nevada Indians in historical times. For example:

1849 - The Manly-Rogers party stayed for nine days at a "hastily abandoned" Native American village "most likely Hugwap at Cane Spring . . . devouring the winter's store of squash which they found there and fattening their oxen in the stubble in the cornfield [Lingenfelter, 1986][Literature Review and Ethnohistory of Native American Occupancy and Use of the Yucca Mountain Area, Science Applications International Corp. January 1990, p70]

Later resistance to accepting a buyout of the Ruby Valley Treaty of 1883 thus becomes understandable. Indian occupation at Ash Meadows and local areas, combined with foraging of the lands surrounding Yucca Mountain, also lead to a certain reluctance to give up all rights to the area. To bring the land ownership issue into court,

Shoshone sisters Mary and Carrie Dann, in 1974 elected to "trespass" on Bureau of Land Management controlled land by grazing their cattle without a permit. The Dann family consistently resisted the takeover of the Yucca Mountain and Test Site area in actions since 1974, one member going to the extreme of dousing himself with gasoline and setting himself on fire.

Other tribal groups in Nevada interested in the Yucca Mountain issue are the Western Shoshone National Council and the Intertribal Council of Nevada. However, these groups are not necessarily the final authority on the best interests of tribe members. Having become locked into alliances with environmental groups with diverging agendas, the Western Shoshone have become ineffective negotiators for the rights of the indigenous people.

# 34. Helen Caldicott & Nuclear Sin

One of the most strident opponents of nuclear technology is an Australian by the name of Dr. Helen Caldicott. On October 14, 1991, she was invited to speak at the University of Nevada Las Vegas by various groups, most notably Citizen Alert, American Peace Test, Boulder City Peace and Justice Committee, Nevada Desert Experience, Nevadans for Peace, Pace Bene Franciscan Center, Rainbow Coalition, the Sociology Department (UNLV), Student Health Services, Student Government and the Women's Studies Department .

Also sponsoring Caldicott was the state of Nevada's Nuclear Waste Task Force led by Judy Treichel (though when pressed she denied providing any monetary support). Caldicott's concerns for the environment bordered on hysterics and the NWTF risked its credibility by offering moral support for such views. If Caldicott's environmental convictions were implemented, it would turn the Las Vegas Valley to a desert ghost town.

Among the many things decried by Caldicott in her speech were gambling, clothes dryers, TV's, air conditioners, advertising, and anything vaguely nuclear. Whether or not Judy Treichel and NWTF sponsored Caldicott directly, the doctor's speech serves as a good example of the apocalyptic environmental mindset enveloping Yucca Mountain. Apparently, representatives of the state of Nevada supported a speech by someone who opposed nearly everything the state stands for.

A question raised by the Caldicott appearance is why the Las Vegas media chose to downplay the outrageous statements contained in Caldicott's speech. Two reasons come to mind: 1) anti-nuclear protest is old news in Las Vegas; or 2) some of the media people who did attend Caldicott's speech were biased in her favor and swept the most outrageous comments under the rug.

It is interesting to note the tone of a report titled "Las Vegans Told To Switch To Solar" filed in the Las Vegas Sun by Mary Manning who covered the Caldicott speech:

**"Southern Nevadans have been warned to close the test site, run Las Vegas on solar power and let California keep its hazardous and nuclear wastes." [Manning, Mary; Las Vegas Sun, Oct 15, 1991, p3A]**

Even reporter Mary Manning, known for her anti-nuclear biases, couldn't keep all of Caldicott's bizarre statements under wraps in her article. Still, some of the Doctor's most damaging statements were conveniently absent from the Sun piece. Specifically, Caldicott hammered gambling, the lifeblood of Las Vegas, as immoral and suggested the Nevada Test Site had made the area uninhabitable, but these potentially explosive statements were left out of the Manning report.

Sherman Frederick of the Las Vegas Review Journal did have a few choice words to say about Treichel's support for Caldicott's speech. In an editorial on October 20, 1991, Frederick wrote:

**"But on Monday, Treichel's Nevada Nuclear Waste Task Force revealed its true colors when it co-sponsored Dr. Helen Caldicott's lecture at UNLV.**

**Among other things, Caldicott gave the audience tips on how to stop nuclear testing, accused President Bush of starting the Persian Gulf War to help the stock market, and generally assailed all things nuclear as well as the military establishment.**

**Treichel defends her organization's co-sponsorship of Caldicott by saying no money was expended. But she admits the Caldicott speech was sanctioned by the Nevada Nuclear Waste Task Force and is the kind of activity that fulfills her lucrative contract with the state."**

Caldicott's statements are useful because the Australian's rhetoric shows in a nutshell that the Yucca Mountain is not principally a local affair. Instead, debate is driven by the dynamics of out-of-state and international anti-nuclear activists with extremist environmental agendas.

*Update*, the faculty newsletter for the UNLV campus, provided some biographical background for the Caldicott visit:

**"Nobel Peace Prize nominee Helen Caldicott . . . is one of the foremost spokespersons for nuclear disarmament, saving the environment, and Americans' taking responsibility for their government. She has been honored with the Gandhi Peace Prize and the University Women Peace Award for her leadership with**

International Physicians for Prevention of Nuclear War. She is founder of Women's Action for Nuclear Disarmament and International Physicians for Saving the Environment." [Update, University News and Publication, Sept. 23, 1991, Vol. 20, No. 39]

Rather than overly analyze Caldicott, we'll let her own words speak for themselves by quoting from her Las Vegas speech.

---

# Helen Caldicott
## Lecture: October 14, 1991

I've never been to Las Vegas, although I've been to almost every other city and town in America. As I drive around it's interesting to see that there are real people actually, who live here, amongst all these pseudo castles, neon lights and parking lots and huge hotels. And in the hotel where we're staying, it's forever twilight with these poor little people, who clearly come from the lower levels of society it seems, particularly in my hotel, where they're pouring their money into these machines, and I just wonder who owns the machines and how much profit they're making and they're skimming. It makes it look like Disneyland or fairyland, but in fact it's ripping their money off. I suppose it's a good place for gamblers and those who are addicted to gambling. - - It sort of makes me feel a bit sick. I have to say though, that's just my first hour's impression of being here."

. . .(I flew in over) beautiful tall mountains, are they pines on them? They look like pines from the air. I just can't believe how we screw things up. Us. White people. . .

[NOTE: there are no pine trees approaching Las Vegas, Caldicott apparently conjectured Nevada's desert scrub are are radiation mutated pine]

I know Uncle Walter (Cronkite) and he told me (that if the country knew what he thought they wouldn't let him past the front door). But I said, God, the country needs you. He said, years ago, that the America should disarm unilaterally. And he said if thousands of people lay down in front of tanks invading Europe from the Soviet Union, that that would stop them. I mean, that's where Walter Cronkite is. Amazing, he's an amazingly wonderful man."

Take it into your souls. Take it into your souls. Here lies the nadir of the second American Revolution. It goes back to the problem of media culture. We all live in a projection room. And those politicians who are able to massage our egos get away with it. The projection room is in the lavatory, the bathroom, the bedroom, the sitting room, the kitchen and everywhere you go the television is blaring out this garbage at everyone, manipulating with subliminal sophisticated psychology. These ads are for manipulation, it's a form of social engineering. I think George Bush is the most deeply unprincipled man in American politics today. . . . . Now I knew Reagan, I spent an hour and a quarter with him in the White House. . . .

My daughter just went to medical school, for five years, and I didn't pay a penny. My government paid for her education. No college education, no university education, should be paid for, it should be paid for by the government, free, with your tax dollars. The government is there to disperse your tax dollars for your benefit. Not to put the money into the coffers of General Electric, who've made every nuclear weapon since time began, into Westinghouse, who makes refrigerators and stoves and nuclear weapons, into AT&T, thank you for calling AT&T (cynical inflection). AT&T runs Sandia Labs, who've designed all the nuclear weapons since time began, I've spoken at Sandia Labs to the scientists, that's what AT&T do (sic), As a side effect, they run a telephone system. But they also run satellites in the air that have wired the world up like a ticking time bomb ready to go off at in a minute in a nuclear war -- first strike nuclear war, which is still the policy of this country. So somewhere along the line, this country's lost its way and the corporation's got hold of you. Hands up, those of you who think they live in a democracy, a true democracy.

So let's get back to Nevada. So they're testing weapons here, ehh? And they keep testing weapons and they keep testing weapons even though there's no enemy anymore. The communists have disappeared. I used to be called a communist. They used to tell me to go to Russian when I said we'll have to learn to live with the Russians or we'll all die together. And Russian flesh burns at the same temperature as American flesh. And they said you're a communist, go and live in Russia. They can't call me that anymore, it's rather nice. I'm actually an Australian.

. . .And Ronald Reagan thought Star Wars was this yellow plastic shield over America, and that weapons came in and they went boink! And they just bounced off. That reminds me of Las Vegas here, with all those medieval castles with all that gambling and crime and stuff. He didn't understand. But Reagan was the one person in this country who really thought Star Wars would work.

. . . Hydrogen bombs are very easy to make and very cheap. And it was his idea (Teller) and Oppenheimer was opposed, because when he saw the first bomb explode in the desert, in New Mexico, and the bomb was called Trinity, after the Father, the Son and the Holy Ghost. And if the bomb didn't work, they were going to send a telegram to the president saying, it's a girl. But the bomb worked, so the telegram read, it's a boy. Note! Note! The revolting sexism in that. I mean, really, that's so offensive. . . .

Analyzing Caldicott's various statements one can discern a certain disconnection from reality. While opposition to nuclear technology can certainly be formed in a rational manner, there is a fine line which Caldicott appears to have crossed into hysteria.

# 35. Greenpeace Meets Pinocchio

Greenpeace and Citizen Alert in April of 1991 attempted to premier a television ad called "Radioactive Nose". Produced by Jim Weisiger, a director at Bruce Dorn Films and his wife, Joyce, executive producer of the company, the ad utilized actor Martin Sheen to stereotype the scientists at Yucca Mountain as lying Pinocchio's. The Greenpeace spot featured an official telling a public hearing that a *"dump site we are proposing is perfectly safe. Trust me."*

As the ad continued, the nose of the official grows and glows green with radioactivity, tipping over a pitcher of water when the speaker turns his head. *"As for our track record,"* the official says, *"accidents will happen. Clean it up."*

The fact that "Radioactive Nose" was devoid of technical information comes as little surprise because the purpose of Greenpeace and Citizen Alert was not to educate the public, but to attack the integrity and credibility of DOE and subcontractor scientists. Applied to an individual or a private concern, this type of character assassination might well have seen its day in court. "Radioactive Nose" was designed to destroy the credibility of the professional men and women working at Yucca Mountain who had little recourse to defend their reputations because of their positions as federal employees.

Of course, there is a fine line between hard-nosed political advocacy and political disinformation. However, the extraordinary amount of effort expended by NWPO, Mountain West, Nevada politicians, and national environmental groups to portray DOE as hopelessly untrustworthy must make us wonder whether the goal of this criticism was to improve DOE or to destroy it. While few would argue that DOE was without sin, the agency made substantial efforts to improve its operations and its interface with the public, notably under the Task Force on Radioactive Waste Management. While calls for DOE to improve its performance are constructive political advocacy, claiming the agency's staff are pathological liars and cheats seems to border on political theater.

Out-takes and commentary about Radioactive Nose were aired on the local news stations, but the ad itself was not viewed by the public. Advertising representatives of the American Nuclear Energy Council made it abundantly clear to the broadcast media that if Radioactive Nose were run for free as a public service announcement, as hoped by the Sierra Club and Citizen Alert, the industry would expect equal free time. This effectively killed the campaign.

Perhaps more remarkable than the ad was a Review Journal Response editorial written by Bill Walker and Jason Salzman of Greenpeace:

**Last week, Greenpeace and Citizen Alert premiered "Radioactive Nose," a 30-second video combining humor and high-tech animation to argue that proponents of the Yucca Mountain nuclear waste dump are, like Pinocchio, "stretching the truth. [Las Vegas Review Journal, Response, May 8, 1992]**

Not everyone saw the humor in this ad.

The two environmentalists argued that local television stations should run the campaign as a free public service. *"Here's why: The time belongs not to the broadcasters, but to the audience."* Further, they complained that *"What's really at stake is the role of a mass media in a democracy"*. Actually, Walker and Salzman were parroting the party line of the Safe Energy Communication Council with whom they were coalition members. The SECC was attempting to revive the Fairness Doctrine which would allow special interest political advocates such as themselves free air time.

If democracy were truly at stake, Walker and Salzman might have suggested free time for ads from the nuclear industry to make up for the sympathetic press advantage enjoyed by the environmental movement, but it is unlikely democracy was their true interest. If ANEC had followed the stylistic precedent set by Greenpeace and Citizen Alert in Radioactive Nose, the industry could have theoretically portrayed the environmentalists as drug-crazed burnt-out hippies on welfare, a caricature based on images of Test Site protesters and not without some humor itself. Yet it is unlikely Walker and Salzman would have supported an industry ad that (to mimic their words) 'combines humor and high-tech animation to argue that opponents of the Yucca Mountain repository are, like drug crazed hippies, protesters looking for free media gratification.'

# Gamble Nevada

Nevada is a great bet for this year's vacation. Come and soak up spring sunshine in the desert. Hike or mountain bike through deep red rock canyons, pinon pines, and flowering cactus. In the summer, water ski at Lake Tahoe, swim at Lake Mead or tour Hoover Dam. Of course, it's always the season for trying your luck at games of chance all across Nevada – home of some of the nation's best and most famous casinos. You've never experienced anything like strolling down the Las Vegas strip, taking in the lights and excitement – and of course the "exploding volcano". Don't wait. Now is the time to visit Nevada.

## GREENPEACE
1021 PEARL STREET, SUITE 200
BOULDER, COLORADO 80302

> **WARNING:** The government has proposed dumping the nation's most deadly radioactive waste under Nevada's Yucca Mountain, located about 100 miles north of Las Vegas. But earthquakes could cause wastes to contaminate ground water and rise to the surface, giving "hot spring" a deadly meaning. If you care about Nevada – and especially if you or your descendants plan on visiting the state sometime during the next quarter million years – help stop the Yucca Mountain dump. Call your senators and representatives today at 202–224–3121.

Of course, neither visual images of green glowing radioactive noses nor of burnt-out hippies dancing in trances are rational arguments for or against Yucca Mountain. Instead, such emotional images pander to base psychological instincts irrelevant to the science being done at Yucca Mountain. As much as Greenpeace and Citizen Alert disliked the pro-dump ads produced by the American Nuclear Energy Council, those ads did at least attempt to address technical questions (Is the waste solid or gas?. How strong are the transportation casks?) and did not represent an attack on the integrity of the environmentalists in the same vein as "Radioactive Nose".

Walker and Salzman's editorial also contended that ". . . the Nevada Nuclear Waste Task Force found numerous examples of distorted or misleading "facts" in the (American Nuclear Energy Council's) ads." However, as we've seen, distorted and misleading facts in NWTF literature were themselves legion and NWTF director Judy Treichel was not a nuclear engineer. The links between Citizen Alert and NWTF (and by inference, the Sierra Club) were long incestuous, making NWTF little more than a Citizen Alert front paid for by nuclear rate payers. Thus, Greenpeace in quoting NWTF as an independent agency was itself using distorted and misleading information.

The Response editorialists then claimed "Radioactive Nose simply -- and accurately -- points out that people pushing the dump are not telling the whole story." Greenpeace's psychological manipulation using imagery of green glowing noses is an appeal to mass nuclear paranoia which is neither simple nor accurate. Neither did Greenpeace nor Citizen Alert tell Nevadans their whole story either in regard to their positions on the repository. Their literature makes it amply clear their purpose in opposing Yucca Mountain is not primarily to protect Nevadans from radiation, but to shut down all nuclear power in America, no matter what the effect might be on global warming or on our economic competitiveness.

Finally, Walker and Salzman claimed poverty as a reason for needing free public air-time: "For 1992, Greenpeace has about $157,000 to fund its work on all nuclear issues, nationwide." This didn't seem to hinder Greenpeace from producing the $400,000 Radioactive Nose ad. Moreover, one of Greenpeace's main interests is anti-nuclear protest. Greenpeace was not poor (According to a Forbes expose, Nov. 11, 1991, Greenpeace U.S.A. raised $64 million dollars in 1990) and it receives generous free publicity from the press. Moreover, its membership in the SECC meant Greenpeace's effective budget for opposing nuclear projects (in pool with numerous other environmental organizations) was anything but small.

If Greenpeace and its allies at Citizen Alert were serious about encouraging democratic discourse at Yucca Mountain, there are some steps they could have taken:

1) Labeling the scientists at Yucca Mountain as liars resembles character assassination more than constructive criticism.

2) Radioactive Nose was devoid of scientific content. Concentrating on providing hard scientific evidence to back up claims of the dangers of radiation would have added to the quality of the debate.

3) Revealing their ties with Citizen Alert, the Nevada Nuclear Waste Project Office, the Nevada Nuclear Waste Task Force and the national coalition of anti-nuclear environmental groups would have helped Greenpeace restore trust in their advocacy. Presenting themselves as an impoverished local grassroots organization hid the non-Nevada origins of this protest and misrepresented local democracy.

Unfortunately, Greenpeace seemed determined to fight dirty in its battle against Yucca Mountain. The year 1993 found them distributing a flier meant to affect the Nevada tourist industry by suggesting gamblers should be worried about the repository. Whether this was constructive political advocacy conducted by environmental activists or a form economic terrorism is open to question.

# 36. Environmental Liberation Theology

Religion played a much larger role in the battle over the nuclear waste repository than most may suspect. This religious involvement is something different than the usual Sunday morning sermons of pastors interpreting local issues in a moral context. At Yucca Mountain and the Nevada Test Site there has been a melding of elements of earth worship with Christian charismatic and evangelical movements, the political Liberation Theology of the Franciscans, Indian tribal spirituality and a multitude of other religious flavors. The resulting hybrid is perhaps best described as environmental liberation theology.

Among the religious coalitions protesting Yucca Mountain, the Nevada Test Site, and nuclear technology in general, the most prominent are as follows:

- Nevada Desert Experience (an outgrowth of Lenten Desert Experience, co-founded by Judy Treichel)
- Pace E Bene (Franciscan organization studying non-violence)
- Western Shoshone Tribal Elders
- People of Faith
- Clergy and Laity Concerned (Nevada chapter founded by Judy Treichel)
- The Catholic Worker

Two main currents underlie the involvement of these religious organizations: 1) a peace movement dedicated to the elimination of nuclear weapons, and 2) a social justice movement which sees land reform through decentralization as the means to fight oppression of the poor. While the religious peace movement elements are familiar, social justice theories and their relation to Yucca Mountain take some explaining. It should be noted that Pope John Paul II opposed many of the more liberal aspects of the peace and justice movement as derived from the offshoot Catholic Liberation Theology of Paolo Friere and others.

Prayers, pilgrimages, vigils, fire and brimstone sermons, retreats and a whole religious ceremonial tradition already were built around the Nevada Test Site and Yucca Mountain. An article by Mary Manning of the Las Vegas Sun gives this substance:

## RADIATION MONUMENT FINDS A HOME

A monument to the world's estimated 18 million radiation victims has found a permanent home.

About 50 people representing atomic veterans, Shoshone Indians and anti-nuclear groups gathered Wednesday at Cactus Springs, an area about 50 miles northwest of Las Vegas where the monument is rising from the desert.

The 4-foot high obelisk had been dedicated on Memorial Day 1984 at Camp Desert Rock near the Test Site.

But U.S. Department of Energy officials confiscated the memorial. Anthony Guarisco of the Alliance of Atomic Veterans credited Nye County sheriff's deputy Lt. Jim Merlino with helping the veterans recover their monument.

The monument is being built on land near Cactus Springs recently returned to the Shoshone Nation. A series of shrines honoring religious figures will be part of the monument. The first one, Our Lady of Guadalupe, was placed there Wednesday.

In all, 22 acres of land was purchased this summer by Genevieve Vaughn, of Texas, and her family and returned to the Shoshone Nation.

Two acres will become a shrine dedicated to those who die from radiation exposure. The other 20 acres was turned over to the tribe, which lays claim to most of the Great Basin, including the Nevada Test Site.

As smoke from a sage fire drifted across the second annual National Veterans Day Celebration, Western Shoshone spiritual leader Corbin Harney chanted and directed the smoke with eagle feathers in each hand.

"Remember, this is our land," Harney said. He urged those forming in a circle in the sagebrush to clean up the land and stop further nuclear testing at the nearby Test Site.

"If you don't remember anything else I say today, I hope that you remember that this is the engine that drives it all," Guarisco said, gesturing towards the Test Site.

**Franciscan Priest Louis Vitale led the litany of names of radiation victims read into the cold wind.**

**Franciscan Sister Rosemary Lynch read a speech by Black Elk, a Sioux Indian chief who had a vision of how people could live together within the "sacred hoop."**

**To honor Veterans Day, formerly Armistice Day, the group observed a moment's silence at the 11th hour of the 11th day of the 11th month. [Manning, Mary; Las Vegas Sun Newspaper, Nov 12, 1992]**

It is hard to imagine a technological priesthood of DOE officials so intent on worshiping Yucca Mountain that they would bother to 'chant and direct smoke with eagle feathers in each hand.' In contrast, anti-nuclear activists gave every sign of having created a budding religious movement dedicated to worshiping a patch of desert.

A shrine for radiation victims, dedicated to Our Lady of Guadalupe, indicates the close relationship between certain religious orders and nuclear protest. The late Corbin Harney of the Western Shoshone, Franciscan Priest Louis Vitale and Sister Rosemary Lynch and Sioux Chief Black Elk were long central to nuclear protest in Nevada. The mystic rituals, religious iconography of a 4-foot obelisk and reference to Armageddon (11th hour of 11th day of 11th month) shows the apocalyptic nature of this movement.

Historically, the main goal of the various religious protest organizations in Nevada was to stop nuclear weapons testing at the Nevada Test Site. This anti-nuclear movement came to include Yucca Mountain in the Test Site protest because they saw an unbreakable chain between nuclear waste and nuclear bombs due to the presence of fissionable plutonium in spent fuel elements. The protesters also saw parallels in the open air weapons tests of the fifties and sixties (which exposed residents of the Western states to varying amounts of radiation) and the possibility Yucca Mountain could also cause mass radiation exposure. As the Nevada Test Site curtailed its activities in response to the end of the Cold War, the anti-war religious coalitions were quick to take up the anti-Yucca Mountain cause in its stead.

The religious roots of organizations like the Nevada Desert Experience explains much of the zealotry of the opposition to Yucca Mountain. In the minds of the anti-nuclear factions, the repository is not a technical problem, but a question of the salvation of mankind from the horrors of radiation and from the Armageddon that nuclear bombs would bring. The counter-argument that the effort required to turn spent fuel into any sort of bomb grade material is both too difficult and dangerous to make this a plausible threat has succumbed to the moral conviction that plutonium is symbolic of warmaking capability. Technical questions can be approached rationally, but religious battles are often only solved (if at all) through protracted holy wars.

In the book of Genesis in the Bible there is a verse we might call the Judeo-Christian environmental manifesto.

**"And God blessed them, and God said to them, Be fruitful and multiply, and replenish the earth, and subdue it: and have dominion over the fish of the sea, and over the fowl of the air, and over every living thing that moveth upon the earth." [Genesis 1:27-28]**

These Biblical lines are perhaps the first environmental laws ever formulated: "replenish the earth, and subdue it." In this biblical context, Yucca Mountain would replenish the earth by isolating dangerous nuclear waste from the biosphere. By completing the nuclear energy fuel cycle on which mankind depends for almost twenty percent of his electrical energy, Yucca Mountain is also obviously an attempt to subdue the earth. In this light, Yucca Mountain satisfies Judeo-Christian ethics (at least in traditional interpretations).

In contrast, environmental religious theory evolving in Nevada casts Yucca Mountain as an evil technology attempting to subdue Mother Earth. In an earth centered theology, (such as practiced by Nevada's Shoshone and Paiute) mankind seeks a passive rather than active role in the environment. Yucca Mountain thus takes on negative religious significance as a symbol of environmentally interventionist technology.

The literature of the anti-nuclear Nevada Desert Experience shows the divergence from traditional Judeo-Christian environmental ethics. Nevada Desert Experience is the outgrowth of Friar Louis Vitale's 1978 tour of the test site which sparked the idea of a Franciscan vigil on the road to Mercury, Nevada. A flier promoting the eleventh Lenten Desert Experience in 1992 shows how that original vigil has expanded into an earth worship movement:

**HEALING THE EARTH**
**HEALING OURSELVES**
    **Lenten Desert Experience XI**
    **March 4 - April 19, 1992**
    **The nuclear bomb has deeply wounded our earth, ourselves. Join the Desert Lenten Experience XI in seeking healing, and expressing prayerful protest at the Nevada Test Site.**
**Schedule**
    **March 4 Ash Wednesday**
    **March 13-15 Franciscan Weekend**
    **March 20-22 Lutheran-Episcopalian Weekend**
    **March 27-29 Friends Weekend**
    **April 9-10 Methodist Witness**
    **April 12 Palm Sunday**
    **April 13-19 Holy Week**
    *"We have only one planet: one water, one air, one land.*
    *If we don't take care of it, it won't take care of us.*
    *I pray for forgiveness from the water, the air, Mother Earth.*
    *I pray for healing for the people."*
    **Corbin Harney, Western Shoshone Spiritual Elder**

Judeo-Christian ethics traditionally have been concerned foremost with the wounding of men's souls rather than injury to the "deeply wounded earth". The above flier, however, represents an entirely new religion, a pantheism dedicated to "earth healing". The fact that tribal elders, Franciscans, Friends, Lutherans, Episcopalians and Methodists were joined together on this one issue should not be viewed as ordinary. Normally, these religious factions would fight like cats and dogs at the prospect of being lumped together on the same program, but on the nuclear issue they suddenly became one happy family.

Further demonstrating the odd mix of religion, pacifism, anarchism, environmentalism and paranoia evident in the Nevada anti-nuclear movement is a quote from Frits ter Kuile, Nevada Desert Experience Las Vegas Coordinator, in an April 1993 NDE flier :

*"When the sun was setting and most of the over 50 buses and hordes of cars had left the NTS, we closed with a reading from 1 Samuel 8 in which the people ask Samuel for a king and Samuel prayed to God. God said to Samuel:*

**Hearken unto the voice of the people in all that they say unto thee: for they have not rejected thee, but me, that I should not reign over them. Since I led them out of slavery, they have forsaken me by serving money, power, lust and a host of other gods, and so they do also unto thee. Therefore, hearken unto their voice: how be it yet protest solemnly unto them, and show them the manner of the government that shall reign over them.**

*And Sam told the people that asked of him a coercive law and order: government will take your sons, and employ them for its own benefit, to pay its deficit, to make you stand in lines and fill out its forms, man its tanks, be its fighter bomber pilots, and some will be marched to ground zero straight after a nuclear explosion: and it will appoint Chiefs of Staff, and Secretaries of State and War and Commerce and Education and send them to each of its colonies, and to reap heavily sprayed grapes, coffee, and cotton, and to make its NAVSTAR satellites, and brainwash your kids into the rat race.*

*It will exploit your daughters to be cleaners, confectioneries, cooks, secretaries, bakers and nurses for the wounded soldiers. And it will munch up your vineyards, orchards and forests - even old growth, and turn them into monstergod: green $ bills capturing the souls of the people. And ye shall cry out in that day because of the government you have elected; and God will not hear you in that day.*

*So now we are born in this world full of freely elected principalities and powers. However, in the silence of the wind, I strongly felt God still does hear us in that, in this day.*

*Let us join our hands together, and work for a test ban so the burros and flowers can live in peace, daddies can keep on reading Winnie the Pooh to the little ones, and we take time to reconsider Sam's words and rid ourselves of the king by obeying the Prince of Peace.*

The implications of this transformation of traditional Judeo-Christian environmental ethics into a form of earth worship should not be underestimated. If Earth is more sacred spiritually than humans, then morality is defined by whatever it takes to minimize human presence on this planet to the bare minimum. Salvation in this new theology would therefore be determined not by the redemption of human souls, but by the number of acres of Yucca Mountain land set aside as wilderness area.

## ROOTS OF LIBERATION THEOLOGY

Much of the ideology of the Nevada Desert Experience is in reality a spinoff of liberation theology. Primarily a Catholic philosophy, liberation theology has boiled and stewed for years among the clergy of Latin America, most notably in Brazil and later Nicaragua and Central America. Marxist in origin but taking on a flavor of its own, liberation theology is now in essence a land reform movement.

Liberation theology's most noted proponents in America are the Catholic Maryknoll orders. The subtle but enormously disruptive power of this movement is evidenced by the number of Maryknoll nuns and priests regularly silenced for their views and whose corpses littered the Central American isthmus over the last decades. Similar views seem to influence the Franciscan founders of the Nevada Desert Experience.

In one of the seminal works of liberation theology Gustavo Gutierrez writes:

**". . . The liberation of our continent means more than overcoming economic, social, and political dependence. It means, in a deeper sense, to see the becoming of humankind as a process of human emancipation in history. . . . Ernesto Che Guevara wrote: "We revolutionaries often lack the knowledge and the intellectual audacity to face the task of the development of a new human being by methods different from the conventional ones, and the conventional methods suffer from the influence of the society that created them. . . . ."**

**". . . one of the most creative and fruitful efforts implemented in Latin America is the experimental work of Paolo Friere, who has sought to establish a "pedagogy of the oppressed." By means of an unalienating and liberating "cultural action," which links theory with praxis, the oppressed perceive - and modify - their relationship with the world and with other persons." [Gutierrez, Gustavo; A Theology of Liberation, Orbis Books, 1968, p56]**

Liberation theology is a revolutionary movement as evidenced by the quotation of noted revolutionary Che Guevara. It views the active education of the masses (pedagogy of the oppressed) as a key to implementing that revolution. To the extent that liberation theology exists within the protest movement at Yucca Mountain, we show (within reason) that opposition to the repository is a religiously motivated revolutionary land reform movement rather than a matter of resolving the actual technical questions involved. There are more than a few tantalizing threads linking liberation theology to the religious and social opposition to Yucca Mountain.

We see evidence of liberation theology in the development of the State of Nevada's equity theory of nuclear waste disposal (i.e., on-site storage is required for reasons of equity, not safety) in Roger Kasperson's earlier cited work, *Participation, Decentralization and Advocacy Planning*. Roger Kasperson, as political geographer and designer of the NWPO socioeconomic studies, wrote admiringly of Paolo Friere, one of the main architects of Brazilian liberation theology. Interpreted this way, Mountain West composed the "concientization" element of Friere's revolutionary theory, collecting social information and acting as an organizing force to influence information channels within the state with their reports and poll results. The revolutionary foot-soldiers are represented by the activists from Citizen Alert, Nevada Desert Experience and other activist groups. The "oppressed" are the Western Shoshone, whose land claims to half the state provide a convenient justification for revolutionizing. This analogy can obviously be taken too far, yet there does appear to be a liberation theology eddy to the main current of protest at Yucca Mountain.

Another subtle link to liberation theology is Judy Treichel. Besides her many other protest activities, Treichel helped found not only Nevada Desert Experience, but also in 1983 the local chapter of Clergy and Laity Concerned, a Catholic advocacy group devoted to "social justice". Clergy and Laity was originally called Clergy and Laymen Concerned About Vietnam, but has lately protested the U.S. invasion of Haiti (although Jeanne-Bertrand Aristide was defrocked as a priest for harboring liberation theology sentiments). Treichel's protest of America's involvement in Nicaragua and Central America through Clergy and Laity in the mid 1980s makes it clear her "social justice" theories have been influenced by liberation theology, though perhaps not consciously. Clergy and Laity Concern's connections to the Maryknollers and liberation theology are well known.

## EARTH WORSHIP FUSES WITH LIBERATION THEOLOGY

Economically, earth worship implies the absence of property rights; mankind exists at the pleasure of the earth god, "Mother Earth". This relates to Yucca Mountain through the tribal land claims of Nevada's Western Shoshone to the Yucca Mountain area, who coincidentally see nuclear testing and disposal of nuclear waste below the surface as an affront to Mother Earth. The Western Shoshone's veneration of the land "We have only one planet: one water, one air, one land.", tugs at the heartfelt romantic notions of the noble savage being at one with nature. Since the most direct supporters of the Western Shoshone are the Franciscans of the Nevada Desert Experience, joined by the activists of Citizen Alert and the environmental movement, it appears there is a natural affinity of opponents of Yucca Mountain to a form of earth worship. This implies some of the opposition to Yucca Mountain is part of a much larger religious and social revolution which seems to be occurring at an almost subliminal level.

THE WESTERN SHOSHONE NATIONAL COUNCIL & THE GLOBAL ANTI-NUCLEAR ALLIANCE CALL FOR

# HEALING GLOBAL WOUNDS

## OCTOBER 2-12, 1992    LAS VEGAS/ NEVADA TEST SITE

# STOP NUCLEAR TESTING
# ON NATIVE LAND

APPROVED FOR POSTING
PULL DATE

Cosponsored
by
KUNV

# END 500 YEARS | OF INJUSTICE
Moyer Student Union    | UNLV

**Oct 2-4:** Indigenous People's Forum, Pow Wow,    Health & Medical Effects of Radiation Conference/Las Vegas

**Oct 5:** Demonstration at Nevada Test Site Opera-  tions Office/Las Vegas

**Oct 5-9:** Join European Peace Pilgrimage and Walk Across America for Mother Earth for the 65 mile, completion of their 8 month 3000+ mile continental walks.

**Oct 7-12:** Nevada Test Site Encampment, workshops, action preparation, etc.

**Oct 10:** Native Led All Nations Healing Ceremony for Mother Earth

**Oct 11:** Multi-Cultural Rally & Mass Nonviolent Action

**Oct 12:** Commemoration of 500 Years of Resistance Since the Arrival of Columbus

For information contact: **HEALING GLOBAL WOUNDS**   Local Contact:
**PO BOX 4082, LAS VEGAS, NV 89127  (702) 386-8696**

ARTWORK BY JACK MALOTTE

173

# PART V: POLITICAL KINGPINS

# 37. Nevada Political Reality

Yucca Mountain has served to harden the political fault lines in Nevada. Popular fears of radiation, amplified by the political climate, led significant numbers of citizens to vote based on their uneasiness that nuclear spent fuel might be transported and stored in the state. Yucca Mountain thus became a litmus test of political orthodoxy among Nevada's incumbent political elite. If a politician failed to claim the Yucca Mountain nuclear waste repository horrifyingly dangerous, they were effectively blackballed from political office.

Clearly, some politicians found it in their self interest to perpetuate the fear of Yucca Mountain. Without Yucca Mountain, Nevada is a political backwater whose greatest problems are competition from Indian gaming and allocation of the few drops of precious water left from the Colorado River. Creating a tragic myth out of Yucca Mountain served the Nevada political establishment in two important ways: 1) it acted as a springboard to the Washington political arena on an important national policy issue and 2) it polarized the electorate in one political party's favor.

## POLITICAL DEMOGRAPHICS OF NEVADA POPULATION

Demographics explain why demonization of Yucca Mountain has proven such a potent political strategy. Nevada has long had a transient population, growing by leaps and bounds, with the majority gravitating towards low-tech gaming and service industries. In contrast to its vibrant service economy, the state has supported a relatively small technological and manufacturing base. The service and technical populations tend to vote at opposite political poles.

The minority of Nevadans are educated in technical areas where they have the scientific or industrial background to objectively judge risk levels of nuclear waste disposal. Engineers and scientists tend to be politically conservative, as are elements of the military stationed in Nevada. Others from the technical community who have experience handling radiation, for example workers from the Test Site, are also conservative. Business groups and construction trade labor unions who see positive economic impact in the construction of a waste site are the traditional backbone of Republican and Democratic boll weevil politics. Consequently, there was a huge potential political windfall for Democratic politicians if they could convince the electorate that the nuclear waste repository was a conspiracy by Republicans to turn the state into a dying wasteland.

Ironically, Yucca Mountain is the type of project Democrats used to take pride in. Similar in national impact to the building of Hoover Dam, Yucca Mountain is a large public works project which would prove an environmental and economic plus for the entire United States and potentially aid Nevada's limited economic repertoire of gaming and mining. It therefore wouldn't normally be an issue dividing Democrats from Republicans.

Indeed, at the national level the most prominent advocates of Yucca Mountain were Sen. J. Bennett Johnston (D. La). author of the so-called "Screw Nevada" bill and Congressman John Dingell (D. Mich.)., leading nuclear proponent in the House, both Democrats. During the Clinton administration, Secretary of Energy, Hazel O'Leary, was a Democrat with a strong nuclear utility background and saw the benefits of completing Yucca Mountain. Even Kent Oram, the public relations expert for the American Nuclear Energy Council, was an insider in Democratic politics, having run former Nevada Governor Miller's campaign. Opposition to Yucca Mountain thus became a litmus test for the Nevada Democratic Party, as much in response to local political demographics as through consideration of the repository's economic impact on the state. Later, the Nevada Republican party moved towards the same lockstep position in a game of political catch-up.

## POLITICAL HISTORY OF YUCCA MOUNTAIN

The negative image that nuclear waste shipments would irradiate school children and result in nuclear holocaust was so successful for politicians who cultivated it during the 80s and 90s that a balloon of nuclear hysteria expanded nearly to the bursting point. Powerbrokers like Senators Richard Bryan and Harry Reid, Mayors Jan Laverty-Jones and Oscar Goodman, former Governors Grant Sawyer and Bob Miller, all benefited from their opposition stance on the nuclear waste issue. However, this posed a vexing problem. If the political elite allowed even the tiniest hint of dissent to their anti-repository position, their power base could be destroyed as the bubble of fear that kept their followers in check burst. It was with increasing stridence that these politicians tried to silence pro-nuclear critics, posing for group solidarity photo-ops to enforce unanimity. Unfortunately, their intimidation tactics seemed to work.

While at the surface the debate was over nuclear waste, at another level a small clique of Las Vegas good-ole-boys were trying to establish a political dynasty using Yucca Mountain as a key issue. Bryan, Reid and Bilbray had all attended Las Vegas High together and their control over Nevada politics eventually became a hammerlock. Those not able to pass the political purity test of being 100% anti-repository were politically tarred and feathered. These punishments served as examples of what could happen to lesser politicians who dared to oppose the ruling political class on other issues that drove the state.

The practical payoff was huge. Since the Yucca Mountain nuclear waste repository is a leading hot button issue in Nevada, the winners in this critical debate could establish a political dynasty well into the next century. The risk was that if your facts weren't straight and your supporting science weak, you could lose everything if the truth came out.

Unfortunately, the Republican establishment in Nevada was afraid of its own shadow and was unable to effectively oppose the Democratic political machine on Yucca Mountain. Ironically, when Republicans finally jumped on the anti-repository band-wagon it was too late to gain political benefit:

"Fed up" with the Department of Energy's handling of the nation's first high-level nuclear waste repository, Nevada Republican congressional members joined Democratic Rep. Harry Reid and Gov. Richard Bryan in vowing to take the issue to court. . . .

**"I congratulate my elected colleagues on their long over-due decision to take a stand in Nevada's behalf," Reid said. "It is unfortunate, however, that their opposition to making Nevada a nuclear garbage dump is three years late in coming." [Manning, Mary; Las Vegas Sun, Aug. 23, 1986, p1A]**

Once forced to play catch-up opposing Yucca Mountain, the Republicans compromised their position repeatedly. For example, in 1987 Republican state attorney general Brian McKay folded under to pressure to pay lobbyists to oppose Yucca Mountain:

**Attorney General Brian McKay raised objections to paying $400,000 for state lobbyists in Washington, D.C.**

**During a Board of Examiners hearing, McKay questioned payments to the Sutherland, Asbill & Brennan firm.**

**The Nevada Nuclear Waste Project Office wants the firm to lobby Congress against placing a high-level nuclear dump at Yucca Mountain.**

**"That's a lot of money," McKay said. "It is like paying four lawyers full time."**

**McKay said that the Nevada congressional delegation is supposed to carry out nuclear dump lobbying.**

**"We have two U.S. senators and two congressmen," he added. "I thought that was their purpose."**

**In the end, however, McKay went along with Gov. Richard Bryan and Secretary of State Frankie Sue Del Papa and approved the contract. [Payment To Lobbyists Questioned, Las Vegas Review Journal, Capitol Bureau, 10/21/87]**

Barbara Vucanovich, the Republican congresswoman representing northern Nevada, attempted to avoid falling in lockstep with her Democratic congressional cohorts, but offered a rather feeble alternative:

**Furious at Congress targeting Nevada as the nation's high-level nuclear repository, Rep. Barbara Vucanovich, R-Nev., said Wednesday she supports efforts by Marshalls Islands officials to be studied as a potential dump site." [Manning, Mary; Vucanovich Wants Nuke Dump Site in Pacific, Las Vegas Sun, 1/14/88, p1A]**

In the 1988 elections, Governor Bryan defeated then Senator Chic Hecht by the slim margin of 1%. While Chic Hecht was known as an honorable businessman, he was never accused of an excess of charisma, so the closeness of the 1988 election indicated Bryan's electoral weakness. What may well have put Bryan over the top was his opposition to nuclear waste storage, a hot-button issue on which Hecht never properly defined his position. Hecht preferred to push reprocessing and seabed disposal as solutions, positions which were too complex for easy analysis by voters:

**Another Hecht bill calls for considering reprocessing of spent nuclear fuel before investing in a deep-earth repository.**

**Ben Rushe, director of the Energy Department's civilian radioactive waste program, told the panel reprocessing to retrieve uranium is uneconomical in the United States, which has plenty of fossil fuels.**

**But Hecht said the government should consider it anyway because it reduces the danger to public health. [Las Vegas Sun, 7/17/87]**

It is interesting to note in the same article that then Las Vegas mayor Ron Lurie, part of the Democratic establishment, in part agreed with Hecht:

**. . . Las Vegas Mayor Ron Lurie said in written testimony the city "strongly endorses" Hecht's Nuclear Energy Waste Policy Act requiring the 50-year storage period.**

**"Once the 50 years has passed, transportation of the waste would be safer," Lurie said. [Las Vegas Sun 7/17/87]**

Later, after Hecht left office, the State of Nevada's position would come to be that a fifty year delay was advisable.

The only bright spot for the Republicans was the short-lived Vision Project started by businessman Bob Gore, who attempted to expose the hypocritical positions of the state political establishment on the Yucca Mountain issue. It was Gore who suggested possible benefits of accepting the repository to a AFL-CIO Union meeting in July of 1991:

**"We want nuclear energy to be to Nevada what oil is to Alaska. Alaskans don't pay any taxes. At the end of the year they get a rebate. . ."**

**"Well I tell you, what I learned on the labor line is next to jobs, everything else is crap!" [Bob Gore speech to AFL-CIO, 7/30/91]**

Gore received a sympathetic response from union members worried about the obvious loss of jobs should Nevada lose not only the Nevada Test Site, but Yucca Mountain as well. This combination might well leave the state's technical community and economy in limbo. Bob Loux was quick to contest Gore's claim that jobs were available at Yucca Mountain, going head to head with Gore on a Channel 3 News interview. Gore's Vision Project conducted an aggressive fax campaign on the Yucca Mountain issue, calling the actions of Bob Loux and Senator Bryan to task, but the revelations were given little heed by the popular press.

Gore ran for the Republican U.S. Senate nomination in 1992 advocating a pragmatic realism on Yucca Mountain, but his campaign was ill funded and lost in the primary to a cattle rancher from the north. The successful Republican nominee, Demar Dahl, ran away from the Yucca Mountain debate and consequently was destroyed in the general election for lack of a key issue to battle incumbent Harry Reid. Reid brutalized Dahl, who was unable to bring himself to fight the down and dirty campaign run against him. In the process, the state reelected a senator whose greatest hope for Nevada's future at that time seemed to be covering the Test Site with solar collectors, no matter how unlikely this technology.


## CAMPAIGN '94

The motivations of key politicians in regard to Yucca Mountain became an issue in the 1994 Nevada elections. If Yucca Mountain was benign and Nevada had sacrificed hundreds of millions of dollars in benefits to keep incumbent politicians in office, those incumbents would have a difficult time explaining their actions. The two key races were for Bob Miller's seat as governor and Richard Bryan's U.S. Senate seat. Both vigorously opposed Yucca Mountain and benefited politically from these positions.

Gov. Bob Miller stood at risk in the 1994 election from an assault from Jan Jones, mayor of Las Vegas and like Miller a Democrat, but won the primary by 40 percentage points. Yet, the bitter primary race this engendered created a natural opening for Republican Jim Gibbons who took the Republican primary from Secretary of State Cheryl Lau. The questionable management of the Nevada Nuclear Waste Project Office became a possible wedge capable of delivering a normally Democratic state into Republican hands. Gibbons claimed to be anti-Yucca Mountain himself based on his education as a mining engineer, so he was able to take both a populist stance against the repository, while also attacking Millers lack of fiscal responsibility.

Miller found himself in the difficult position of having to deny that he had already negotiated with the nuclear industry through his former campaign consultant Kent Oram (who also conveniently was the chief representative of the American Nuclear Energy Council). It was also inconveniently found that another ANEC consultant, Don Williams of Altamira Communications, had stood to gain $7 million in a BLM landswap arranged by Democratic Congressman Jim Bilbray of Nevada. It thus began to look like there was a network of good-ole-boy politicians who railed against Yucca Mountain in public, but dealt with nuclear insiders behind the public back. Miller was the least insulated from these charges.

The Republican frontrunner in the senate race against Richard Bryan was Hal Furman. Furman, a former Laxalt aide, was immediately accused by Bryan of being a patsy for the nuclear industry:

**A pro-nuclear lobbyist sent out 100 invitations for a fund raiser this week for Nevada GOP Senate candidate Hal Furman, a longtime friend of the lobbyist and a challenger to Sen. Richard Bryan, D-Nev. . . .**

**Furman said he doesn't want to take money from ANEC, because he doesn't want to imply he has prejudged what scientists might determine about Yucca Mountain. ["Barbs Abound in D.C. Bryan foe helped by nuke lobbyist", Las Vegas Sun, 10/1/93]**

**Furman, however, sees the lack of success by Senator Bryan on the Yucca Mountain issue as a political opening. While not supporting the t imposition of the repository on Nevada (no politician does), Furman is willing to negotiate for benefits and holds Bryan accountable for the managing of NWPO.**

**"He (Bryan) campaigned for and was elected to the Senate in large measure because of a promise to prevent Nevada from becoming the nation's high-level nuclear waste site," Furman said of Bryan.**

**"But in the Senate, Mr. Bryan has been outmaneuvered, and he's been outfoxed. And today, Nevada is closer to receiving the nuclear waste than when we hired Senator Bryan nearly five years ago." [Kanigher, Steve; Las Vegas Sun 10/20/93, p2B]**

### THE FALLING HAMMER

What Nevadans were unaware of as the 1994 elections approached was that the Nuclear Waste Policy Act was to be revisited in 1995. The 1987 Amendments to NWPA was known as "Screw Nevada" in the popular vocabulary, but the 1995 amendments threatened to be an even larger "Screw Nevada II"

First there was the question of siting of a Monitored Retrievable Storage facility, which by the 1982 Act had been limited to states not being sited for the main geologic repository. Common sense, however, suggested that the MRS facility should be sited as close to Yucca Mountain as possible. The need for an immediate solution to this problem was in part driven by Congresses' NWPA commitment to begin accepting nuclear waste by 1998. Thus it was likely Nevada would have the MRS facility shoved down its throat because although there was strong opposition within the state to accepting such a site, acceptance elsewhere would be equally if not more difficult.

Consumers Power Co. Chairman William McCormick was quoted in July of 1994 on his feelings concerning the need of the industry to press forward on the NWPA revisions:

**"I think that everyone realizes that it is very important that we get a federal interim storage site authorized as soon as possible." To do that, McCormick said, the industry has to define the issue early. "I think there is a consensus developing that we need to get this done as quickly as possible," he said. [Newman, Pamela; "Nuclear waste strategy Splits Industry", Tuesday, July 26, 1994, p1]**

Among the areas likely to be addressed in the 1995 amendments were:

1) The need to establish DOE's obligation to accept spent fuel some time in 1998.

2) The need to grant the nuclear industry full access to the nuclear waste fund.

3) The need to provide benefits to the high-level repository host state.

The Nuke Lite campaign in 1994 thus worked to neutralize the negotiations issue as part of the democratic process within the state. This effectively allowed incumbents politicians who had vowed not to negotiate under any circumstances to slide into the elections without ever having had to discuss what their position would be in 1995 when national legislation might force Nevada to accept the worst of a bad deal.

# 38. Senator Bryan

Nevada politics were dominated during the 80's and early 90's by Richard Bryan, who rose from lawyer to State Attorney General, to Governor, to Senator. The consummate political animal, Bryan spent the majority of his life in politics. Much of his later career was built out of opposition to Yucca Mountain and he went on to chair the Nevada Commission on Nuclear Projects.

A graduate of the University of Nevada, Reno, Bryan received his law degree from the University of California, Hastings College of Law in San Francisco. Public service began in 1964 when he served as Deputy District Attorney, later becoming Clark County's first public defender. He was elected to the State Assembly in 1968, where he served until 1972 when he traded jobs for the State Senate. Elected Nevada Attorney General in 1978, he served in that position until his election in 1982 as Governor of Nevada, defeating incumbent Robert List. Re-elected governor in 1986, Bryan served until 1988 when he defeated Chic Hecht by 1% to become Senator from Nevada.

In 1983 it was Governor Bryan who appointed Robert Loux, a former football star at the University of Nevada Reno and high school history teacher to head the Nevada Nuclear Waste Project Office. Despite a lack of management or technical credentials, Loux became an overnight expert in nuclear engineering. He also became Richard Bryan's frontman for the ensuing decade of debate over Yucca Mountain. Bryan's successor, Governor Miller, retained Loux after Bryan was elected to the senate (as did Governors Guinn and Gibbons). However, the following governors didn't actively involved themselves in NWPO policy and the state's opposition to Yucca Mountain is still mostly shaped by Richard Bryan's agenda.

Ironically, Bryan seemed to support Yucca Mountain when he was first elected governor in 1982. What brought about his change of heart is unclear, however, he since exploited anti-nuclear hysteria for political advantage. For example, some Nevadans remember the shipments of radioactive dirt from New Jersey that Governor Bryan blocked in 1985. New Jersey had proposed to ship landfill with slightly elevated levels of radon to Nevada for disposal and Bryan ran an aggressive campaign to stop the radioactive dirt from coming west. His efforts proved successful, although the health of Nevadans never seemed threatened.

In 1987, congress passed the amendment to the Nuclear Waste Policy Act which designated Nevada as the sole study site. While the protests of the entire Nevada Congressional delegation were strident, then Governor Bryan was especially incensed. Bryan blustered that Senator Bennett Johnston's bill to offer $100 million-a-year in compensation 'nuclear blackmail':

**"My reaction is Nevada is not for sale," Bryan said. "This would be the equivalent of a federal flash roll that's being dangled before Nevadans with the hope that we will close our eyes, ignore the safety of our citizens, swallow hard and accept the nuclear waste dump." [Bryan Labels $100 Million-a-Year Deal 'Nuclear Blackmail', Las Vegas Sun, 3/26/87]**

Curiously, Bryan later claimed that no benefits were available from the Federal government in exchange for Nevada's acceptance of the repository. Yet, Bryan was not necessarily against radioactive materials and hazardous waste if there was a political or financial incentive:

**Bryan said there is a "critical distinction" between his fight two years ago to stop a train loaded with low-level radioactive dirt from New Jersey coming to Nevada and his support now for the Superconducting Super Collider which will generate 100 tons of low-level radioactive waste each year . . . .**

**The governor in an interview, said the "critical distinction is there is a helluva difference between taking care of a waste byproduct that is generating technology and in acting as a dump."**

**Now the state will submit an application next month to the Energy Department to be considered for the $4.4 billion giant atom smasher research project which, in addition to the radioactive waste, will also result in 10,000 gallons of hazardous waste annually. Bryan is strongly backing the application. [Cy Ryan, Bryan sees 'distinction' between collider waste, NJ dirt, Las Vegas Review Journal, 8/11/87]**

One of the benefits of acceptance of Yucca Mountain might have been a nuclear studies institute capable of "generating technology", so the Supercollider and the repository both promised to buttress the local technical community. Yet, unlike the Yucca Mountain project, Bryan was prepared to accept the Super Collider, and the state was prepared to spend $300 million to woo the project and its low-level radioactive waste to Nevada:

**The state says it is prepared to provide at no cost to the Energy Department the 15,850 acres, an adjacent new community, water and sewer facilities, an upgraded airport capable of handling jets and construction of highways to serve the site. [Ryan, Cy; Nevada Woos Super Collider, Las Vegas Review Journal, 8/29/87, p1A]**

Governor Bryan rode the populist anti-nuclear bandwagon into the senate by defeating former senator Chic Hecht in 1988 by the narrow margin of 1%. Many feel Bryan squeaked out his slim victory on the back of the Yucca Mountain issue:

**Bryan told a group of national reporters that Hecht had failed to help Nevada on a series of issues, including nuclear waste, and the state would be better off with a new senator even though it would lose Hecht's six years of seniority.**

**"If you can't find the key to the men's room no matter how long you've been back there, there isn't much you can do," Bryan said. [Koenig, Dave; Bryan: Hecht ineffective, 'can't find key to men's room', Las Vegas Review Journal, 5/19/88, p1B]**

Bryan's stinging portrayal of Hecht as soft on opposing nuclear waste seemed to bear fruit. One of Bryan's ads showed low-level radioactive dirt from New Jersey in a truck passing a school bus, implying the governor had saved the children of the state from being irradiated. However, the New Jersey dirt is an alpha emitter, a type of radiation which can't penetrate a piece of paper, much less the walls of a truck or even the dirt that holds it. Alpha emitters are only dangerous if inhaled or ingested, events unlikely to result from packaged containers whose concentration of alpha emitters was admittedly small. Sources close to this issue suggest Bryan was told the New Jersey dirt was absolutely safe to transport, but ignored the information.

## CORPORATE AUTO FUEL EMISSIONS

The question posed by Bryan's handling of Yucca Mountain and other nuclear issues is whether his interests lay in promoting Nevada or his own political fortunes. An example was Bryan's attempt to pass the Corporate Auto Fuel Emissions (CAFE) bill of 1992 which would have imposed a forty miles per gallon fuel efficiency standard on American cars. Bryan's sponsorship of this bill rubbed powerful congressional delegations from the rust belt auto manufacturing states the wrong way (especially Dingell of Michigan), seriously damaging Nevada's leverage in Congress.

Nevada is a vast, mountainous, desert state where much of the population finds occasion to traverse long stretches of roadway on a regular basis. This is road hog country, meant for big-engined, air-conditioned, 90 mph behemoths of the road that people can stretch out in. Nevada is so sparsely populated that pollution is only a worry in Las Vegas proper, and not nearly to the same extent as L.A. or most large metropolitan areas. CAFE would in effect mandate small, crash vulnerable vehicles, although Nevada is one of the least likely states to benefit from such a regulation.

One can argue the benefits of CAFE nationally, however, Senator Bryan chose to make CAFE a priority even though Nevadans were more interested in issues like water rights and competition from legalized gambling in other states. Bryan was perhaps influenced by Colorado Senator Timothy Wirth, a cosponsor of Bryan's bill who had long been associated with the Safe Energy Communication Council and other Naderite groups. Some of the allies of the Safe Energy Communication Council had long pushed for 100 mile per gallon downsized cars (notably Amory Lovins of the Rocky Mountain Institute).

While Bryan was quick to point out the faults and risks of Yucca Mountain, he was been notably shy about allowing CAFE to receive criticism. The Corporate Average Fuel Emissions bill was meant to promote a cleaner environment and ensure the security of America by reducing our foreign oil needs, but has serious safety ramifications. A report in the Journal of Law and Economics puts Senator Bryan's priorities in perspective:

## CONCLUSION

**Earlier analyses of the effects of fuel-economy regulation have missed an important point. Fuel economy regulation inevitably leads to smaller, lighter cars that are inherently less safe than the cars that**

would be produced without a binding fuel economy constraint. We have shown that even if the pursuit of fuel economy were costless to producers, the cost of the added life and serious injury from traffic fatalities would more than offset its benefits in reductions of gasoline consumption for 1989 model year cars. We estimate that these 1989 model year cars will be responsible for 2,000 - 3,900 additional fatalities over the next ten years because of CAFE. Thus, when any discussion of energy conservation focuses upon externalities in energy consumption, we would suggest that all such externalities be included. When safety considerations are included, CAFE appears to be a very costly social policy. [Crandall, R.W. and Graham, J.D.; "The Effect of Fuel Economy Standards on Automobile Safety", Journal of Law & Economics, April 1989, p97]

The Crandall and Graham analysis studied the 500 pound vehicle weight reduction which followed the original introduction of CAFE standards in the early 80s. Senator Bryan's CAFE standards would have raised the fuel efficiency to 40 miles per gallon, a nearly fifty percent increase in mileage. However, most of the straightforward fixes to improve mileage (such as changes in fuel injection, more valves, better oils, computerization, streamlining, etc.) had already been tried. Consequently, the most obvious place to affect a change in mileage is in vehicle weight, compromising structural integrity.

Putting this in perspective, Senator Bryan was willing to accept 2,000 to 3,000 more deaths a year to promote automobile energy conservation, while objecting to an estimated risk of death of one-tenth person per year from Yucca Mountain. Over the ten thousand years Yucca Mountain is designed to contain the waste, this amounts to 20,000,000 deaths attributable to Senator Bryan's CAFE bill, versus 1,000 for Yucca Mountain, a ratio of 20,000 to 1.

## BRYAN VERSUS BENEFITS

Part of Senator Bryan's tactic to defeat Yucca Mountain was to claim that Nevada would never see any benefits if the waste repository came to the state. So tied was Bryan to promoting this position that he went to extremes to prevent Nevadans from believing benefits were even possible. Bryan's efforts to preempt benefits was not limited to calling them 'nuclear blackmail'.

The end result of Bryan's having alienated the Senate over the CAFE bill and other issues was that he lost all influence over the Energy Bill of 1992. This bill effectively took away Nevada's last bit of leverage regarding Yucca Mountain. Not content to accept the 84 to 6 defeat of his filibuster of the energy bill in the last hours of the '92 session, Bryan attempted to torpedo the parallel funding bill for the New Mexico Waste Isolation Pilot Project (WIPP).

The Bullfrog County Times (the admittedly pro-nuclear fax attack produced by Altamira Advertising) couldn't help but rub salt in the wound when Bryan's filibuster was defeated:

### WHY WHIP WIPP?
As usual, what we heard from Washington isn't what really happened. And as usual, the truly important facts were left out of the story. Yes, we all know that Sen. Bryan fought to the end, not only against the energy bill, but against a bill that dealt with the WIPP program in New Mexico. WIPP (which stands for Waste Isolation Pilot Project) is designed to find out whether nuclear waste generated in defense programs can be safely stored in salt deposits. Dick Bryan fought like a wildman against the WIPP bill, but not because it had anything to do with Nevada. It didn't. In fact, the energy bill, with its provisions concerning Yucca Mountain, had already passed when Bryan attacked the WIPP bill.
So what was the motive? Well, the WIPP bill contained a provision which will pay the state of New Mexico $431 million over the next 15 years. The payments start at $20 million per year, plus five percent per year for inflation, which would raise the payments to nearly $40 million a year. That's a lot of money for a small state. But WIPP is a small program compared to the Yucca project. Dick Bryan fought tooth and nail against WIPP. He didn't want New Mexico to get $431 million. And he didn't want Nevadans to even know about these payments. Why?
### BENEFITS ARE (WERE) REAL
Bryan and his pals have been telling Nevadans for years that promises of benefits are illusory. The anti-nuke crowd insists that the government can't be trusted, that there is no money for any benefits, and that even if the money existed, we will never see a penny of it. New Mexico will get $431 million but Nevada's

**deficit-plagued state government won't get a cent. We have Dick Bryan to thank for that. [Bullfrog County Times, Volume 1, no. 17, October 1992]**

SAVING THE NEVADA TEST SITE

An issue closely related to Yucca Mountain has the future of the Nevada Test Site, which shares land with the repository. Long on the forefront of atomic weapons research, the NTS is now threatened by moratoriums on the testing of nuclear weapons. Future nuclear projects which could take advantage of the Test Site technical expertise and provide it a long term mission include not only Yucca Mountain, but nuclear weapons disassembly and a possible nuclear studies institute. Most nuclear projects have met resistance from Senator Bryan, at times causing a rift with his senate colleague from Nevada:

**BRYAN REAFFIRMS OPPOSITION TO NUKE ASSEMBLY PLANT**

**Escalating the debate over the future of the Nevada Test Site, Sen. Richard Bryan on Wednesday seized upon a nuclear accident in Russia this week to press his argument that the state should not bid for a government factory to disassemble nuclear weapons.**

**Citing reports that a tank of uranium waste exploded at a nuclear fuel plant in the Siberian city of Tomsk-7, Bryan urged "caution before Nevadans offer the state as a plutonium storage and reprocessing location as part of a planned reconfiguration of the country's nuclear program."**

**"This simply is not my vision of the future of Nevada," Bryan said.**

**Bryan's remarks . . . underscored his disagreement with Sen. Harry Reid, D-Nev., who is considering whether to propose the Nevada Test Site as the home for a nuclear weapons disassembly plant. [Batt, Tony; Las Vegas Review Journal, 4/8/93]**

Having ruled out most nuclear technologies as part of his vision for Nevada, and calling compensation for Yucca Mountain nuclear blackmail, Senator Bryan was faced with convincing the electorate he was capable of saving jobs at the Nevada Test Site.

Belatedly following Senator Reid's lead, in 1994 Senator Bryan proposed that the Nevada Test Site be converted into a giant solar farm. He claimed this would make Nevada the Saudi Arabia of energy, creating thousands of jobs for the Silver State. In reality, Senator Bryan's solar legislation may have only been an attempt to avoid responsibility for not having helped reconfigure the Nevada Test Site years ago. In fact, solar energy is unlikely to be funded long term for Nevada, in part because Senator Bryan has little political clout in Washington and in part because even a profligate Congress recognizes the huge cost escalation likely from solar research Bryan is proposing.

Senator Bryan in conjunction with the DOE's solar research group (Office of Solar Energy Conservation) in pushing a solar makeover for the NTS held a conference on this subject June 2-5 of 1994. However, George Sterzinger, a consultant hired by DOE at Bryan's request to prepare a pro-solar report noted a number of key hurdles standing in the way of bringing solar projects to the Nevada Test Site:

**\* Water is a major physical constraint on solar energy development.**

**\* An alternative habitat for the Desert Tortoise, a threatened species, will have to be developed to mitigate the impacts of solar development at NTS.**

**\* Transmission service to the site is inadequate for large-scale solar energy development. To accommodate 1000 Megawatts of solar energy the lines would have to be upgraded to 500kV, at an approximate cost of $33 million, assuming $500,000 per mile over a 65 mile line ranging to $55 million for upgrading 110 miles of line.**

**\* Natural gas service to the site is a roadblock to hybrid solar electric development. Bringing gas to the site for solar thermal trough development would cost roughly $60 million.**

**\* Water: Groundwater is the only local source of water at the NTS. Total existing water use on the NTS totals 9900 acre-feet per year drawn from 14 active wells. Roughly 56 per cent of annual consumption is drawn from nine wells on the Ash Meadows subbasin. . . .**

**Before relying on groundwater supplies at the NTS, more detailed analysis of groundwater reserves and the impacts of extraction, especially in Area 25, will be required. The wells in Area 25 may be connected to groundwater supplies for the Devil's Hole spring. Devil's Hole is habitat for the endangered pupfish, and therefore water use that could affect either water quality or quantity at Devil's Hole would trigger the**

**Endangered Species Act protections. [Nevada Test Site Solar Feasibility Study, April 1994, prepared by George Sterzinger and DynCorp for U.S. Department of Energy]**

Thus not only would the Nevada Test Site require hundreds of millions of dollars in upgrades in transmission lines and gas outlets merely to make it possible to support a relatively small 1,000 MW solar facility, but the site also threatens the Desert Tortoise, threatens the Devil's Hole pupfish, and severely taxes the local groundwater supply. Other countless hundreds of billions of dollars would need to be spent building a water pipeline from the ocean and high pressure hydrogen gas lines to fulfill Senator Bryan's complete solar utopia for the Test Site.

Curiously, Senator Bryan said for decades that you can't trust the Department of Energy, and that you can't trust the federal government to provide funds for the state of Nevada (specifically in regard to Yucca Mountain). Yet, when he proposed a dubious solar utopia requiring hundreds of millions of federal dollars, Bryan suddenly found the federal government and the DOE a likely donor.

## ANALYZING BRYAN'S MOTIVATIONS

Bryan's motivations are as important to the reader as his actual pronouncements regarding Yucca Mountain. There are a number of disturbing trends in this broader record:

1) Bryan is willing to promote other technical solutions (CAFE and the Super-Collider) that have apparent risk levels tens of thousands of times greater than Yucca Mountain.

2) Bryan's only counter to Yucca Mountain and nuclear technology was based on under-researched solar proposals whose technical and financial merits were undemonstrated even by his own analysts.

3) Bryan was willing to deny his state benefits for the study of Yucca Mountain, and even filibuster another state's benefits, while claiming benefits from solar technology. This appears to be inconsistent.

## THE NUCLEAR INDUSTRY'S RESPONSE

Given Senator Bryan's apparent lack of consistency and vulnerability on these issues, one would have expected the nuclear industry to attack Bryan head-on in the 1994 elections. However, apparently a deal (called 'Nuke Lite') was struck between nuclear trade insiders, Hazel O'Leary, Bennett Johnston and others within the Clinton administration that no direct attacks on Bryan would occur in a bid to help save his seat in the 1994 election for the Democrats. In return, Bryan would be expected to not attack the nuclear industry during 1994, at least not as vehemently as before.

The reason such a deal was particularly inviting to Bryan and to Governor Miller at the State level, was that it literally freed up millions of dollars in campaign funds which otherwise would have been needed to block the nuclear industry's campaign. As cited earlier, Bryan's Republican opponent, Hal Furman, made this charge in the opening salvos of the 1994 election. It would reappear frequently during the election.

It would seem Senator Bryan was willing to play both sides of the fence on technological issues ranging from the Super-Collider, to NTS, to CAFE to Yucca Mountain. The unfortunate side effect of the nuclear coalition's decision not to directly confront Bryan in open debate, attempting to win its battles in Washington's backrooms, is that they may well have lost the long term battle this way. While political opposition to Yucca Mountain was been neutralized in the short term, no positive constituency has been created within Nevada, or the nation, for continuation of nuclear power.

# 39. The Governors

After Richard Bryan stepped down as governor in 1988 to take a seat in the U.S. Senate, Lieutenant Governor Bob Miller took over the reins of Nevada state government. Although a political hegemony had developed for Democrats in Nevada, Bob Miller inherited a host of problems in his new post that threatened his power. As lieutenant governor, he'd acted the good cop to Bryan's bad cop, but now he had to make some tough decisions.

One decision Miller mishandled (though this became a pattern for future governors as well) was not replacing Bob Loux as head of the Nevada Nuclear Waste Project Office. As volatile as the Yucca Mountain political situation was, leaving Loux in position meant Miller was forced to carry the millstone of Bryan's anti-nuclear campaign without any realistic hope of political benefit for himself. Miller couldn't negotiate with the feds for benefits for Nevada without repudiating both Bryan and Bryan's appointed director of NWPO.

Miller must have felt he couldn't fire the mercurial Loux without being perceived by Nevada Democrats as having to give in to pressure from the American Nuclear Energy Council and their representatives from OIZ Advertising and Altamira Communications. Miller had relied on campaign advisers Kent Oram from OIZ and Don Williams from Altamira in his 1990 gubernatorial campaign (in fact the close ties of these political consultants to the governor was why they were viewed as so valuable by the nuclear industry), and these advisers were in a bitter war with Loux. As the 1994 elections approached, Miller's political situation became even more difficult because Republican Secretary of State Cheryl Lau was running a hard hitting race and threatened to control the Yucca Mountain issue by questioning Loux's competency.

As governor, Miller lost control of the most controversial agency under his administration. If Bob Loux answered to anyone (at times a question in itself), it was to Senator Bryan, leaving Miller to take the political heat for Loux's indiscretions. According to the Elko Daily Free Press:

**Yesterday, we received one letter for the dump, from a private group called Bullfrog County Times, and one against the dump, from the state's "Agency for Nuclear Projects." Why they don't come right out and call it the agency "against" nuclear projects, we don't know, but we suspect it has something to do with the fact Gov. Miller is involved, and he doesn't even know the difference between a budget cut and a spending hike." [Steninger, Dan; Elko Daily Free Press, Elko, Nevada, July 1992]**

Being tied to the random policies of Loux and the staff at the Nuclear Waste Project Office forced Miller into an inconsistent political stance. While opposing Yucca Mountain, he heartily supported continued testing at the Nevada Test Site. Also perplexing were Miller's statements in support of a Nuclear Rocket study planned by the Air Force for the test site. In reality, the risks of radiation contamination from such tests, though quite low, were still substantially higher than that of Yucca Mountain (the Challenger explosion demonstrates why). In contrast, Miller opposed reopening of the Beatty low-level radioactive waste dump in the 1993 legislative session, even though $20 million a year in fees were offered. Miller's nuclear position, in short, seemed dictated by political expediency rather than a cost benefit analysis.

If Miller's first mistake in office was not firing Loux, his second was giving in to the special interests in the 1991 legislative session. Between the teacher's union, a recessionary fall in gaming revenues and Nevadan's innate antipathy to a business tax, Miller created a fiscal vise. In 1992, the vise closed tight.

With a projected $136 million budget shortfall (approximately 13% of the budget) and no economic recovery in sight, Miller was forced to make draconian cuts across the board in everything from mental health benefits to teacher's wages, antagonizing almost everyone and pleasing no one. There were even rumors that a mental patient committed suicide as a result of the cuts in the mental health budget. While these rumors may have been unfair and unfounded, they nevertheless indicate the political difficulties Miller faced in the 1994 political campaign in the form of a compassion issue.

To put the governor and the state of Nevada's plight in perspective, there was more than the 1992 recession howling at the state's door. A whole pack of devastating fiscal wolves were circling the Silver State at the same time.

First, there was the threat of Indian gaming and startup casino operations in New Orleans, Chicago and elsewhere to worry about. Miller recognized that the State's income was almost totally derived from the discretionary spending of California vacationers and from the federal government (sources that could dry up at any moment). Moreover, Nevada had been unsuccessful in trying to lure major industries to take up residence. It didn't help when the governor proposed a new business tax in 1990, and then complained that the tax that resulted wasn't large enough. The tax was supposedly to cover the state's shortfall but in part covered Miller's concessions to the special interests. By the 1993 legislature, Nevada was faced with substantial recessionary loses in gaming revenues at the same time it had little concept how to attract new light-industries to stabilize the tax base.

Secondly, the Nevada Test Site was about to be substantially downsized, if not eliminated completely. One of the benefits of the end of the Cold War was the downsizing of the American military and defense in general. Nevada was vulnerable to this downsizing at the Nevada Test Site, at Nellis Air Force Base and the Tonopah Test Range. The combined job loss from these sites threatened to number tens of thousands, a huge hit on a low population state like Nevada. And these weren't just any jobs, they were high-tech, high-paying jobs with higher wages than low paying casino positions.

A third threat was the collapse of the economic miracle in the neighboring state of California. The bloom was off the rose in the Golden State and they were reduced to using IOUs to pay employees in mid 1992. Miller knew very well that California's weakened economy was suffering a malaise worse than the rest of the U.S. and this might hurt Nevada. The Silver State is in many ways only a province of its bigger neighbor, dependent on a bloodline of vacationers with discretionary income.

Finally, bills before congress prompted by Secretary of the Interior Bruce Babbitt threatened to wreck mining interests in Nevada by shifting taxation from the production to the prospecting end of the mining cycle and to attack cattle ranching by raise grazing fees. Though perhaps well intentioned by eastern politicians, in a state with little other viable manufacturing or industrial base than mining and ranching, these bills promised economic havoc.

## MISSED OPPORTUNITY

The fifty to a hundred million dollars in yearly benefits suggested in Bennett Johnston's 1987 amendment to the Nuclear Waste Policy Act would have saved Miller politically from all these headaches. Kent Oram of OIZ Advertising, who was handling the public relations for the American Nuclear Energy Council, made Miller aware from the beginning with frequent phone calls that such sums were available and not mere Washington rhetoric. Oram had run Miller's previous political campaigns and so the governor knew very well what Oram could do for him (or against him) in regard to obtaining benefits for the State. The simple political solution would have been for the governor to accept money from the Nuclear Waste Fund for studying Yucca Mountain while still opposing the final siting of the repository in Nevada. Thus, Miller was given plenty of opportunity to make a graceful separation from the the anti-Yucca Mountain policies of Richard Bryan before the 1994 campaign.

Mayor Jan Jones of Las Vegas, a Democrat, took the unprecedented step of opposing the sitting governor in the 1994 primaries. Jones made a bold political move to encircle Miller on nuclear issues by voting to allow the transportation of radioactive landfill to the Nevada Test Site. As noted by political columnist John Ralston:

**At first, it was only a momentary tingling when I noticed the Las Vegas City Council, led by Mayor Jan Jones, refused last week to join a state law suit designed to prohibit 16,000 truckloads of radioactive dirt from being trucked onto the Nevada Test Site from Fernald, Ohio. Union officials insisted the lawsuit would shut down the test site while an environmental impact statement was conducted.**

**Now, though, I feel an overdose coming on as the last couple of days have seen virtually every Nevada pol from Las Vegas to Washington become involved. And even better, this lawsuit seems to have caused a percolation of a problematic issue - the test site jobs vs. anti-nuclear dump rhetoric - which has complicated a debate that once seemed so one sided.**

**"This rips asunder the thinly welded alliance on the nuclear issue," said one veteran of the slippery slope. And what it also has done is transformed what was a short-lived off-off Broadway play showing at City Hall into A Chorus Line of political performers. . . . . .**

**Governor Bob Miller: Mr. FOB No. 1 {Friend of Bill Clinton}, with the unions mobilizing and Jones already with them, said he feels consideration needs to be given to amending the language. [Review Journal, July 14, 1994, p7B, John Ralston]**

## NUKE LITE

The reason John Ralston, the usual insider on Nevada's political dirt, had heard so little about the nuclear waste issue in 1994 was because of the secret agreement between Miller and the nuclear forces to not conduct a campaign of mutual destruction in Nevada in the 1994 elections. Called "Nuke Lite" by insiders, Senator Bryan was also included in the 'hands-off' political deal which canceled a million-dollar-plus campaign against Miller and Bryan by the nuclear industry. In exchange, Miller apparently made a backroom deal with his friend Kent Oram to negotiate for benefits after the election and to take care of the difficult Loux by finally moving him to the side.

Unfortunately for Miller, the nuclear issue did not die in the 1994 campaign with the defeat of Jan Jones. Republican candidate Jim Gibbons attacked Miller claiming the Governor had allowed Bob Loux at NWPO to deliver million dollar plus contracts without Requests for Proposals and without competitive bids. Gibbons argued that if Miller were so lax in his oversight of this agency because of cronyism, the governor couldn't be trusted to run other agencies as well.

Gibbon's also opposed Yucca Mountain, so his argument that Miller had sold out to the nuclear industry and Kent Oram in a crass deal to throw the election was particularly biting. With the upcoming revision of the Nuclear Waste Policy Act in 1995, Nevada risked not only having a Monitored Retrievable Storage site shoved down its throat, but also risked having its future potential NWPA benefits erased. Miller's credentials as a Friend of Bill Clinton had not significantly slowed the repository study and now with a Republican Congress set to be elected Gibbon's attacked Miller's ability to keep the site out of the state.

The nuclear industry thus once again shot itself in the foot by taking the low road on Yucca Mountain with Machiavellian backroom deals. The decision to forward nuclear industry interests within the Washington Beltway and through scratch-your-back politics with Governor Miller instead of through of Nevada voter education boomeranged and became a primary issue in the 1994 campaign.

# 40. Mayors Jones & Goodman

Despite the glitter and flash of Las Vegas, even after rapid growth it is in many ways still a small town, a puddle in which the biggest fish all know one another and comprise the small, politically correct "A" list of Nevada. Among the big fish in this small pond are Jan Laverty-Jones and Oscar Goodman.

## THE JONES YEARS

Jan Laverty-Jones , mayor of Las Vegas and her shark attacks on Yucca Mountain were none too subtle .

A native of Santa Monica, Jan Laverty's grandfather founded the Southern California grocery chain Thriftimart, making her part of that city's rich-and-famous in-crowd. A Stanford graduate with a B.A. in English, Jan had twenty years of marketing experience and was a millionaire in her own right before becoming mayor of Las Vegas. In the early 1980s Jan Laverty married her high school sweetheart, Ted Jones of the Fletcher Jones car-dealership patriarchy of Las Vegas (both had ended earlier marriages). Jan had long been known on the West Coast along with her brother Biff for the zany ads they did for the family owned food chain. Now she carried into her new marriage a flair for commercial making for the Fletcher Jones car dealership.

The Fletcher Jones dealership was the most extensive in Nevada; they are wealthy people. The aggressive advertising campaigns run by the Jones dealership became the political springboard to the mayor's seat for Jan by giving her the name recognition it takes others decades to achieve. The tongue-in-cheek ads were effective not only for the dealership, but also in giving Jan Laverty-Jones the credibility to run for and be elected mayor on the basis of her image as a competent business woman.

It is after coming to power in city government that the line between wacky commercial making and media manipulation began to blur, especially in regard to Yucca Mountain. In a July 8, 1991 article appearing in the L.A. Times shortly after she was elected mayor titled "A Whole New Image", Laverty-Jones is quoted as saying "Perception is much more important than reality, in fact it's everything." The mayor seems to have applied this motto to her handling of Yucca Mountain.

Good friends of Mayor Jones were Ken Johnson and Jim Tofte, two local disc jockeys for radio KKLZ, whose irreverent humor had attracted sizable following (they once threatened to kill a puppy if they didn't increase their listening audience, a sort of black humor not without its attraction). The friendship with Johnson and Tofte led to the making of some quite funny, but also deceptive ads meant to criticize the Yucca Mountain repository. Johnson and Tofte fused on screen to become the two-headed Yucca Mountain man, supposedly the result of an encounter with a green radioactive slime. The two disc jockeys were squeezed into a single set of giant bib overalls and did car commercials for the Fletcher-Jones dealership complete with references to radioactive pellets that caused Biff Laverty to glow green.

The effectiveness of these ads points up the inherent difficulty in educating the public about the Yucca Mountain debate. While a seemingly innocuous simulated pellet in an ad done by OIZ Advertising for the American Nuclear Energy Council in late 1991 drew 10 months of heated debate, two technologically unsophisticated disk jockeys dressed in what could only be described as a clown suit were allowed to take the moral high ground unchallenged.

The fact that the mayor of Las Vegas was behind these commercials made the situation even more bizarre. Had ANEC (which was also a private entity) composed humorous ads using crash dummies which claimed the Fletcher Jones dealership was selling defective cars that kill babies, they would have created a firestorm of protest. But the Yucca Mountain Man ads, which in essence claimed that the nuclear industry was out to kill Nevadans with radioactive pellets, went unscathed because they were politically correct.

Even more important was the way Yucca Mountain Man tied into the national anti-repository campaign run by NWPO and their surrogates from Mountain West. In a Wall Street Journal Editorial, James Flynn of Decision Research wrote an attack on the advertising campaign run by OIZ Advertising entitled "How Not To Sell a Nuclear Waste Dump." According to Flynn:

"Perhaps the most devastating rejoinders to the American Nuclear Energy Council campaign came from a pair of Las Vegas disc jockeys, who launched a parody of each TV ad as it appeared. The main character in these skits bore the mock name "Ron Ditto," whose rather simple-minded pronouncements were heaped with ridicule. "Hi! This is Ron Ditto, your formerly respected sportscaster, trading in your respect for much needed dollars." Local businesses joined in.

A TV advertisement showed the disc jockeys climbing into a huge pair of overalls to create a two-headed mutant, "Yucca Mountain Man," as part of a commercial for an auto dealership in Las Vegas. . . . . .

The Yucca Mountain advertising campaign failed because Nevadans thought the depository unfair and because they did not trust the experts who told them it was safe. Neither trust nor a sense of fairness can be built by an advertising campaign that assumes public ignorance is the cause of opposition to a nuclear waste repository." [Flynn, James; "How Not To Sell A Nuclear Waste Dump", Wall Street Journal, Wednesday, April 15, 1992, B8]

The Flynn analysis is curious. First, Flynn works for Decision Research and was previously a vice-president of Mountain West, so his article was in essence a hit paid for by the State of Nevada's Nuclear Waste Project Office, hardly an unbiased observer. Second, Flynn tried to imply that Yucca Mountain Man was a spontaneous outpouring of grass roots sentiment when in fact it was for all practical purposes bought and paid for by the Mayor of Las Vegas, Jan Laverty-Jones, who also sits on the Nevada Commission on Nuclear Projects.

Third, Flynn claimed the Yucca Mountain advertising campaign failed because the public distrusted the experts. Obviously, little public trust could be expected when Robert Loux had spent much of $50 million over eight years thinking of every way possible to discredit DOE, a campaign in which Flynn had himself taken part.

Finally, if the ANEC ad campaign assumed the ignorance of the public, Yucca Mountain Man used two disk jockeys squeezed into a clown suit to play on fears straight out of the nineteen-fifties nuclear horror movies. If there was any disrespect for the intelligence of Nevada's citizen's, it may well have been shown by Mayor Jones, Johnson and Tofte.

In fact, Mayor Jones had long played on the idiosyncrasies of the Las Vegas public. According to the 1991 L.A. Times article:

"In five years of funky television spots, Jones cavorted as everything from Little Red Riding Hood to a glitzy casino moll stroking the hindquarters of a sleek new automobile. She crinkled her nose and cleverly teased her niggardly father-in-law, the company founder, with the famous tag line "Nobody's cheaper than Fletcher Jones." And in her most outrageous mode, she ad-libbed soap-opera vignettes with a female impersonator who plays Las Vegas lounges."["A Whole New Image", L.A. Times, July 8, 1991]

Obviously, Mayor Jones career as a free-swinging car-saleswoman suggested her attention to facts in regard to Yucca Mountain took a backseat to promotion. The mayor's lack of technical schooling (B.A. in English from Stanford), didn't stop her from making pronouncements about the dangers of Yucca Mountain, both in the form of Yucca Mountain Man ads and off-the-cuff statements. In reference to ads produced by the American Nuclear Energy Council which attempted to demonstrate that nuclear waste is a solid and not a liquid or gas, Jones stated:

"Showing an individual holding a pellet of nuclear waste that in reality would kill that person within a matter of minutes and then saying well this is just imagery and the public understands, well to me that just goes beyond borderline misinformation." [Chan 3 News, 6:00 PM, 10/24/91. Interview by Tonya Ellis]

While the nuclear pellets are dangerous, death would not occur within minutes as Jones stated, but perhaps over the course of a day. Neither the mayor's office nor Channel 3 reporter Tonya Ellis attempted to ask the makers of the ANEC advertisement, OIZ Advertising, whether there was any intent to mislead on the subject of the radiation dangers of radioactive pellets. In fact, the advertisements were made on the basis of polls which showed the public feared nuclear waste could be released as a liquid or gas.

Mayor Jones involvement didn't stop at Yucca Mountain Man or occasional interviews, urging Nevada politicians to become involved in CAN-WIN (Citizens Against Nuclear Waste In Nevada) in 1992. Perhaps most disquieting was Mayor Jones inclusion on the Sawyer Commission. It is difficult to imagine that the mayor was able to keep an objective view of the data being presented to the Commission while at the same time financially supporting a series of parodies on the airwaves.

## JONES CAMPAIGN 1994

Mayor Jones' involvement in the Yucca Mountain debate took an important turn when in late 1993 she announced her candidacy for the Nevada Governor's seat. The 1994 primary campaign would prove to be brutal and divisive for the Democratic Party, leading to her defeat. Interestingly, Judy Treichel, employee of Governor Miller, backed Jones and developed ties with the mayor's campaign. Miller had ungrudgingly carried the Nuclear Waste Project Office as a political liability during his years as a governor yet he ignored claims that Treichel had sold him out.

Jones' position on Yucca Mountain evolved substantially as a candidate. While not backing the repository study, the mayor at least saw the advisability of keeping open negotiation options. Jones' answers to key questions regarding negotiations were addressed in the Nevada Monitor:

> **QUESTION 1: . . . If you could be assured that Nevada would not forfeit its legal right to oppose a repository by engaging in a negotiation process, would you, as governor, engage the federal government in negotiations for benefits to Nevada?**

> **JAN JONES: I believe that if we receive iron clad guarantee that a discussion of benefits does not preclude or prejudice our right as a state to object, and our congressional delegation's right to oppose the siting at Yucca Mountain, we would be neglecting our fiduciary duty to Nevada's citizens if we didn't consider discussing benefits.**

> **QUESTION 2: . . . what issues would you, as governor, place on the table?**

> **JAN JONES: Before I answer this question, I want to say that it is my belief scientific advances may well preclude the necessity of burying nuclear waste under Yucca Mountain. However, the fact is that a lot of scientific studies and construction activity are currently taking place at Yucca Mountain. I strongly believe that since this work is proceeding, and will continue to proceed, with or without our approval, the State should be compensated. This compensation could, and should, take many forms, some of which are listed below:**

1) **The federal government currently owns 86% of the land in Nevada. I would like to see some land turned over to the State, or the public, soon after negotiations begin, if and when, Yucca Mountain begins receiving nuclear waste. . . .**

2) **The Nuclear Waste Policy Act offers certain incentives to the State of Nevada if we agree to negotiate over Yucca Mountain. I believe if we start negotiating with the federal government prior to January 1, 1995 we can still capitalize on this provision of the Act. This may trigger annual federal assistance in the millions of dollars, which would be a great help to our state treasury. I would also suggest that compensation be calculated retroactively from the time preliminary scientific or construction activities begin at Yucca Mountain.**

3) **. . . . Whether or not a repository is ultimately located at Yucca Mountain, I would press hard to see that the plant to build these shipment containers is located in Nevada.**

4) **We should be able to bargain for expanded scientific and industrial use of the Test Site. This would include solar and hydrogen projects. Also, I would like to see an international research facility built on the Test Site. . .**

5) **Our university system should participate in any scientific or research work done at Yucca Mountain or other areas of the Test Site . . .**

> **Question 5: Some people in Nevada support continued work at the Nevada Test Site, including the transporting, and storage of special nuclear materials, yet they strongly oppose Yucca Mountain, which would be a storage place for high-level nuclear materials if a repository were actually built there. How do you feel about this dilemma? As a governor, how would you deal with it?**

> **JAN JONES: This question brings out the hypocrisy concerning Yucca Mountain. For many years nuclear devices and/or waste have been transported, handled, stored and detonated on the Test Site.**

Area 5 of the Test Site is currently being used to store nuclear waste. Bob Miller and other public officials have consistently favored nuclear testing and nuclear operations at NTS. However, when a few years ago the political winds temporarily shifted, Mr. Miller jumped on the bandwagon and started to oppose any studies of Yucca Mountain.

➢  [Nevada Monitor, September 1994, p1]

# 41. The Sawyer Commission

In 1985 the state legislature created the Nevada Commission on Nuclear Projects (NCNP). Then Governor Richard Bryan appointed Grant Sawyer, former governor of Nevada (1959-1966) to chair the commission, which became known as the Sawyer Commission. Composed of a number of leading citizens, the Commission unfortunately took on the role of kangaroo court. Grant Sawyer and the other members met ritualistically to hear testimony from Bob Loux and the Nuclear Waste Project Office staff, who inevitably touted the failures of the Department of Energy. Then the Commission would pass out judgment against the repository. According to the Commission's 1992 summary report:

**The Commission has been actively studying the federal nuclear waste repository program for over six years. Looking back over that time, it becomes apparent that current national waste management policy and the implementation of that policy by DOE exhibit certain characteristics that reflect serious problems for the country's nuclear waste disposal efforts unless fundamental changes are made. This observation derives from the Commission's experience with various aspects of the federal program and from findings and experience generated through Nevada-specific oversight and research efforts. [Report of the State of Nevada Commission on Nuclear Projects, September, 1992, pix]**

If the oversight and research efforts reviewed by the Sawyer Commission had actually been generated in good faith by NWPO, alarm over DOE's reported lack of credibility would have been warranted. Unfortunately, the Commission was misled by the psychologists of Mountain West and by NWPO's political agenda. Under the guidance of Grant Sawyer, the commission overlooked NWPO's shortcomings in favor of DOE bashing.

Richard Bryan formally announced the appointment of the original members of the Sawyer Commission in November, 1985. The Commission was composed of former governor Grant Sawyer, then Las Vegas mayor Ron Lurie, Commissioner Thalia Dondero, James Cashman III, labor representative Frank Caine, Michdon Mackedon and Ann Pierce. During 1992, Mayor Jan Laverty Jones, County Commissioner Don Schlesinger, and Peter Thomas replaced Lurie, Dondero and Cashman.

Frank Caine appears to be the only one on the board who was not adamantly opposed to Yucca Mountain. Peter Thomas leaned against the repository, but seemed disturbed at the kangaroo court atmosphere of the Commission meetings. Any dissent to the official position is not, however, evident in the reports issued by the Sawyer Commission, which are stridently defamatory of DOE. In fact, so eager were the powers running the Commission to squelch all support for the Yucca Mountain project, that Frank Caine, the labor leader represented on the commission, was not allowed to see the 1992 summary report before it was issued, much less offer a minority opinion.

A chance street-corner conversation between this author, Grant Sawyer and Don Schlesinger (county commissioner and Commission member) after the the Sawyer Commission meeting of July 2, 1992 suggests just how biased the proceedings are. During the conversation, I suggested that some of the material I was researching for this book was highly indicative of problems at NWPO, both in regards to the unregulated flow of money to out-of-state socioeconomic studies, as well to the lack of appropriate scientific rigor of those socioeconomic studies. I pointed out, as an example, that the Commission had just received two hours of testimony from William Freudenburg, a rural sociologist, who spoke on the subject of engineering risk analysis and Murphy's Law. These were fields in which Freudenburg's qualifications are suspect, yet no one voiced any concern.

I was told by Grant Sawyer that I should be spending my time investigating DOE rather than NWPO. The implication was that in Sawyer's eyes, NWPO was incapable of sin. This struck me as an inappropriate attitude for a public servant to be taking, and even Mr. Schlesinger, who has exceptionally close ties to Governor Sawyer, seemed taken aback.

Other statements from the 1992 Sawyer Commission report are also enlightening:

**Likewise, the Commission considers research on socioeconomic and environmental risks associated with a Yucca Mountain repository to be compelling. Nevada's unique tourism-dependent economy is perhaps more vulnerable to disruption by the impacts associated with a nuclear waste repository than that of any other state . . . . [Report of the State of Nevada Commission on Nuclear Projects, September, 1992, x]**

Of course, the theory that socioeconomic doom awaits Nevada due to negative nuclear imagery associated with the repository has been forwarded most forcefully by Paul Slovic of Decision Research, who had been a researcher with NWPO since 1985 as part of the Mountain West socioeconomic studies. Slovic made a career out of getting his nuclear imagery theories published in multiple venues, including NWPO reports, articles in the journal Science, articles in the Journal of Risk Analysis, etc. until his views seemed to predominate. Of course, the fact that Slovic worked for NWPO and was funded with millions of dollars tells us that his subsequent publication in various journals are not necessarily independent works, nor are they therefore "compelling" as indicated by the Sawyer Commission. Our earlier discussions of Slovic's analysis of negative imagery with rebutting arguments from Basset and Hemphill are given in the earlier Chapter 19, Fear Sells.

It is interesting to note that the Commission report of 1992 cites two articles in its Attachment IV, and Attachment V which supposedly voice independent academic concern over Yucca Mountain and its effects on Nevada. Attachment IV, Time To Rethink Nuclear Waste Storage (Issues in Science and Technology), is by James Flynn, Roger Kasperson and Paul Slovic, all contractors to Mountain West who we've met before in various guises. Attachment V, Perceived Risk, Trust, and the Politics of Nuclear Waste (Science, 13 December 1991, Vol 254, pp 1603-1607), is by Paul Slovic, James Flynn, and Mark Layman, also part of the original Mountain West studies. Consequently, the state was citing its own paid contractors as independent academic opposition. It seems deceptive for the State of Nevada to imply that there is nationally academic outrage over Yucca Mountain when these academic papers are entirely self generated.

The Sawyer Commission may have come to believe its duty was to reshape national energy policy:

**The experience with the current flawed waste program in Nevada has contributed to the emerging realization that a policy reassessment and a change in direction are needed. The lessons learned in Nevada, by Nevada researchers, policy makers and others, can help shape a more successful and publicly acceptable waste disposal strategy for the future. The State has sought to communicate those lessons to national policy makers and others who will determine the course of the country's nuclear waste efforts after Yucca Mountain. Likewise, researchers and technical experts working within the State's oversight program have promulgated their findings and insights through the professional literature, where it is beginning to be noticed and used for informing incipient policy decisions. [Report of the State of Nevada Commission on Nuclear Projects, September, 1992, xi]**

**The Commission is correct that notice has been paid the State's researchers and their inordinate impact on policy making. NWPO's undisciplined scientific oversight of Yucca Mountain makes it unlikely such an unguided policy missile will again be allowed to dominate America's energy policy. The movement to avoid a repeat of the Nevada experience is already apparent in the structure of New Mexico's approach to the Waste Isolation Pilot Project, which takes advantage of in-state talent to conduct socioeconomic impact assessment.**

**The Waste Isolation Pilot Plant bill (HR 2637) that passed Congress in October of 1992 was strongly influenced to not follow the example of the Nevada Commission on Nuclear Projects. The funding of socioeconomic studies in New Mexico is being done through the University of New Mexico system. The Socio-Economic Impact Assessment Group (SEIAG) within the Waste-management Education and Research Consortium (WERC) in New Mexico is designed to guarantee researchers will be predominantly in-state professionals with a stake in the outcome.**

**Unlike Nevada, New Mexico has accepted benefits for accepting the WIPP project, amounting to over $400 million over the next twenty years. Fears of nuclear waste disasters and negative nuclear imagery have not been inordinate as predicted by Slovic et. al. from NWPO and New Mexico appears to be well on its way to adjusting to the WIPP project.**

**The Sawyer Commission made three policy recommendations in 1992 for the following two year period:**

**The commission urges the legislature, governor, local elected officials and the state's congressional delegation to continue and maintain strong and unified opposition to the Yucca Mountain Nuclear Waste Project. Further, that no actions be taken that could be interpreted to imply that state policy is in any way open to change.**

The Commission urges the governor, legislature, congressional delegation and others to actively pursue a reassessment and redirection of national nuclear waste policy.

The Commission concludes that the agency for Nuclear Projects is effectively carrying out statutory responsibilities for independent oversight, state and local government coordination, and public information. Further, the Commission recommends that the governor and legislature continue to provide state funding for important agency objectives.

[Report of the State of Nevada Commission on Nuclear Projects, September, 1992, xiii]

In a nutshell, this is a call for Nevada to refuse benefits, obstruct the study of Yucca Mountain through the efforts of NWPO and obstruct national energy policy by imposing Nevada's views on the country. These are far from modest objectives the Sawyer Commission has set for itself.

## GRANT SAWYER'S AGENDA

Perhaps Grant Sawyer's main contribution to his Commission was his lawyerly ability to act as inquisitor of anyone who testified in a positive manner about Yucca Mountain. As an example, an exchange at the July 1992 Commission meeting between the former Governor and Bill Andrews (previously of Science Applications International Corporation and later of the Harry Reid center for the environment) shows the prosecutorial flavor:

➢ Chairman Sawyer: "You are involved with a contractor to the DOE?"
➢ Mr. Andrews: That is correct.
➢ Chairman Sawyer: Are you at all concerned about the credibility of the DOE.
➢ Mr. Andrews: Yes, Sir - I am.
➢ Chairman Sawyer: Have you told the DOE your concerns about their credibility?
➢ Mr. Andrews: That is right, I have.
➢ Chairman Sawyer: How have you done it?
➢ Mr. Andrews: And in the area where I deal with them - I don't always agree with them - I'm sort of a straight-on-person, you know. If the Emperor has no clothes, he has no clothes.
➢ Chairman Sawyer: I can see that and I appreciate it. Have you given the DOE the sort of material you are giving us today? Questioning the credibility of the Nevada Office (NWPO)?
➢ Mr. Andrews: Yes I have.
➢ Chairman Sawyer: You have? Could we have a copy of that?
➢ Mr. Andrews: Uh - I guess - state your question again - maybe I misunderstood?
➢ Chairman Sawyer: Well today you have given us written material (interrupted by Mr. Andrews)
➢ Mr. Andrews: Questioning the credibility of DOE?
➢ Chairman Sawyer: Questioning the credibility of the Nevada Project Office - you said you had also expressed the same concerns to DOE, who is your employer. . . (interrupted by Mr. Andrews)
➢ Mr. Andrews: Not in writing - I have, I have asked - I have and it may appear self-serving since I was managing the umm, the rail access development, but I recommended to them that would continue - for several reasons - one there's a lot of local governments involved now and it's a job that needs to be done. SAIC no longer has responsibility for that work, so I can say that.
➢ Chairman Sawyer: Okay. You're concerned about the credibility of DOE - you say you orally have told them that you're concerned about the credibility?
➢ Mr. Andrews: That is correct. Because I want them, in my area, which is transportation, I make sure that the words come out right. So I get after them in the same way, if I think somethin's wrong.
➢ Chairman Sawyer: Okay, well we appreciate your input Mr. Andrews, very much. We will file your remarks.
[Minutes of the Meeting of the Commission on Nuclear Projects, July 2, 1992]


This was the only verbatim section of the July 2, 1992 Commission on Nuclear Projects meeting transcribed in the minutes, omitting Mr. Andrews other more telling and coherent remarks about the failure of NWPO to conduct adequate oversight. This particular testimony was transcribed because of the Sawyer Commission's preoccupation with trying to compromise proponents of Yucca Mountain as either stooges of the DOE or perhaps even liars. Grant Sawyer, as a lawyer, was adept at using prosecutorial techniques to confuse anyone speaking before the Commission in favor of Yucca Mountain. Through a combination of interruption, irrelevancies, limitations on speaking time and by questioning the character of the speaker, Mr. Sawyer was able to confuse speakers and divert attention from accusations that NWPO may not be perfect. Sawyer used similar techniques to disrupt the September 1992 presentation of Carl Gertz, head of DOE's Yucca Mountain Project Office.

In the spring of 1994, Gov. Sawyer suffered a stroke and the Sawyer Commission was not afterwards active a participant in the battle over Yucca Mountain. However, the release of Sawyer's autobiography, *Hang Tough*, in 1994 gives some further insight into his views on Yucca Mountain.

**Nevada produces no nuclear waste of its own, yet the most expensive political campaign in the history of the state is going on now to assure us that radioactive waste is nice and safe, and that we should relax and enjoy it." [laughter] But the bottom line is, if nuclear waste is all that safe, why don't the states where it comes from keep it? We are talking about a ten thousand year situation here, longer than the history of civilization! I have asked them, "Well, tell me, what was Yucca Mountain like ten thousand years ago? How can you assure us that it will be safe for ten thousand years?" The DOE-paid scientist's position is absurd: as Senator Kerry of Nebraska said, "They lie a lot, the Department of Energy." Of course they do. It's all very interesting.**

**It finally dawned on us that the federal government was fronting for an extravagantly profitable private industry! It isn't tax dollars, but utility dollars that pay for every bit of what's going on. Congress is permitting itself to be used and financed by a private industry to promote what they call a "federal" project. Well, Congress may let itself be used in this fashion, but Nevada doesn't necessarily have to be its victim. Unfortunately, some people see all this money floating around, and they get in line. No vote our commission has ever taken on the dump has been unanimous - everyone has been five to two against it, and two members who are for it wonder about the rest of us. They think Nevada should deal with the Department of Energy, with the utilities; they recommend that we accept the dump and get what we can.**

**Our commission began recommending to the governor and the legislature certain actions to protect Nevada. Governor Miller has been terrific; and so have Senators Bryan and Reid, and Congressmen Bilbray and Vucanovich. The state legislature has enacted legislation that our commission proposed, virtually every county in the state has passed a resolution against the dump, and most (not all) elected officials have been opposed to it. We have a pretty strongly unified voice in Nevada. If we can sustain that unity, it will be very hard for the government to force this project on a state that just will not take it.**

**Nevada may be a wilderness, as Senator Johnson would have it, but there are a million people here who have a lot of pride, and we're not going to be patsies for the federal government; nor will we suffer for the mistakes that the utility companies have made over the years. . . .**

**[Sawyer, Grant; Hang Tough, University of Nevada Oral History Program, 1993, p214]**

The key element to note about Sawyer's views is the strong strain of anti-big-business and anti-DOE populism that ties the argument together. This is at least as strong a motivation in Sawyer's opposition to the repository as the possible dangers of radiation. Ironically, if private industry had been allowed to pursue its own solution to the nuclear waste problem without Congressional intervention, Nevada might very well still have been chosen as a site for a repository. In that case, the state might have had even less input into the siting process than it presently enjoys.

# 42. City & County

It has been clear for decades that the actual socioeconomic impact of Yucca Mountain on Nevada (specifically Las Vegas, Clark County, Nye County and the other counties surrounding the site), will in fact be slight or perhaps even positive. The peak number of workers to be assimilated is in the range of 11,000, a number easily absorbed by the Las Vegas community now approaching two million. (See accompanying Table) The load is also minimized by the downsizing of the Nevada Test Site. The minimal impact has actually been softpeddled by the DOE and its contractors as a goodwill gesture, since the benefits that could be obtained by the Southern Nevada community in part depend on their being a demonstrated economic loss.

For local governments, Yucca Mountain is at heart a battle over property rights; i.e., the Federal Government's right to put a repository on government owned land over the objections of local government. Traditional constitutional interpretations would give local government no right to limit the use of Yucca Mountain as a repository, unless it could be shown that some unusual and permanent damage would result for which there was no compensation. However, considering the actual dangers of Yucca Mountain versus the perceived dangers, it is apparent that the health and economic impact of the repository are trivial. In fact, Yucca Mountain may even prove economically beneficial to the area even without subsidies.

The small size of standard socioeconomic effects, like population displacement, led NWPO researchers to emphasize special socioeconomic effects, like adverse risk perception, in their studies. Instead of the actual impact of Yucca Mountain on Southern Nevada, an emphasis was placed on proving that fear of Yucca Mountain would devastate the area. The counties surrounding Yucca Mountain found this approach little help in determining the actual impacts of transportation routes and demographic shifts necessary to calculating the cost of mitigation. Consequently, the counties have conducted their own socioeconomic studies. Rather than being forced to deal with Yucca Mountain through the state, the counties have been relatively independent of NWPO interference, interfacing with DOE through its Affected Units of Government program.

Two prominent political opponents of Yucca Mountain within Las Vegas and Clark County were County commissioner Don Schlesinger and Las Vegas Mayor Jan Laverty Jones. The two became influential in the Yucca Mountain debate because of their inclusion on the Nevada Commission on Nuclear Projects, the Sawyer Commission, overseeing the Nuclear Waste Project Office. Schlesinger lost his primary bid in 1994 and his clout diminished within the Commission. Mayor Jones, in her unsuccessful run for the Democratic gubernatorial nomination in 1994, was persuaded to take a much more practical look at the issues and while remaining opposed to the repository, seemed willing to negotiate for benefits. Within the City Council, Frank Hawkins and Arny Adamson became counterweights to the anti-repository position.

There are a number of counties and cities who receive funds from the Nuclear Waste Fund to study and mitigate the socioeconomic impact of Yucca Mountain. The use of these funds has led to mixed results.


## LAS VEGAS

The original problem for the city of Las Vegas in relation to Yucca Mountain was not planning for the repository, but figuring out how to spend the money it received from NWPO. Funds started flowing shortly after 1983 when Bob Loux informed the city that grants were available. Although the original grants were small, in the hundreds of thousands of dollars per year, the money lacked a purpose and much was poorly spent.

Since there was little real work for the Las Vegas Nuclear Waste Division to do, the funding glut at first went to purchases of extraneous supplies. Furniture and computers which went unused sat in the corridors of city hall, or were frequently borrowed by other departments in the city. This didn't absorb the entire surplus, so one of the prime means of soaking up funds was to take trips to conventions. A lack of controls meant that after representatives of the city returned from these fact finding missions, few reports were made to update other members of the staff. Little was learned that could help the city should a repository someday show up on the Southern Nevada doorstep. Perhaps the most irritating aspect is that there is now a gag order on the Las Vegas repository technical staff which prevents them from making comments at forums on siting issues, neutralizing their effectiveness.

Extravagance with nuclear ratepayer funds isn't unusual, in fact the City of Las Vegas has a history of ill spent money. Locals will recall the High-Speed Mag-Lev Train fiasco, the Minami Towers hole, the Main Street Station disaster, potentially the Fremont street renovation and many other planning disasters. Interestingly, there was a time when the city entertained the possibility that benefits might be available from Yucca Mountain:

### LURIE RETURNS FROM WASHINGTON WITH VISIONS OF SUGARPLUMS

**Las Vegas Mayor Ron Lurie said Friday that it is time for Nevada to begin putting together a "Christmas tree list" of things they want from the federal government if the nuclear waste dump is placed in Nevada.**

**Lurie, who met with several members of Congress in Washington earlier this week, said he made it clear that he strenuously opposes placement of the repository at Yucca Mountain, 110 miles northwest of Las Vegas.**

**But he said he also discussed compensation he would want local governments to receive if the dump is built at Yucca Mountain.**

**"On the one hand you don't want it,' Lurie said. "You're going to do everything you can to prevent it. But at the same time, they say, 'You better tell us what you want or you're not going to get anything.'"**

**City Manager Ashley Hall, who accompanied Lurie to the meetings, agreed that it is time to begin discussing compensation. "The worst possible position we could have would be to fight it tooth and nail to the bitter end and then have nothing when it does come," Hall said.**
**RJ March 25, 1988 p1c Diane Russell**

Later, when Jan Laverty-Jones became mayor and a member of the Sawyer Commission, the City's position swayed even further against the repository. Her Yucca Mountain Man television advertisements and support for Citizens Against Nuclear Waste In Nevada set the tone of the city's opposition. Only later in 1994 did the mayor take a pro-negotiations stance, likely in search of union endorsements for her bid for the Democratic gubernatorial nomination. Yet, others on the city council were not quite ready to adopt the mayor's position wholesale. In a city council meeting December 16, 1992, a committee was put together by council members Arnie Adamson and Frank Hawkins to study Yucca Mountain. The fact that this sidestepped a city office already devoted to this issue indicates the councilmen realized it was time to begin preparing for the inevitability of nuclear waste repository despite the mayor's opposition.

## CLARK COUNTY

One person in local government who did an admirable job in regards to preparing for Yucca Mountain Dennis Bechtel, Clark County Nuclear Waste Division Coordinator. Clark County officially opposed the siting of Yucca Mountain, through a resolution sponsored by Paul Christiansen in the mid 1980s. Bechtel was "Supportive of City Council, but tried to maintain objectivity." [Interview, 12/27/93]

This objectivity led Bechtel to diverge from the state in the key area of developing in-house socioeconomic expertise rather than relying on out-of-state consultants like Mountain West. Consequently, Clark County had an experienced staff on hand with ties to the local community to "do future planning, reacting to new situations gracefully."

Bechtel's opinion was that "Standard effects don't seem to be that great in the valley. However, Indian Springs and Good Springs might see disproportionate growth due to the railroad spur." His concern was that "DOE might not build the railroad spur," to route shipments around Las Vegas.

Bechtel had raised objections to Mountain West's Section 175 report in 1988, which he didn't feel addressed the needs of the community for solid socioeconomic analysis. Given the mindset of State and city officials, Bechtel's objectivity was no small feat. Bechtel had been skeptical of the "special effects" studies on risk perception done by the state, but a case in New Mexico which awarded damages on the basis of risk perception regarding WIPP changed his feelings somewhat, although he continued "Sorting out the issues."

## NYE COUNTY AND AFFECTED UNITS OF GOVERNMENT

Clark County is the most populous affected county, but Yucca Mountain actually sits in Nye County and transportation routes would also affect Mineral County and others. DOE conducted meetings with the counties through the so-called Affected Units of Government giving the counties input into the siting process. The counties also ran their own divisions devoted to repository issues.

While the counties received preliminary funding to allow them to conduct studies into the impact of Yucca Mountain, the critical question has been whether they could receive tangible benefits. Part of the process of determining such benefits were the Grants Equal to Taxes studies which evaluated Yucca Mountain as a lost taxable resource. Since these studies were highly objective, in the end the Grants Equal to Taxes issue was resolved with the Department of Energy through a process more akin to horse-trading than definitive economic analysis. In early 1994, Nye County was awarded $35 million in benefits, answering at least one small part of the compensation question.

For other counties, transportation and emergency preparedness are the critical issues. These questions may only converge on solutions when final decisions are made on the size of transportation casks to be used. If the nuclear industry moves to large Multi Purpose Units transportation casks, conceivably weighing as much as 125 tons, this will force the building of a railroad spur and better define the impacts on other rural areas of Nevada.

# 43. Nevada Legislature

Although Nevada's legislature is bicameral and operates fairly close to the federal model, it differs from most states in that it meets only every two years. Biannual sessions combined with the independent streak of Nevada's population and the state's low population density give these sessions a unique populist flavor. In Nevada's citizen legislature, debate over Yucca Mountain has taken many unexpected turns, events which serve as windows into the dominant political power structure within the state.

The 1993 legislative session saw the introduction of Senate Concurrent Resolution 57, which illustrates the legislative dynamics in play on the Yucca Mountain issue. A relatively innocuous statement, the resolution said:

**"We support the efforts of the legislative committee on nuclear waste to clarify the compensation and benefits due to the host state for accepting the repository."** [ S.C.R. 57]

Adamant opposition to this resolution in the legislature led to its lopsided defeat at the end of the session. However, the attempt to stifle all debate on Yucca Mountain speaks volumes about the tactics and goals of the opposition armies.

To understand what happened at the 1993 legislature, it's necessary to backtrack to 1989, two sessions previous. That year, the legislature passed two resolutions, AJR-4 and AJR-6. AJR-4 was a resolution that expressed adamant opposition to the repository coming to Yucca Mountain. The vote on that resolution was 19 to 2 in the senate and on the assembly side, the vote was 36 to 5 with 1 abstention. The companion resolution was AJR-6, which said that Nevada would not approve of the withdrawal of land by the federal government for placing a repository there.

Over the next years a number of things convinced the nuclear industry that they might attempt a counter resolution in the 1993 legislature. First, the Nevada Initiative, the public relations and political campaign sponsored by energy industry trade organizations beginning in 1991, had muted at least some of the opposition to negotiating for benefits from Yucca Mountain.

Lobbyists hired to represent the nuclear industry as part of the Nevada Initiative beginning in 1991 had tremendous impact on legislators. The industry was precise in who they hired and picked Nevadans known to be very powerful lobbyists. Ed Allison, representing the American Nuclear Energy Council, had old family ties in Northern Nevada and had lobbied for a number of different interest groups in the state, splitting his time between Washington D.C. and Carson City. Two former legislative leaders also lobbied for ANEC: Jack Jeffries who previously was majority leader of the assembly, representing labor and Bob Berengo, a former speaker of the assembly and a powerful lobbyist.

One reason the industry thought they had a chance to pass a resolution in the 1993 session was that the Nevada budget was stretched to the limit. Rather than raise taxes to cover a $130 million deficit, Governor Miller made retroactive cuts in the 91-93 budget, chopping everything from mental health services, to university funds, to K through 12 teacher loads. In an atmosphere in which the electorate was saying no to taxes, Miller was certainly under pressure to provide alternative revenue sources. The nuclear power industry's lobbyists pushed the position that there were millions of dollars that the state would receive in return for being the host to the repository, perhaps the entire shortfall. This began to have a lot of appeal for a number of politicians who didn't want to raise taxes, but who also didn't want to have to cut popular social programs.

Labor unions in the state had come out wholeheartedly for the repository, seeing it as a source of jobs, especially in light of the looming closing of the Nevada Test Site. Legislators opposed to Yucca Mountain argued that the two issues were not comparable and that the nine thousand jobs lost at the Test Site would not be replaced by the few thousand jobs at Yucca Mountain for study and construction. They also argued that only a few workers would be needed to watch nuclear waste after the site was completed.

Labor is traditionally tied to Democratic candidates, so unlike in the 1988 legislature, this began to weaken the opposition to the repository that was held so strongly by Democratic elected officials. Consequently, some politicians who before may not have been for the repository began to be less afraid to come out and say they were willing to accept the site. Because the political heat had been so hot previously, proponents had remained quiet in the background, but now they began to come forward on the front line.

In the senate, opposition to the resolution was led by Ernie Adler, from Carson City, and Matt Callister and Dina Titus from Southern Nevada.

Dina Titus, Nevada senate minority leader and UNLV political science professor, gave a short recap of the legislative battle against Resolution 57 to members of Citizen Alert, at a UNLV meeting September 21, 1993. The speech contained significant admissions by Titus that ANEC's campaign and the work of its legislative lobbyist, Ed Allison, had been effective. This contradicted the party line of Bob Loux, the governor and Senator Bryan that the political alliance was holding firm. According to Titus:

**"So anyway, this has gone on since '89 up until the current session. And you can feel the difference. You can see all those that are out there, people aren't quite as excited about it. More people are being swung over to the other side. So they figured, well, let's give it another shot. Let's see if we can turn around those resolutions, that were passed two sessions ago, and that's exactly what they tried to do during this session." [Nev. Sen. Dina Titus; speech to Citizen Alert, 9/21/93]**

Titus' speech also pointed to the effectiveness of Citizen Alert's mail and telephone campaigns against the pro-Yucca Mountain resolution and the lack of a countering force from proponents of Yucca Mountain:

**". . . .And we got busy when we knew it was coming trying to round up votes. To vote no on whatever resolution they came with, regardless of what it said. Now we had a lot of help in doing this that came from y'all. Now it came in the form of these awful green cards. Whoever chose this color, it was a smart choice because you can't not see these. And all up and down the halls, and in everybody's offices, and all the mailboxes you saw these constantly. I mean, these are all the ones I got (about two hundred), and they told them not to bother to send them to me, because that was preaching to the choir, I was already on their side, send them to people where they might have a chance. But these make a tremendous difference, so I thank you for doing that and I compliment you on that because it was very effective. . . . . " [Titus; 9/4/93]**

Interestingly, the anti-Yucca Mountain forces tried their best to crush all debate on this issue. When the resolution was introduced, Titus made the motion to table the measure, feeling that would kill it. Senator Bill Raggio, the Republican Senate Majority Leader, called in a caucus of Republicans with three Democrats and the motion to table failed.

Speaking for the resolution were Joe Neal, Ray Schaeffer, Lawrence Jacobsen, Tom Hickey, Ray Rawson, and Bill O'Donnell. Senator Neal, a gruff representative from North Las Vegas and himself a test site worker, led the charge:

**". . . Today, I am reminded of Shakespeare's Julius Caesar when he was accused of political ambition, which was a crime among the Romans at that time. Marc Anthony was allowed to read Caesar's will, after his assassination, which stated that he had left all of his property to the people of Rome, which in the eyes of the people of Rome made Caesar noble again. Brutus, who took over the reins of government, became the enemy of the people and found himself on the battlefield with Marc Anthony's and Octavius' armies marching toward him. And knowing he had missed his opportunity for greatness said these words: "There'll come a tide in the affairs of men which taken at the crest would lead to good fortune, omit it and all the voyages of our lives will be left wallowing in the shadows." This is what this debate, involving the Yucca Mountain repository is to determine, whether or not we have missed the tide of benefits.**

**We have allowed certain individuals to take center stage and tell us what benefits we could receive. We thought for a while that they were correct, but now we are beginning to find that it is possible that there were greater benefits than those of which we have been told. . . ." [Senate Daily Journal, July 1, 1993, Remarks from the floor regarding S.C.R. 57]**

Speaking against the resolution were Dina Titus, Ernie Adler and Matt Callister. The negative hyperbole offered by Senator Titus rivaled the literary eloquence of Senator Neal:

"My final and most important reason for opposing this resolution is that we must not take any action which might be interpreted as implied consent. . . . . Well this is one of those issues where you can't have it both ways. You either take the thirty pieces of silver or you do not. You either barter away the future of this state or you fight to save it.

"I strongly urge you to do the latter. Because if you don't and this resolution passes, we might as well build a statue on the border of Nevada, similar to one that stands in New York Harbor, welcoming thousands of immigrants to this country. But, instead of a lady with a beacon who cries out "Give me your tired, your poor, your huddled masses longing to breathe free, we should make ours a giant garbage can and inscribe on its base "Give me your vile, your poisonous and deadly toxins, yearning to roam free". Because if you pass this resolution, that is the message you will be sending forth from Nevada for generations to come. Please, please think long and hard before you condemn Nevada to such a fate." [Senate Daily Journal, July 1, 1993, Remarks from the floor regarding S.C.R. 57]

Obviously, this was a very emotional debate. The house was packed, with members from the assembly side, lobbyists and the press. When the vote was taken, the resolution was defeated fourteen to seven. This was rightly viewed as a tremendous victory for the opponents of Yucca Mountain. ANEC attempted to claim a minor victory, arguing that at least the resolution was heard. This was not much of a victory, for it demonstrated that the political opponents of Yucca Mountain were, in the words of Dina Titus, going to "vote no on the resolution they came with, regardless of what it said."

Indeed, the need to defeat the nuclear lobby at all costs was so intense that even members of Senator Bryan's staff entered the fray. According to Titus in her speech to Citizen Alert:

"Maureen Lockhart from Senator Bryan's office. He just kind of put her on assignment in Carson City. And she was down there, we were tracking everything, every possibility, what was going to happen, she was running around with a little vote count trying to see who was going to vote how." [Titus; 9/4/93]

The propriety of Senator Bryan's staff member lobbying at the state level was dubious and questions were raised about whether Maureen Lockhart ever registered as a lobbyist. This does not diminish the fact that an extensive grass roots and savvy political effort was expended by those opposing Yucca Mountain to make sure they crushed the nuclear industry-backed resolution. Unfortunately, Titus et..al. succeeded in crushing the nuclear resolution not by winning the substantive debate over Yucca Mountain, but by closing off discussion of the issues. Not only did they legislatively defeat resolution S.R.57 which asked for an investigation of possible benefits, but they also succeeded in cutting off debate in public forums. Titus acknowledged her fear that the industry's education campaign, especially its open tours of the repository site, might actually work:

"You've had a tremendous public relations effort by the industry, at trying to shape public opinion, trying to screen public opinion, to be more favorable and thereby affect the legislature to be more favorable. And you see this manifested in a number of ways. In all the forms of the media; you've heard the ads on the radio, you saw those terrible ads on TV where the person is holding the nuclear pellets saying this won't hurt. Remember those? The newspaper ads, big ads, full page ads placed in all the newspapers by ANEC. Tremendous amounts of money spent on this kind of advertising. You've also seen in the stepped up number of visits that DOE and contractors are taking to the test site. Now just about anybody can go out there and catch a bus and ride out to Yucca Mountain and tour the place. This is a PR effort to familiarize you with it, so you'll see how desolate it is. To capture you on a bus and give you their side of the story. That's all that is, and they're available to any group, anytime you want to go. " [Titus; 9/4/93]

Apparently, Titus and others in the opposition hoped that citizens would not take the tour of the repository, perhaps believing ignorance is bliss. Stifling debate is not particularly fair or democratic, though it is an effective and legitimate political tactic. Indeed, Titus and her allies against Yucca Mountain targeted Bill O'Donnell, Joe Neal, Lawrence Jacobsen, Tom Hickey and others who supported the ANEC resolution and campaigned against them on this issue. The Yucca Mountain debate has remained in the center of the political stage in the legislative elections.

# 44. Senator Harry Reid

Senator Harry Mason Reid early on made a decision to support solar energy at any cost, and that included the cost of dismantling Yucca Mountain. In part this was because Reid has always followed the polls where the votes are, but that analysis is simplistic. Financial self interest could also play a part since his former properties in Searchlight, Nevada, benefited from being on a solar energy corridor. Reid's devotion to solar energy, no matter its origins, became so all encompassing that it significantly hurt the energy future of Nevada.

Of course, poll following can lead to poor legislation. An early example was when Reid jumped on the Alar pesticide bandwagon and helped lead the scare campaign against apples. The Natural Resources Defense Council decided in 1989 to demonize apples by claiming school children were being given carcinogens in the form of Alar, a preserving agent that was used to prolong the shelf life of fruits. After the scare hit the television show, 60 Minutes, it was suddenly politically correct to be an apple scaremonger. After sniffing the political winds, Reid began denouncing Alar to the press, even cosponsoring a bill to have Alar banned.

**Nevada Sen. Harry Reid was among a bipartisan group of senators who introduced legislation Wednesday to ban the apple ripening agent Alar. . . The bill would ban the sale and use of Alar, also known as daminozide, on domestic and imported food products.**

**"We should not feel sorry for Uniroyal," Reid said, saying the sole manufacturer of Alar can continue to sell the product for use on non-food products such as flowers. [ Las Vegas Sun (UPI) 1989]**

The only problem was, Alar wasn't dangerous and the economic losses to the apple industry because of the reaction of activists like Reid rose to more than $150 million. Actually, Alar is a carcinogen, just as is nearly every other chemical at some dose, but apples already contain a number of natural carcinogens more potent than Alar and Europe continues to use the chemical without ill effect. Reid's populist inclination to take sides on environmental issues in search of votes is not thus benign, but has potentially devastating economic side effects.

So what do apples, Alar, Harry Reid and the Natural Resources Defense Council (NRDC) have to do with Yucca Mountain? The NRDC, key to the Alar scare campaign, was one of the organizations opposing Yucca Mountain, specifically under the umbrella of the Safe Energy Communication Council. The NRDC is also an active opponent of the Nevada Test Site, having at one time helped one Charles Archambeau with a million dollar contract to study seismic events from nuclear weapons testing. Archambeau's company, Technology Resource Assessment Corporation, later became the recipient of a $1.3 million, non-competitive contract issued without Request-For-Proposal by Bob Loux and NWPO.

Robert Pollard of the NRDC are long time anti-nuclear activists with ties to Marvin Resnikoff and Ralph Nader's groups and showed up frequently at various nuclear hearings. These relationships demonstrate that opposition to Yucca Mountain is not an isolated grass roots movement, but part of a larger activist environmental movement spearheaded by professional alarmists such as those at the NRDC.

The tenacious hold of the Washington based apocalyptic environmental groups over Nevada's Senators very much affected the political resolution of Yucca Mountain. Harry Reid was caught up in the Alar issue by following the lead of the NRDC. Washington based environmental groups affiliated with the Safe Energy Communication Council, influenced Richard Bryan's proposed Corporate Average Fuel Emission bill and certainly helped fuel their opposition to Yucca Mountain. Both Reid and Bryan are lawyers lacking technical expertise to decide on their own what constitutes legitimate environmental threats. Without a scientific education, Reid and Bryan were forced to rely on other lawyers and non-technicians among the Green movement for critical analysis of environmental hazards. Consequently, the hazards Reid and Bryan attacked (including Yucca Mountain) were often exaggerated and their solutions Utopian.

The largest perceived hazard in Nevada has for a long time been Yucca Mountain. As it became clear to Reid that the Nevada Test Site would be phased out, taking with it many jobs, the senator was forced to scramble to show concern for his constituents by replacing the lost revenues. Bryan had locked Reid into the position of refusing all benefits for Yucca Mountain, leaving little room to maneuver as the 1992 senatorial election approached (Reid need not have worried, his Republican challenger, Demar Dahl, was politically inept). From a basket of dubious make-work suggestions for revitalizing the Test Site, Reid chose to champion a solar solution resurrected by the Green movement. The magical idea was that the Nevada Test Site could be transformed into a giant solar farm and Nevada could be at the forefront of a new age of renewable energy.

During 1992 Harry Reid suggested alternative uses for the Nevada Test Site:

**"His (Reid's) favorite is a hydrogen research center that he said "could revolutionize the world."**

**"I've talked to (Democratic vice-presidential candidate) Al Gore about this. We know it works. This could be one of the biggest programs in the history of the country." The program, aimed at producing hydrogen fuel that could power most vehicles that now rely on dwindling petroleum supplies, ties in with Nevada's vast potential for solar energy.**

**Rows of solar power dishes, much like satellite television dishes, would line barren areas at the test site, 65 miles northwest of Las Vegas. Using electricity that could be generated from the dishes, water could be electrolyzed, splitting its components -- hydrogen and oxygen -- to draw off hydrogen, a clean burning fuel.**

**So where does Nevada get the water for the hydrogen?**

**"Maps have been drawn up to get water from the ocean up here," Reid said.**

**"We know it works," he said about hydrogen production process. "They can produce enough solar energy at the test site for the world using solar energy. It is so vitally important for our country."[Review Journal, September 14, 1992]**

A small problem with Harry Reid's solar-hydrogen plan was that after a hundred years of research hydrogen technology still has a number of high hurdles to resolve, despite Al Gore's claim. Among the many problems:

- There is a process called hydrogen-embrittlement which causes pipelines and storage tanks for hydrogen to crack. Also, although hydrogen has a high energy content per mass, it has a low energy content per volume, making it bulky, difficult to store and hard to use. Making a hydrogen energy system prototype is consequently far removed from making the whole country run on hydrogen technology.

- The environmental impact on Nevada would be huge, much greater than Yucca Mountain. Solar projects are extremely resource intensive and making the Nevada Test Site a solar site would be similar in effort to paving 250 square miles of the earth's surface with concrete. The surface ecology of the Nevada Test Site would be totally ruined by a solar project, much more so than even from the nuclear testing done there.

- Millions of tons of materials would need to be mined to provide the necessary aluminum, concrete (cement and gravel), copper wires, piping, conduits, support structures, etc. necessary to build a hydrogen solar project. Thus, the national environmental impact is significant.

- Many of the mirror silvering compounds, anticorrosion paints, lubricants, etc. are toxic. On a scale this large, the chemical pollution would be substantial.

- The mirrors would create a microclimate by changing the albedo (reflectiveness) of the earth's surface.

- Hydrogen tanker trucks are dangerously explosive (unlike trucks carrying radioactive waste). There would be large numbers of these tankers on Nevada's roads.

- Large quantities of water would be needed. If derived from salt water, it would require expensive pumping stations and huge mountains of sea salt would need to be disposed of. If local water was used, it would cut into the already parched allotments for Nevada development. This would dwarf the socioeconomic impact of Yucca Mountain.

- Disruption of the environment from construction could be substantial due to dust.

- Desert tortoises would be displaced. The desert tortoise is endangered, and such a large solar facility would threaten their habitat.

- Radioactive soils left over from bomb testing at the Nevada Test Site would be disrupted. This might cause a greater radiation danger than from Yucca Mountain.

- Solar is relatively low-tech. Most jobs created would be low tech, like washing mirrors. For the whole site to be economical, wages would need to verge on stoop-labor.

- There is no guarantee that the federal government will compensate Nevada for socioeconomic impacts.

- Last but not least, this solar project is incredibly expensive, requiring hundreds of billions, if not trillions, of dollars in investment

Reid and Bryan had failed to prepare for the closing of the Nevada Test Site and had opposed benefits from Yucca Mountain. Reid tried to transform this fiasco into a political plus using the lure of solar energy jobs, a policy position later adopted by Bryan in the 1994 election campaign. In a Review Journal article, FUEL COULD BRIGHTEN TEST SITE OUTLOOK, December 20, 1992, Reid was taken to task for his motives:

**Reid bristles at the suggestion his enthusiasm for hydrogen fuel is a desperate attempt to find another use for the test site so it will not be closed.**

**On June 19, Reid introduced a $550 million bill aimed at providing retraining and job search services for test site employees who lose their jobs. The bill went nowhere. Some critics said Reid, who was running for reelection, had stubbornly believed there would not be a nuclear testing moratorium and had acted too late.**

**"That's just not true," Reid said. "For lack of a better word, I've been a leader in developing alternate uses at the test site. Everybody knows that." [Las Vegas Review Journal, "Fuel Could Brighten Test Site Outlook", December 20, 1992**

Actually, everybody does not know that, Reid and Bryan had long ignored the possible closing of the Nevada Test Site and the potentially devastating impact. One of the prime alternative uses for the test site and its workers is Yucca Mountain and associated nuclear technologies, not solar energy, but Reid has steadfastly opposed this option. As for the feasibility of hydrogen technology, we quote from the Review Journal Article:

**Dr. Harry Linden of the Illinois Institute of Technology, who serves on a technical advisory panel to the Energy Department, said the consensus in the scientific community is that hydrogen fuel is a feasible energy source.**

**"The big argument is whether it's time to get our feet wet as far as investment is concerned," Linden said. "Some scientists say we should begin investing in hydrogen fuel now. I am among those who think we can wait."**

**Linden said natural gas, which is much cheaper than hydrogen, probably will be the primary energy source for the near future. But he said hydrogen's time will come "within the next 50 to 100 years." [Las Vegas Review Journal, December 20, 1992]**

The relatively long time horizon for hydrogen technology development made prospects for federal political support unlikely, however that didn't dampen Reid's support for an all solar option. Reid further damaged his credibility by promoting the unproven solar-hydrogen system on the basis of newly elected vice-president Al Gore's expertise:

**"We know it works," Reid said. "We sent a man to the moon with hydrogen fuel. . . ."**

**Reid's trump card may be his close relationship to Vice President Al Gore, who seems certain to be the Clinton administration's point man on environmental issues. Reid said he has talked to Gore twice about the hydrogen fuel concept, once before the election and once after.**

**"He's interested, and his science adviser has been briefed by Lockheed, McDonnell Douglas, EG&G and our staff," Reid said. [Las Vegas Review Journal, December 20, 1992]**

Citing Al Gore, a former journalist, divinity student and politician as an expert in hydrogen energy technology was not particularly confidence inspiring and indeed, the hydrogen economy has been quietly dropped as a talking point. What Reid conveniently forgot to give the consumer (i.e., the taxpayers) the tab for this solar-hydrogen utopia -- three or four trillion dollars! In short, Reid is quick to jump on the newest alternative energy fads, but forgets the ruin his ideas would have produced if followed.

The Obama administration, with the backing of Harry Reid, has continued to push a return to the sorts of energy policies developed during the Carter administration when many of the same Green policy activists were in power. Projects begun in the Carter years like synfuels and the windmills of Altamont, California have been allowed to quietly die over the last decades because they simply weren't cost effective. Unfortunately, solar projects have nine lives and have been resurrected as political payoff to senators like Reid.

## SHUTTING DOWN YUCCA MOUNTAIN

Senator Reid took specially delight that in 2010 Yucca Mountain was theoretically canceled. From his 2011 website:

**I am proud that after over two decades of fighting the proposed Yucca Mountain nuclear waste dump, the project is finally being terminated.**
**The proposal to dump nuclear waste at Yucca Mountain threatened the health and safety of Nevadans and people across our nation. Yucca Mountain, which is 90 miles northwest of Las Vegas, is simply not a safe or secure site to store nuclear waste.**

Absent from the statement was a solution to the nuclear waste problem, which will now have to go through the same multi-decade process to come to yet again a similar dead end. Reid and the Obama Administration thus sloughed off their problem on yet another Blue Ribbon Commission.

### Finding Alternatives to a Flawed Proposal

**I have long worked with the Nevada delegation and other Nevada leaders to put a stop to this flawed plan.**

**In 1982, the United States Congress passed the Nuclear Waste Policy Act instructing the DOE to identify possible sites to build and operate an underground disposal facility for the nation's spent nuclear fuel. In 1984 the DOE chose ten sites to study as potential locations, but after only three years, Congress prematurely instructed the DOE to study only Yucca Mountain. In 2002, Congress recklessly approved President Bush's decision that Yucca Mountain was suitable for nuclear waste.**

**In 2008, the DOE announced that it was raising Yucca Mountain's estimated price tag from $57.5 billion to over $96 billion. Beyond its bloated budget, the Yucca Mountain project faced a laundry list of scientific, technical, public health, legal, and safety problems. The skyrocketing price tag, the steadfast opposition of Nevadans and their congressional delegation, and the growing understanding that Yucca was a mortally flawed proposal have led to the project's demise.**

**The time is long overdue for America to find a new approach for solving the nation's nuclear waste problem. That is why I was joined by Senator John Ensign in proposing the creation of a Blue Ribbon Commission of experts to make credible, scientifically sound recommendations for a new approach to nuclear waste.**

**I am pleased that President Obama and Secretary Chu agree with this approach, and on March 3, 2010, announced the creation of the Blue Ribbon Commission on America's Nuclear Future. The commission includes distinguished nuclear energy experts, geologists, policymakers, and environmental policy experts. The panel will present their final report on the best alternatives to Yucca in early 2012. While this commission prepares its report, I will ensure that Nevada's health and safety are never again threatened by nuclear waste.**

While the cost overruns at Yucca Mountain are to be deplored, some of that extra expenditure is due to the actions of Reid himself. If the true environmental and societal costs of solar energy were known, it is likely those technologies would also require a Blue ribbon Commission.

# 45. The Secretary of Energy

## CLINTON, O'LEARY & RADIATION VICTIMS

When Bill Clinton was elected president of the United States, political insiders in the Nevada Democratic Party thought they had gone to heaven. Now they would be rid of Admiral James Watkins, Bush's head of the Department of Energy, hopefully to be replaced by someone who would staunchly oppose Yucca Mountain and nuclear energy in general.

The betting was that former Colorado Senator Timothy Wirth would take the position of Secretary of Energy considering Wirth's close ties to environmentalists and especially to the Safe Energy Communication Council. Wirth had also been a co-sponsor of Richard Bryan's ill fated Corporate Average Fuel Emissions bill, so Nevada's Senators and Governor Miller believed Wirth would be get the position because of their collective clout with newly elected vice-president Al Gore. After all, Reid had been one of the first sponsors of Gore's aborted 1988 presidential campaign.

In December of 1992, Bill Clinton's choice of Hazel O'Leary as secretary of energy over the Green former senator Timothy Wirth sent chills up the spine of Nevada's congressional delegation. An editorial in the Review Journal stated the political situation tartly:

**Nevada leaders (Democrats) promised us that a vote for Bill Clinton was a vote against the Yucca Mountain nuclear repository. Clinton, the Big Lie went, would be more sympathetic to the good fight than those big-business Republicans, Ronald Reagan and George Bush.**

**Remember these quotes?**

**- If he's elected, Clinton will "take another look" at plans to build a nuclear waste repository at Yucca Mountain, Sen. Richard Bryan said last July.**

**- "Clinton was very attentive and very sympathetic with Nevada' position (on Yucca Mountain)," Bryan said in August after meeting with the Arkansas governor.**

**- Clinton's victory could give Nevadans more influence on the nuke waste issue, Bryan said Nov. 4. "We could wind up with a secretary of energy who does not believe the world has to glow in the dark to make progress on energy dependence."**

**- Clinton's campaign platform calling for a decreasing reliance on nuclear energy is a strong sign that Nevada will be treated more fairly on the Yucca Mountain issue, Gov. Bob Miller said last week.**

**But it was all fertilizer spread by opportunistic politicians and gullible Friends of Bill. [Las Vegas Review Journal, December 22, 1992, Opinion]**

The Review Journal's characterization of local politician's statements promising better times under Clinton as a Big Lie was probably too severe. Nevada's powerful Democratic political machine no doubt believed Tim Wirth, a staunch opponent of nuclear energy, would be selected to head the Department of Energy.

The fact that Clinton picked Hazel O'Leary to head DOE was thus no less than a kick in the teeth to Nevada's top politicians, despite their best spin control to the contrary. Hazel Rollins O'Leary at the time of her nomination was Executive Vice President of Northern States Power Company (NSP) and was in the process of being promoted to President of NSP's Gas Utility. Northern States Power provides energy to five contiguous states in the northern Midwest, a fact made important because of its dependence on nuclear power. As O'Leary stated in favor of nuclear energy before the Senate Committee on Energy and Natural Resources March 31, 1992:

**"NSP relies on the low cost environmentally sound production of nuclear power to meet 35% of the electrical energy needs of its customers." [Statement of Hazel O'Leary Before the Committee on Energy and Natural Resources, U.S. Senate March 31, 1992]**

O'Leary went on to remark about the need for Yucca Mountain in her testimony to the Senate:

**Some believe licensing reform alone is needed to encourage investment in new nuclear facilities. I believe they are wrong. If the waste issue is not resolved, there will be no new investment in nuclear power under present circumstances. It is not reasonable to assume responsible business people will risk billions of dollars to invest in new nuclear plants when there is no place to store spent fuel." [Statement of O'Leary to Senate, March 31, 1992]**

Whatever Clinton's reasons were for choosing O'Leary over Wirth, this represented an abrupt sea change in politics in the newly elected administration. During the campaign, Clinton had spoken of a need to phase out nuclear power, and his vice-president, Al Gore, has long been caught up in the search for a solar utopia (along with his friend Harry Reid).

The supposedly close ties of Harry Reid, Richard Bryan and Governor Bob Miller with President Bill Clinton and Vice-president Al Gore were indicative of just how far environmentalists had burrowed into the political bureaucracy. After snubbing Nevada's loyal delegation, especially with the O'Leary appointment to head DOE, Clinton needed to regain their trust. However in 1994, Clinton further antagonized the Nevada delegation by considering a sin tax on gambling, Nevada's lifeblood. Interior Secretary Bruce Babbitt also proposed large increases in mining taxes and grazing fees. The prickly question this raised was how large a political payoff Nevada's politicians needed to stay on board the Clinton juggernaut until 1996 and the next round of presidential elections? A trillion dollar solar-hydrogen project at the Nevada Test Site may have been the unfortunate answer.

The O'Leary appointment also sent shock waves of a different sort through the pro-nuclear community. O'Leary was ostensibly on the side of the nuclear industry, a lone ally in a sea of environmentally activist Clintonites. However, O'Leary's appointment in many ways acted against the interests of the nuclear coalition, especially on the Yucca Mountain issue. So afraid were the trade lobbyists that O'Leary would be scuttled by the administration, that they gave O'Leary a free ride on a number of issues rather than rock the political boat. Because O'Leary didn't like the acrimonious debate generated by ANEC advertising in Nevada, numerous advertising spots were canceled and there was essentially no public relations campaign in 1993 and 1994. This meant the activists at the Nuclear Waste Project Office and in the anti-nuclear lobby took the propaganda field virtually unopposed.

A series of ads that would have shown DOE scientists explaining technical aspects of Yucca Mountain and inviting the public to tour the site never saw the light of day. The Secretary thought it would look like DOE was endorsing the repository. More troubling for the Yucca Mountain study was O'Leary's decision in late 1993 to make public past radiation tests conducted by the D.O.E. and its predecessors, some of which did not pass modern ethical standards for informed consent.

O'Leary's resurrection of the nuclear radiation victim story was in part a payoff to a number of Democratic constituencies, including Senators Bryan and Reid of Nevada. Actually, the story was already old:

## A LONE VOICE IN THE NUCLEAR DEBATE
### Massachusetts Dem Sounded Alarm In '86

Ed Markey knows what it's like to be a lone voice in the wilderness.

In 1986, his house energy subcommittee released a report detailing 31 U.S. government radiation tests on 695 Americans from the mid-1940s through the 1970's. It drew scant notice at a time the nation was rebuilding its military strength.

"Like casting seeds upon the pavement," recalls the Massachusetts Democrat. "The story dies after the first day."

But now - after disclosures of more nuclear tests from declassified files and a pledge from the White House to investigate - Markey is out front on an issue that has exploded into the national consciousness. [ USA Today, 1/11/94]

One contention was that the radiation tests had been conducted without the informed consent of the human "guinea pigs "and violated the Nuremberg Conventions. This resembles the arguments used by NWPO consultant and science philosopher Kristen Schrader-Freschette to paint Yucca Mountain as an illegal experiment conducted by the DOE at the expense of Nevadans. As old as the issue was and given the lack of solid evidence of harm (some of the victims lived into their eighties), the suspicion was that a large element of politics involved in O'Leary's revelations. Even O'Leary seemed somewhat afraid that the revelations would create a counterproductive reaction:

". . . I believe recent disclosures about this dark and narrow corner of our history will result in a full accounting to those wronged by the experiments, a more open and trustworthy government, and stronger ethical guidelines for today's research.

If I am right, our efforts to "come clean" with the American people will have had a positive effect.

If, however, a distrusting American public points to past government abuses as a reason to oppose medical research, the search for an AIDS vaccine, a cure for cancer or treatments for hundreds of other diseases could be set back terribly. [ USA Today, 2/8/1994, p 11A]

A number of environmentalists, notably NWPO consultant Marvin Resnikoff, have repeatedly warned about minute levels of plutonium from Yucca Mountain. Yet, one of the tests conducted in the 1940's involved injection of plutonium into subjects:

**The researchers believed that all 18 people had terminal illnesses - including Mousso and his Addison's disease - and wouldn't survive beyond 10 years, but the last of the group survived until 1991. Mousso survived until 1984, when he died at age 82; the funeral home called it a "natural death." [" People Injected With Plutonium Apparently Kept Tests Quiet", Las Vegas Review Journal, 2/8/94, p7A, Assoc. Press]**

Thus, radiation testing revelations actually support the position that Yucca Mountain is unlikely to be dangerous to local populations, a least in respect to trace releases of plutonium. The political reason for the radiation victim story may well be more complex than a question of public health, as suggested in an L.A. Times editorial:

**. . . the instant-doom day aspects of the scandal appear largely hyperbole. For years the environmental community has advanced the notion that even small amounts of radiation represent a mystical mega-danger of which society should live in utter dread. Why do they so vehemently warn against what is, at worst, a speculative threat? Because nuclear power plants cannot explode but can emit tiny amounts of radiation. Continuing public dread of even tiny levels of radiation is the enviros' main argument against nuclear power - which has many problems but which, on ecological grounds, may actually be good for society. [Easterbrook, Gregg; "News On Tests Not All Bad", Las Vegas Review Journal, 1/12/94, special to L.A. Times]**

According to an editorial in the Wall Street Journal, it appears O'Leary may have been led to resurrect the radiation victim story through the efforts of Dan Reicher of the Natural Resources Defense Council, one of the professional anti-nuclear activists to whom we have by now become accustomed. Nevertheless, O'Leary showed remarkable resilience to the effort to circumvent her advocacy on nuclear issues (namely the attempt to create a Blue Ribbon Commission). Environmentalists and Nevada political leaders found little comfort in statements of the secretary:

**Responding to questions from (Nevada Lt. Gov. Sue) Wagner, O'Leary said she will oppose an independent review of the department's work at Yucca Mountain if it would mean halting site studies. Last week, Sen. Paul Wellstone, D-Minn., and nine other senators asked President Clinton to appoint an independent group to review all the department's nuclear programs.**
**"I have to go forward with it," she said of the Yucca Mountain work. "I will throw my body in front of The White House or anybody who tries to stop site characterization.**
**"If the request (for review) came unattending that effort to stop the characterization, I would be strongly in favor of it, but I am not in favor of something that's used to simply stop the process" on studies that merely seek to determine Yucca Mountain's suitability as a repository.**
**"It doesn't do a bloody thing else," O'Leary said of the site characterization.**
**"I'm far more impassioned about this than I should be, but you have to understand why. I hear from 34 states," she said, referring to states where nuclear utilities are running out of storage for radioactive spent fuel rods. [Las Vegas Review Journal, Friday, March 25, 1994, p 1B]**

## SECRETARY CHU & OBAMA ADMINISTRATION
Steven Chu was appointed Secretary of Energy by the Obama administration. While toeing the administration line that yet another Blue Ribbon Commission was needed to evaluate the storage problem, he was short on answers as seen in this interview with TechnologyReview.com:

- *Technology Review*: **There's some 50,000 metric tons of nuclear waste scattered among 130 sites across the country. What are you going to do with that waste now?**

- Steven Chu: Yucca Mountain as a repository is off the table. What we're going to be doing is saying, let's step back. We realize that we know a lot more today than we did 25 or 30 years ago. The NRC [Nuclear Regulatory Commission] is saying that the dry cask storage at current sites would be safe for many decades, so that gives us time to figure out what we should do for a long-term strategy. We will be assembling a blue-ribbon panel to look at the issue.
- [We're] looking at reactors that have a high-energy neutron spectrum that can actually allow you to burn down the long-lived actinide waste. [Editor's note: Actinides include plutonium, which can be dangerous for 100,000 years.] These are fast neutron reactors. There's others: a resurgence of hybrid solutions of fusion fission where the fusion would impart not only energy, but again creates high-energy neutrons that can burn down the long-lived actinides.
- TR: Is this to burn up existing waste? Or to deal with waste in future reactors?
- SC: It could be for existing, but mostly for future waste. So we're looking at, instead of the way we do it today, where you're using 10 percent or less of the energy content of fuel, can you actually reduce the amount of waste and the lifetime of the waste.
- TR: What about the existing waste?
- SC: Some of the waste is already vitrified. There is, in my mind, no economical reason why you would ever think of pulling it back into a potential fuel cycle. So one could well imagine--again, it depends on what the blue-ribbon panel says--one could well imagine that for a certain classification for a certain type of waste, you don't want to have access to it anymore, so that means you could use different sites than Yucca Mountain, [such as] salt domes. Once you put it in there, the [salt] oozes around it. These are geologically stable for a 50 to 100 million year time scale. The trouble with those type of places for repositories is you don't have access to it anymore. But say for certain types of waste you don't want to have access to it anymore--that's good. It's a very natural containment.
- TR: Waste you know you don't want to reprocess.
- SC: Yes, whereas there would be other waste where you say it has some inherent value, let's keep it around for a hundred years, two hundred years, because there's a high likelihood we'll come back to it and want to recover that.
- So the real thing is, let's get some really wise heads together and figure out how you want to deal with the interim and long-term storage. Yucca was supposed to be everything to everybody, and I think, knowing what we know today, there's going to have to be several regional areas.
- TR: That will deal with some of the transportation problems.
- SC: Right. It makes it less of a problem.
- TR: I know you've come out in favor of nuclear power. It's been decades since any new plants have been constructed. What progress has been made so far in getting some new plants built?
- SC: We're now going to a two-step licensing. You license the generic plant, and then there's a separate license for the site. And this helps speed along the process. Before, the way we did it is every plant was a new one.

# 46. Al Gore, God & Energy Policy

Al Gore in accepting the vice-presidency during the Clinton Administration also claimed to be the environmental torchbearer for the nation and set the tone of nuclear opposition for decades to come. Gore's motivating force appears to be a religious environmental pantheism similar to that developed in Nevada within the peace movement and anti-Yucca Mountain protesters. His beliefs have had substantial impact at Yucca Mountain because he was viewed by the environmental movement as the likely leader of a Blue Ribbon Commission to review all nuclear technology.

Gore visited Nevada with his ally Harry Reid during his unsuccessful 1988 run for the Democratic presidential nomination. Reid had been one of the first to endorse Gore and the candidate appeared ready to repay the favor by reconsidering the Yucca Mountain site:

**Gore criticized the government's handling of the nuclear waste dump siting process, which culminated last year with the congressional decision to put the high-level repository at Yucca Mountain, a portion of which is at the test site.**

**"I did not believe it was handled correctly," said Gore, who will be forty next month. It's obvious to me that they weren't following procedures in the law."**

**Gore, whose home state once was threatened with an interim waste storage facility, said he would work as president to "look at the process from start to finish from geologic and scientific principles rather than political principles." [Ralston, John; Las Vegas Review Journal, 2/21/88 p1A]**

Although former Vice President Al Gore has been one of the leaders of the anti-Global Warming movement, his thoughts on environmental issues have been less than meticulous.  Gore has no technical credentials other than having run numerous symposia with environmental philosophers and global warming enthusiasts during his tenure in the senate and as vice president  Yet his impact on our lives has been substantial.

Gore's undergraduate training in journalism and graduate work as a divinity student predisposed him to scientific voyeurism rather than concise analysis, as demonstrated in his well known documentary "An Inconvenient Truth". This lack of formal scientific training did not stop him from earlier writing what became an environmental best seller, *Earth In The Balance*, which outlined an ecological utopia achieved through a theological interpretation of technology:

**In my own religious experience and training - I am a Baptist - the duty to care for the earth is rooted in the fundamental relationship between God, creation, and humankind. In the Book of Genesis, Judaism first taught that after God created the earth, He "saw that it was good." In the Twenty-fourth Psalm, we learn that "the earth is the Lord's and the fullness thereof." In other words, God is pleased with his creation and "dominion" does not mean the earth belongs to humankind; on the contrary, whatever is done to the earth must be done with an awareness that it belongs to God. [Gore, Al; Earth In The Balance, Plume / Penguin, 1993, p244]**

This is no minor restatement of orthodox religious doctrine, but a major rift with traditional Judeo-Christian views. What Gore was suggesting was that man is not the principle creation on earth, in a theological sense, but is second fiddle to a "holistic" earth. Since every technology man uses, from making toothpicks to building structures like Yucca Mountain, is an effort to stake dominion over a wild earth, what Gore is suggesting is that every activity humans engage in becomes a moral question. If Gore were solely a lone Baptist philosopher in the wilderness, this view would not be controversial, but as as former vice-president this environmental theology has real consequences.

For example, the vice-president also expresses his view of nuclear energy in his book:
**In my own view, the present generation of nuclear technology, light water-pressurized reactors, seems now rather obviously at a technological dead end. The research and development of alternative approaches should focus on discovering, first, how to build a passively safe design (whose safety does not depend on the constant attention of bleary-eyed technicians) that eliminates the risks of current reactors, and second,**

**whether there is a scientifically and politically acceptable means of disposing of - in fact, isolating - nuclear waste.**

**In any event, the proportion of world energy use that could be practically derived from nuclear power is fairly small and likely to remain so. It is a mistake, therefore, to argue that nuclear power holds the key to global warming. Nevertheless, research should continue vigorously . . . . the emphasis in the short term should be on conservation and efficiency . . .[Gore, Al; Earth In The Balance, Plume / Penguin, 1993, p328]**

Gore was right that later generations of nuclear reactor designs needed to be pursued. Thus it was a non sequiter for Gore to see minimal hope for nuclear energy worldwide while pinning its future not on its scientific practicality but to whether a political solution can be reached. At least this was an agnostic stand on nuclear compared to Secretary of Energy Hazel O'Leary's decidedly pro-nuclear background. Because the anti-nuclear forces in Nevada and nationally realized that O'Leary could not be easily neutralized within the Clinton Department of Energy, they attempted to sidestep the Secretary. Calling for a "Blue Ribbon Commission" to independently investigate Yucca Mountain, military waste, and in essence the entire nuclear issue, the environmentalist's true intention was to cut the pro-nuclear O'Leary completely out of the process, creating an omnipotent commission they hoped would derail all nuclear technology.

The environmental and NWPO lobby wanted Gore to run the Blue Ribbon Commission because they viewed the vice-president malleable on the nuclear issue. Gore's belief that science is a religious question rather than an objective scientific debate is similar to the religious perspective of the Nevada Desert Experience. Al Gore's religious philosophy thus bears on how he might lead a review of Yucca Mountain, merging religion and science.

In Earth In The Balance, Gore further hypothesizes a Heisenberg uncertainty principle between science and religion that gives some indication of where his line of thought might lead:

**Yet science itself offers a new way to understand - and perhaps begin healing - the long schism between science and religion. Earlier in this century, the Heisenberg [Uncertainty] Principle established that the very act of observing a natural phenomenon can change what is being observed. Although the initial theory was limited in practice to special cases in subatomic physics, the philosophical implications were and are staggering. It is now apparent that since Descartes reestablished the Platonic notion [of separation of mind from matter] and began the scientific revolution, human civilization has been experiencing a kind of Heisenberg Principle writ large. The very act of intellectually separating oneself from the world in order to observe it changes the world that is being observed - simply because it is no longer connected to the observer in the same way. This is not a mere word game; the consequences are all too real. The detached observer feels free to engage in a range of experiments and manipulations that might never spring to mind except for the intellectual separation. In the final analysis, all discussions of morality and ethics in science are practically pointless as long as the world of the intellect is assumed to be separate from the physical world. That first separation led inevitably to the separation of mind and body, thinking and feeling, power and wisdom; as a consequence, the scientific method changed our relationship to nature and is now, perhaps irrevocably, changing nature itself. [Gore, Al; Earth In The Balance, Plume / Penguin, 1993, p253]**

The Heisenberg Principle applies to subatomic particles at the limits of the physical laws exhibiting quantum effects that have no application to the human condition. Thus, Gore has the unnerving ability to create pseudo-science by mixing religion, philosophy and vaguely understood scientific principles. A Blue Ribbon Commission on nuclear technology and Yucca Mountain chaired by vice-president Gore would likely be driven by similar metaphysical musings.

# PART VI:  BACKROOM POLITICS

# 47. Washington Elite

The legislative road to constructing a national nuclear waste disposal policy has been long and torturous, pitting federal bureaucracies, local governments and citizen activists against the U.S. Congress. These efforts first produced a comprehensive policy in the form of the Nuclear Waste Policy Act of 1982, followed by the 1987 Amendments Act known as "Screw Nevada."

Reviewing the history of the Washington elite who have shaped policy so far thus gives insights into the probable direction revisions will take in 1995.

## DEVELOPMENT OF NUCLEAR WASTE POLICY

The Energy Research and Development Administration (ERDA), successor to the Atomic Energy Commission, proposed a plan for disposal of high-level and long-lived commercial nuclear fuel wastes in 1975 and 1976. The plan proposed six sites be built by the 1990s, based on the then projected needs of several hundred reactors in service. However, by mid 1977 with nuclear reactor orders declining and with President Carter's decision to defer reprocessing of nuclear fuel based in part on nuclear deterrence issues, the waste program began to receive an overhaul.

White House energy adviser James Schlesinger came to feel that the ERDA siting process was too wide spread and a magnet for un-needed political opposition. Colin Heath, then director of ERDA, in turn prepared a policy memorandum which limited the siting to six states with salt formations plus Hanford (basalt) and Yucca Mountain (tuff). The Hanford and Yucca Mountain sites already were seen to have some advantages over the salt sites: the plasticity of the salt effectively ruled out future retrievability of the waste compared to the hard rocks of Nevada and Washington and migration of brine created corrosive conditions for canisters.

The Interagency Review Group (IRG) was set up by then Secretary of Energy Schlesinger to review the current options. The group concluded the first repository could be built by the 1990s, but suggested that even using multiple natural and engineering barriers some uncertainty and risk of release of radionuclides would always remain. The IRG made a positive statement that the nuclear waste problem should not be left for future generations. President Carter first made a policy statement based on the IRG findings in February of 1980; however, his weakened political standing led to his policy suggestions being ignored. Nevertheless, Carter's existing policy became the de facto policy of the incoming Reagan administration.

## THE NUCLEAR WASTE POLICY ACT OF 1992

The Nuclear Waste Policy Act grew out of efforts of the 96th and 97th Congresses between 1979 and 1982. The American Nuclear Energy Council was at the forefront in representing the industry's concern that interim storage be addressed and that geologic storage provided the best long term solution to the waste problem. The environmental lobby, spearheaded by Ralph Nader's Public Citizen and joined by the Sierra Club, Friends of the Earth and other Greens pushed their agenda through friendly members of Congress, notably Morris Udall of Arizona. The Department of Energy was eager for a bill because they believed this would take some of the pressure off their management of the siting process.

A number of issues confronted the Ninety-sixth Congress in its writing of the Nuclear Waste Policy Act. One question was whether a moratorium on further licensing and operation of reactors would be imposed until the waste problem had been addressed. Senator Gary Hart of Colorado had proposed that continued use of nuclear power be subject to Nuclear Regulatory Commission findings in 1985.

Another critical issue was the question of whether host states would be given veto power over a siting decision, because an unlimited veto would have likely deadlocked the entire process. As a first compromise, a state's veto was to be overridden only if sustained by at least one house of Congress. Late in the process, under threat of a filibuster by Senator Proxmire, Congress approved a veto provision sponsored by Congressman Broyhill which required both houses of Congress to override a state's veto.

Other issues included the siting and integration of a Monitored Retrievable Storage facility with the geologic waste storage program, the creation of a nuclear waste fund to pay for research and development costs, a "horserace" process for studying parallel sites, Indian issues, host state compensation and a variety of subtle concerns. After significant compromises between the concerns of environmentalists (designed to halt the nuclear industry) and nuclear interests (perhaps overly eager to build a repository without the burden of environmental regulations) the Nuclear Waste Policy Act was passed December 20, 1982 and signed by President Reagan January 7, 1983. Interestingly, the environmentalists deserted the bill as it reached its final conclusion despite having obtained significant concessions on environmental regulations and licensing.

## THE 1987 AMENDMENTS

Implementation of the parallel study requirements of the 1982 NWPA proved problematic because the cost of conducting nine studies simultaneously was cost prohibitive. After the sites were winnowed to three in 1984 (Deaf Smith, Texas; Yucca Mountain, Nevada; and Hanford, Washington) environmental opposition began to focus on these three sites. Intense lobbying efforts by environmental groups preceded the designation of Yucca Mountain as the sole site for study in the 1987 amendments to the Nuclear Waste Policy Act. According to Caroline Petti, now of EPA but then a lobbyist for the National Nuclear waste Task Force, environmental opposition may have been doomed by the the 1987 designation of Yucca Mountain as the sole site because now the problem was Nevada's and national activists no longer saw a threat to their home states.

J. Bennett Johnston was the main architect of the 1987 amendments to the Nuclear Waste Policy Act, the "Screw Nevada" bill. The effect of this bill was to narrow the search for a nuclear repository from three sites (Deaf Smith, Texas; Hanford, Washington; and Yucca Mountain, Nevada) to Yucca Mountain alone. Had this bill only set Nevada aside as the site for a nuclear waste repository, it would certainly have deserved its nickname of Screw Nevada for its forced siting. However, concurrent amendments sponsored by Bennett Johnston contained provisions for compensation to Nevada to the tune of $50 to $100 million dollars per year, one-tenth the Nevada State budget.

In the Bush years, the most important national politicians to watch were J. Bennett Johnston (D. La.), Chairmen of the Senate Energy Committee, John Dingell, Congressman from Michigan and Admiral James Watkins (a former submariner with a Ph.D. in engineering) who was Secretary of Energy. The most important of Nevada's national politicians was of course then Governor Richard Bryan, who in part owed his Senate seat to previous Senator Chic Hecht's waffling on the 1987 Amendments. In contrast, Nevada's three other representatives; Sen. Harry Reid, Rep. Jim Bilbray and Rep. Barbara Vucanovich opposed Yucca Mountain but seemed open to suggestion given the possibility of reasonable political favors in return.

Bryan's three colleagues rarely ventured out front on the anti-Yucca Mountain side. Bryan, on the other hand, was a powder keg, repeatedly antagonizing Bennett Johnson and John Dingell, two of the most powerful Congressmen in Washington D.C. Harry Reid took a different path courting Bennett Johnston's friendship (though this is not something Reid publicized to the extent he does his friendship with Al Gore). Yet, for better or worse, Bryan acted as the crucial linchpin in Nevada's official position towards Yucca Mountain during his senate tenure.

## 1992 ENERGY BILL

Where the Washington story next played out most conspicuously was in the passage of the Energy Bill of 1992. Among the objectives of this omnibus bill was the streamlining of the process for licensing nuclear reactors and provisions for accelerating the Yucca Mountain project. Obviously, there was a certain pressure to pass this bill before the rapidly dying embers of the Bush presidency totally faded. Senator Bryan bitterly opposed the 1992 Energy Bill, going so far as to attempt to filibuster the act. In the end, Bryan was defeated 86 to 4.

A different approach towards nuclear waste and compensation had been taken in New Mexico by Senator DeConcini who helped draft legislation that emphasized the role of New Mexico's university system in technical oversight and especially in doing socioeconomic impact assessment. While socioeconomic studies done within a state's academic community may carry their own political baggage, at least that baggage is subject to the influence of local political and social currents.

As described previously, Senator Bryan attempted to scuttle the benefits for New Mexico as compensation for WIPP at the end of the 1992 session. This may in part have been in retaliation for having his filibuster of the 1992 Energy Bill defeated, and may also have been an attempt to hide grants of compensation to New Mexico from Nevada voters. Bryan had made a continual point within Nevada that no compensation or benefits could be expected, even though Senator Bennett Johnston had made a commitment to a schedule of $50 and $100 million dollar payments parallel to the 1987 Nuclear Waste Policy Act Amendments. Bryan antagonized Johnston, Dingell and others with his obstruction of the 1987 Amendments Act to the extent that Johnston was heard to claim he would never again unilaterally offer benefits to Nevada.

Early in 2002 the Secretary of Energy recommended Yucca Mountain for the only repository and President Bush approved the recommendation. Nevada exercised its state veto in April 2002 but the veto was overridden by both houses of Congress by mid-July 2002. In 2004, the U.S. Court of Appeals for the District of Columbia Circuit upheld a challenge by Nevada, ruling that EPA's 10,000-year compliance period for isolation of radioactive waste was not consistent with National Academy of Sciences (NAS) recommendations and was too short. The NAS report had recommended standards be set for the time of peak risk, which might approach a period of one million years. By limiting the compliance time to 10,000 years, EPA did not respect a statutory requirement that it develop standards consistent with NAS recommendations. The EPA subsequently revised the standards to extend out to 1 million years. A license application was submitted in the summer of 2008 and is presently under review by the Nuclear Regulatory Commission.

The Obama Administration rejected use of the site in the 2010 United States federal budget, which eliminated all funding except that needed to answer inquiries from the Nuclear Regulatory Commission, "while the Administration devises a new strategy toward nuclear waste disposal." However, the NWPA is still the federal law and is not a project of the President and could not be canceled by either President Obama or the energy secretary. In late 2013, a federal court ruled that the Department of Energy must stop collecting fees for nuclear waste disposal until provisions are made to collect nuclear waste.

# 48. Department of Energy

The Department of Energy held the unenviable position of being both the problem, and the solution to the Yucca Mountain war. On the one hand, DOE's cleanup problems at Rocky Flats, Hanford and elsewhere, which it inherited from other older agencies such as the Atomic Energy Commission, were naturally tied to the Yucca Mountain project and raised legitimate concerns about their ability to manage the repository study and design. On the other hand, because of these past indiscretions, DOE bent over backwards to make the science at Yucca Mountain air tight and spared no expense trying to calm the fears of the public through in depth science.

## STRUCTURE AND PERSONNEL

DOE runs the Yucca Mountain project through its Office of Civilian Radioactive Waste Management (OCRWM), established by the Nuclear Waste Policy Act. The organizational chart which governs OCRWM is presently as in Figure 10. OCRWM's responsibilities include the following objectives:

- **site evaluation: characterize and evaluate the Yucca Mountain site as required by NWPA and 10 CFR Part 960 to assess its suitability for a geologic repository.**

- **NEPA compliance: initiate the National Environmental Protection Act process and prepare an environmental impact statement (EIS) in accordance with NEPA and NWPA, to assess environmental impacts of a geologic repository at Yucca Mountain and whether these impacts are adverse.**

- **pre-licensing interactions: conduct pre-licensing interactions with the Nuclear Regulatory Commission as required by NWPA and 10 CFR Part 60; prepare a license application and, if the site recommendation is approved, submit the license application to the NRC as required by NWPA and 10 CFR Part 60.**

- **repository/EBS design: develop designs for the repository and the engineered barrier system (EBS) appropriate to support DOE's site suitability evaluations; if the site is suitable, a license application; the NEPA process; and pre-licensing interactions with the NRC.**

- **site recommendation: if the site is suitable, initiate the site recommendation process as required by NWPA;**

  **[Proposed Framework To Facilitate Planning and Integration of the Geologic Repository Program, prepared for the Director OCRWM/DOE]**

The first significant interim milestone was reached in December 1988 when the Site Characterization Plan (SCP) for the Yucca Mountain site was completed and sent to the Nuclear Regulatory Commission and the State of Nevada for review and comment.

The Yucca Mountain project saw three key managers in its early years. Don Vieth headed the program in the embryonic years during the early 1980s until succeeded by Carl Gertz. Gertz was well respected by his community, but if he had any flaw it was that he had an engineer's outlook and style and was ill at ease in the confrontational politics that had grown around the repository issue. Dr. Daniel Dreyfus was nominated by President Clinton and became the Director of OCRWM in late 1993.

## NEVADA OPPOSITION

NWPO, Senator Bryan and Nevada's politicians spent the decade of the 80s painting DOE as irredeemably mismanaged and unable to conduct science at Yucca Mountain. For example, in the 1993 position paper titled

"Why Nevada Opposes Yucca Mountain" prepared by NWPO the agency referred to DOE's trust problem numerous times.

**Nationally, the Department of Energy has major credibility problems due to mismanagement of waste and contamination at almost all defense facilities.**

**The GAO has documented radioactive and hazardous waste contamination of groundwater, soil and air at 124 of the 127 nuclear facilities managed by DOE.**

**The Department of Energy's credibility is so low, especially with respect to waste issues, that it is probably not capable of carrying out a program like the repository.**

NWPO cited DOE problems at Hanford, Rocky Flats, Fernald Ohio and elsewhere and the GAO's estimate of a $200 billion dollar cleanup as evidence of the department's recalcitrance. One problem with this assessment is that many of the cleanup problems DOE faced were the result of defense projects which were carried out under the pressures of the Cold War and many of the projects date back to the infancy of the nuclear age when environmental concerns were not considered as pressing. At Yucca Mountain, the openness of the civilian project and multiple layers of oversight argued that it is unlikely to suffer from hidden contamination hotspots as older defense facilities operated under clouds of secrecy.

To further address the problem of trust, Secretary of Energy Watkins under the Bush Administration, created the Task Force on Radioactive Waste Management which specifically addressed this issue of trust and ways to improve it [see "Earning Public Trust and Confidence: Requisites for Managing Radioactive Waste", Task Force on Radioactive Waste Management, 1993]. Among the many Task Forces' recommendations were:

- Requiring local residence for all employee, contractor personnel, and National Laboratory scientists who spend the majority of their time working at the site.
- Involving stakeholders in the process of selecting external peer reviewers.
- Favor local industries and firms as sources for supplying goods and services to the program.
- Aim to design a repository system whose predictable performance exceeds by a substantial margin the standards set up by regulators.
- Develop contingency plans should Yucca Mountain prove unsuitable for a repository.

The Task Force on Radioactive Waste Management has been a good faith effort to address DOE's credibility problems. An interesting question is why the Nevada Nuclear Waste Project Office should not be subject to these same trust enhancing rules.

## WHISTLEBLOWERS

While DOE seems as a whole to be carrying out a safe program of characterization at Yucca Mountain, this does not mean it is without critics. Indeed, it is possible to argue that large government agencies are structurally incapable of conducting efficient science and industry. One notable whistleblower was Quality Assurance specialist Don Brown who claimed quality assurance measures at Yucca Mountain were often overlooked (Brown was dismissed from DOE but later settled for an undisclosed sum. Another prominent critic was General Joel Hall:

**"There has been no objective evaluation of this mountain. This is an analytical scam," said retired Air Force Brigadier General Joel T. Hall, a former deputy commander of NATO's 5th Allied Tactical Air Force who worked for DOE as one of its major contractor employees. . . . In letters to Watkins, Hall said, "What is happening here is the development of tailored data, data files, computer models, and analysis to support and validate the selected the selected conclusion that Yucca Mountain is suitable for a repository.**

**It is wholly unrealistic to expect the Department of Energy to spend $6.5 billion characterizing" Yucca Mountain and then simply walk away after serious flaws are found.**

**Study plans and engineering designs strongly suggest that the DOE's site characterization program is, in fact, geared towards building the first portion of the dump, not merely "studying" the site as it claims.**

**[Lecture at UNLV, May 7, 1993]**

General Hall's attack on the Yucca Mountain project was hard hitting and while disagreement exists about the validity of his analysis, his views were nevertheless aired in public and were an independent evaluation of DOE.

## DOE'S REAL PROBLEMS

Opponents often accuse DOE of being "schedule driven" in its study of Yucca Mountain. However, it should be pointed out that all private engineering projects are also schedule driven and that they nevertheless seem to be able to build skyscrapers, airports and other large projects without compromising safety. In part, the Nuclear Waste Policy Act drives the DOE schedule, but a case can also be made that schedules are part of good engineering practice because they force conclusions to be finalized.

Further, it may be argued that DOE's real problem is its bureaucratic structure which allows it to design Yucca Mountain as no private firm would ever do. A private consortium building Yucca Mountain would not only be under time constraints, but would also be accountable for bringing the project in under cost (so they could maintain a profit) and within safety specifications (as part of their contract fulfillment). Part of the strategy that would be employed by a private firm would be to minimize the risk that the entire project could be derailed by design flaws made in the investigative stage and therefore they would tend to proceed in an evolutionary fashion. In contrast, DOE tends to design in an all-or-nothing mode, not licensing until the entire project meets specifications. Attempts to make the project more like a private enterprise can thus have significant positive effects.

For example, in order to acquire subsurface data on the repository level, the DOE has two choices, to either drill numerous test holes into the area (as recommended by NWPO) or to conduct tunneling operations. Test borings from the surface to collect data unfortunately compromise the integrity of the mountain by allowing for possible escape routes for nuclear waste and they also delay final decisions on construction, perhaps indefinitely. Fortunately, DOE began full scale tunnel boring program despite the inevitable claim that the construction of tunnels looks like the building of the site rather than characterization. Nevertheless, tunneling is the best way of investigating the properties of the mountain without compromising its integrity and leads to immediate practical data on site suitability.

DOE has been saddled by law with designing a complete finished project when what is really needed is an evolutionary process in which technical problems are resolved over a period of time through decisive action not encumbered by bureaucratic red tape. Removing the roadblocks to quality science and practical site engineering is more critical than restoring levels of trust than public relations window washing. Freedom to take decisive action is important because DOE needs to resolve a number of technical questions which impact the final design

1) What should the repository look like if it is never to be closed?
2) Will test tunnels be followed by immediate emplacement of test canisters?
3) Will the Multi Purpose Cask concept be accepted before committing to the entire program.
4) Will an MRS facility be part of the project?
5) How might engineered barriers enhance the safety of the site?
6) What provisions need to be taken to implement long term monitoring?

# 49. Nuclear Organizations

In 1953, the nuclear industry created the Atomic Industrial Forum (AIF) to focus on the beneficial uses of nuclear energy. This was two years before the international "Atoms for Peace" conference held in Geneva in 1955, marking the dawn of the nuclear age.

In 1987, the industry divided AIF to create NUMARC and USCEA. USCEA originated in 1979 as the U.S. Committee for Energy Awareness.

The Nuclear Energy Institute (NEI) was founded in 1994 from the merger of several nuclear energy industry organizations, the oldest of which was created in 1953. The 1994 merger included:

- Nuclear Utility Management and Resources Council (NUMARC), which addressed generic regulatory and technical issues

- U.S. Council for Energy Awareness (USCEA), which conducted a national communications program

- American Nuclear Energy Council (ANEC), which handled governmental affairs

- The nuclear division of the Edison Electric Institute, which handled issues involving used nuclear fuel management, nuclear fuel supply and the economics of nuclear energy.

ANEC and USCEA were the most critical advocacy groups/trade organizations for Yucca Mountain during the early years of the project: USCEA for the energy industry as a whole and ANEC concentrated on the nuclear issues. These organizations were reorganized under the National Energy Institute (NEI), but many of their original activities will continue.

The most prominent citizen's group with pro-repository sentiments was the Nevada Nuclear Waste Study Committee, whose operation was subsidized by the USCEA. Although there is a monetary linkage to the nuclear industry, members of the Study Committee have interests that are independent of the industry. In fact, the Study Committee predates the presence of industry lobbyists by a number of years and was funded out-of-pocket by citizens concerned over the safety of Yucca Mountain.

## THE NEVADA NUCLEAR WASTE STUDY COMMITTEE

The first lobbying presence in Nevada on behalf of the Yucca Mountain repository was the Nevada Nuclear Waste Study Committee (NNWSC). Officially neutral on Yucca Mountain, the members of NNWSC are for the most part citizens who support geologic disposal in Nevada with the dual provisos that the site be studied thoroughly and the state be given sufficient compensation.

The Study Committee started by Jack Regan, the late Bob Dickinson, Dave Cooper, Hugh Andersen and a small core of Las Vegas businessman who first met in a garage. The Yucca Mountain safety issue was first raised by Jack Regan at the North Las Vegas Chamber of Commerce in 1984. This evolved into an independent group which only later made connections to the USCEA in 1985 through the efforts of Dave Cooper.

Bob Dickinson, Jack Regan and other Study Committee members traveled the state addressing what they viewed was obstructionism by Bob Loux and the Nevada Nuclear Waste Project Office. The two issues that drove the Study Committee members was the feeling that NWPO was conducting biased science and that their refusal to negotiate for benefits left the state vulnerable.

Bob Dickinson, first director of the NWSC brushed against Judy Treichel's lockstep opposition to Yucca Mountain as far back as May, 1985, at a debate titled "The Nuclear Waste Forum" at Clark County Community College. Representing the PRO side of the issue were Dr. Donald Vieth, then Director of the DOE's Waste Management Project Office and Robert C. Dickinson, for the Nevada Nuclear Waste Study Committee. The CON side was represented by Fran Polk of the Franciscan Center for Peace and Justice, Judy Treichel, coordinator of Clergy and Laity Concerned, and William Vincent, staff member of Citizen Alert. Obviously, battle lines formed early in the Yucca Mountain debate, with anti-nuclear protest and religious groups arrayed against the business interests of the Nuclear Waste Study Committee.

It was Bob Dickinson who later in 1988 provided the most opposition to the curious way in which the Nevada Nuclear Waste Task Force was created out of thin air by Judy Treichel and Citizen Alert. The Nuclear Waste Study Committee had submitted a similar educational plan to the state in 1986 but had been rejected out of hand. Even after having shown such an interest, NNWSC was not even given a Request For Proposal from the State to submit a bid to act as the informational source for the State. The Nevada Monitor reported on this in an article titled "Committee Proposes Sweeping Public Education Program to Nuclear Projects Commission."

**At its February 7, 1986 meeting in Las Vegas, the Nevada Commission on Nuclear Projects (Sawyer Commission) received a proposal prepared by the Nevada Nuclear Waste Study Committee for a comprehensive program of public information and education on the Yucca Mountain repository issue. Committee spokesman Bob Dickinson urged the Commission to take the lead in implementing such a program with its resources. . . .**

**The special collections should be promoted by the state, said Dickinson, "as a resource people can readily access to get the objective and technically accurate information they must have to form a sensible opinion on this vital decision we all might soon be facing.**

**An important role was envisioned for the University of Nevada in creating instructional materials not only for its own faculty and students, but for the general public as well. Community colleges, business and labor organizations and such public-interest organizations as the League of Women Voters would be enlisted in the marshaling of an effective statewide information system. ["Committee Proposes Sweeping Public Education Program to Nuclear Projects Commission", Nevada Monitor, Spring 1986]**

The Nevada Nuclear Waste Study Committee built a membership list of 13,000 and began to flex its muscle under new director "Ace" Robison. Nevertheless, the state and its allies attempted to portray the Study Committee as a stooge for the nuclear industry:

## CITIZEN GROUP A FRONT FOR NUCLEAR POWER
**Who and what is the "citizen" group called the "Nevada Nuclear Waste Study Committee" (NNWSC)? Follow the money trail.**

**In politics we clearly see the link between politicians and money. It is well recognized that finding the source of their funds reveals who influences and controls them. We commonly condemn them as being bought. And so it should be.**

**Whose interests are represented by the NNWSC? Look to who funds the NNWSC. They are funded by the nuclear industry, as recently acknowledged by their northern Nevada information advocate, Patti Smith.**

**The NNWSC is simply a front group for the nuclear power industry. The nuclear power industry's paid representatives on the "Study Committee" state that the "Study Committee" doesn't take sides on the Yucca Mountain nuclear waste dump. The reality is that the "Study Committee" is nothing more than a part of that public relations campaign to sell the dump. The evidence is overwhelming. Follow the money trail. [DeWitt, Dennis; Letters to the Editor, Reno Gazette-Journal, May 31, 1994, p 5A]**

Missing from this analysis was the observation that the Nevada Nuclear Waste Project Office, the Nevada Nuclear Waste Task Force and to some extent Citizens Against Nuclear Waste In Nevada, Rural Alliance and Citizen Alert are all either funded by nuclear waste funds or backed by state politicians. In any event, the Study Committee's most important objective was assuring that Nevada was not eliminated from consideration for compensation in the likely event that Yucca Mountain is eventually commissioned. In fact, the Study Committee's Citizen Advisory board began to develop a white paper on the benefits issue in 1994 as preparation for the expected rewriting of the Nuclear waste Policy Act in 1995. Thus, while the Study Committee obviously served some of the purposes of the nuclear industry, this was more of a symbiotic relation with Nevadans based on making sure the repository study was conducted without while assuring the state compensation.

## AMERICAN NUCLEAR ENERGY COUNCIL INVOLVEMENT
While the USCEA had a strong presence in Nevada from the start of the Yucca Mountain campaign, it wasn't until 1991 that the American Nuclear Energy Council was forced to throw its cards into the pile as well with the start of the Nevada Initiative. ANEC was forced to enter the public relations wars and wage an aggressive advertising campaign to counter the adverse image of Yucca Mountain created by former Governor and now Senator Richard Bryan, NWPO director Bob Loux and the protest establishment. The risk was that endless delays

due to political manipulation could lead to possible cancellation of the site. Public sentiment had been raised to such a fever pitch that the political situation simply couldn't be ignored.

ANEC approached Kent Oram of OIZ Advertising and the lobbyist Ed Allison with the prospect of trying to turn public opinion around. OIZ Advertising was long a premier political campaign consulting agency in the State of Nevada and their choice to lead a pro-nuclear charge was inspired though not necessarily a natural match. Interestingly, Kent Oram was originally against Yucca Mountain and had even helped the city of Las Vegas draft a statement against the repository in the early 1980s.

As one of the most successful political campaigners in Nevada, Oram was an obvious choice for ANEC, especially because of his close ties to Governor Miller. However, Oram wouldn't accept the Yucca Mountain job without first being convinced 100% that he was both fighting on the right side of the war and secondly that he had a chance to win.

Oram's most difficult task with ANEC and USCEA was to turn their tactics around. Engineers, lawyers and accountants are notoriously shy (if not inept) at confrontation politics, especially the kind waged by a lawyer like Richard Bryan. However, in the public's mind if a misstatement isn't vigorously challenged in front page media, it is accepted as fact. ANEC and USCEA, composed of traditional lobbyists, weren't used to this confrontational approach. Bryan, Loux and the environmental groups had consequentially gotten away with murder in the press.

NWPO spared no chance to paint the nuclear industry's position as one entirely opposed to the interests of Nevada. In fact, the only reason there was a public relations battle and a plan called "The Nevada Initiative" was because of the industry's desire to settle accounts with Nevada in a favorable and fair way. Had they wished to merely "screw Nevada" as often was claimed, there would not have been a media campaign nor any need for public relations consultants, the industry would merely have crammed the repository down Nevada's throat through its lobbying clout in Congress. Nevertheless, NWPO claimed in its position paper "Why Nevada Opposes Yucca Mountain" that:

**The American Nuclear Energy Council also is funding lobbying efforts in the State to influence unions, business readers, of the media, legislators and others to support the Yucca Mountain Project.**

**Part of this effort has involved offers to pay Nevada millions of dollars to give up its right to disapprove of DOE activities and its right to mitigation and compensation.**

**Nevada's Governor Bob Miller called this approach "a cynical new strategy to try and buy Nevada's surrender." ["Why Nevada Opposes Yucca Mountain", NWPO, 1993]**

In fact, the nuclear industry supports both benefits and the right of Nevada to retain veto rights. In a letter sent from Ed Davis, President of the American Nuclear Energy Council to Nevada Governor Bob Miller and dated April 19, 1993, Davis made the industry position clear:

**At an April 7 press conference held in Carson City to discuss the Yucca Mountain study, it was stated that the industry has misled Nevada state officials by urging negotiations to accept benefits during the study, because acceptance of such benefits would lead to the state's "implied consent" to allow the repository to be sited in Nevada. This statement fundamentally misstates the industry's position.**

**Rather, the industry believes that the state should be allowed to receive benefits during site characterization as well as retain its right to disapprove the repository following completion of the study and that such right could be made an essential part of the state's negotiations. Moreover, the industry stands ready to support a legislative provision to protect the state's right for disapproval as part of a negotiated benefits package during characterization.**

**[Ed Davis, president American Nuclear Energy Council to Governor Miller, April 19, 1993]**

# 50. Nevada Initiative

The nuclear industry realized by 1990 that their efforts to convince Nevadans to accept the nuclear waste repository were in danger of failing. Even though the study of Yucca Mountain was forging ahead, lack of public support could put the entire project at risk. At this time, Kent Oram and Ed Allison, long time Nevada political consultants and lobbyists, developed what came to be known as The Nevada Initiative which they presented to the American Nuclear Energy Council. The Initiative proposed a multi-level political and media campaign to move the sympathies of Nevadans towards acceptance of the site.

Because funding for the Initiative had to come from a number of utilities, not all of whom were dependent on nuclear power, it was perhaps inevitable that the Initiative was leaked to the press and became a political hot-button. Politicians played the media with Nevada Initiative to create fears of outside interests invading the state. Similarly, the staff of NWPO rarely failed to bring up the supposedly nefarious nature of the Initiative, which they liked to portray as the "Secret Plan". Of course, we have pointed out that the environmental movement also had numerous secret plans to shut down the nuclear industry (see Shutdown Strategies from Public Citizen, 1987), and the fact that ANEC and USCEA might finally be building a political counter attack should have come as little surprise to anyone.

Nevertheless, there is something less than forthright about the existence of any secret plans and the public deserves to know what went into the Nevada Initiative. Space limiting, we have included some of the pertinent sections, as published in the 1992 Report of the Nevada Commission on Nuclear Projects :

**The Nevada Initiative**
**The Long Term Program**
**An Overview**
**Proposal by Kent Oram and Ed Allison**
**To The American Nuclear Energy Council, Washington D.C.**
**September 1991**
### CAMPAIGN BACKGROUND
**The most critical priority at hand for the nuclear industry is ensuring that the process of characterizing the Yucca Mountain, Nevada site continues to move forward without further slippage in the deadline for locating a national high-level waste repository. Since 1986, the industry has seen the repository deadline slide 12 years - to the point where the very future of a national repository is in question and the siting process established by Congress is at risk.**

**However, as a result of a stepped up industry effort during the past year, tangible progress has been made to halt this erosion and keep the program on track. A political beachhead has been established in Nevada, and the campaign to bring pressure on the state to issue permits to allow scientific study at Yucca Mountain is making substantial inroads. The industry message has been focused, influential Nevadans have been recruited to help advance the industry's objectives and a working political alliance has been established with the Department of Energy, natural allies and other key decision makers. Aggressive coalition building is under way, an in-house scientific response team has been recruited, an industry boiler room operation is functioning in Nevada and a dialogue has been developed with the media. A paid advertising campaign will begin this month. . . . . . . .**

### OVERALL CAMPAIGN OBJECTIVES
**1. In the short term, create the necessary political and public climate to allow further site characterization of Yucca Mountain to proceed.**

**2. Within the next three years, build the framework for political, media and public awareness that will allow successful site characterization to be completed and accepted.**

**3. Secure a negative agreement with the states' political and elected leaders that provides for the states' cooperation in the site characterization process in exchange for specified benefits.**

### KEY GOALS

* To manage and meet the significant political challenges posed by Nevada's forthcoming three year state and federal election cycle.

* To successfully take the industry's message to the public to reduce concern over the transportation and storage of nuclear waste and elevate the merits of the repository.

* To lower public and political anxieties, thereby giving Nevada political officials "air cover" to negotiate.

* To finish laying the public and political framework over the next 18 months that will allow a negotiated settlement in 1993, a non-election year during which the Nevada legislature will face a projected state deficit.

* To win acceptance by the Nevada Resort Association, the states' most important source of tax revenues.

* To convince prominent business and union leaders who are quietly supportive, to publicly endorse the repository.

Whether one views the goals of the Nevada Initiative as sinister or not depends on whether one believes the industry was acting altruistically in its fight to continue with nuclear technology, or whether it was callously promoting a failed technology to save its own jobs. Thus portions of the Nevada Initiative take on different meaning depending on one's perspective. Perhaps one of the most controversial suggestions in the Nevada Initiative was for the creation of scientific truth squads:

## SCIENTIFIC TRUTH RESPONSE TEAM

A group of scientists was trained by Kent Oram to function as an effective, expert in-house accuracy response team. This team served both as a proponent of the repository and as a truth squad in responding to scientifically inaccurate, misleading or untrue allegations that were published or aired about nuclear issues or Yucca Mountain.

The use of DOE scientists to publicly address questions of safety about the repository became so controversial that Secretary of Energy Hazel O'Leary put pressure on the nuclear industry to cancel all involvement of DOE spokespeople in their media efforts. While involvement of these scientists in media releases designed by consultants of the Nevada Initiative certainly deserved scrutiny, without access to DOE scientists the end result was that no media information was available without resorting to second-tier scientists unattached to the Yucca Mountain project.

Other portions of the Nevada Initiative were similarly devoted to an aggressive, proactive campaign to convince Nevadans that Yucca Mountain, and nuclear energy in general, were safe and necessary. Viewed by some as a campaign to trick Nevadans into accepting the Yucca Mountain repository, proponents of the Nevada Initiative viewed their efforts as educational. The hoped for end results:

With positive movement in the polls and a more informed media that is less susceptible to hyperbole, the anti-repository movement will find fewer elected officials willing to even sanction their cause, much less give them credibility. Consequently, the opposition will dramatically lose numbers and effectiveness as the domino theory falls into place.

Perhaps the greatest political windfall gleaned by the state and opposition from the release of the Nevada Initiative was the figure of $8,600,000 budgeted over three years to run the ad campaign. In reality, reliable sources suggest only a much smaller sum materialized, around $5,000,000 spread over the three years. As a comparison, the Fletcher Jones car dealership spent about two million dollars per year on television advertising in Southern Nevada alone, and Richard Bryan's political warchest was about $3.5 million for the 1994 elections. Consequently, the actual funds spent on the Nevada Initiative were not grossly out of line with the local market.

In contrast, the effective public relations effort of anti-nuclear proponents may be gauged to have cost a million dollars plus per year. Adding in the donated time of members of organizations such as Citizen Alert, the Safe Energy Communications Council and other environmentalists and the fact that the media, public and political establishment were substantially biased against the project, the costs of the Nevada Initiative do not appear to be extreme. Debate over the ethics of the nuclear special interests in running the Nevada Initiative are somewhat mitigated by similar questions about the funding and motives of the environmental special interests who had also invaded Nevada. After all, Ralph Nader's Public Citizen Critical Mass Energy Project, the Safe Energy

Communication Council, the Sierra Club, Friends of the Earth, Citizen Alert, Southwest Research, Nuclear Information Research Service, ad infinitum could not exist for free and represent an advocacy industry whose anonymous backers' motivations also remain obscure.

It is certainly true that the Nevada Initiative was an attempt by an outside special interest, the nuclear industry, to sway Nevadans to a pro or neutral stance to the repository. Given that nuclear industry outsiders were attempting to mold public opinion in Nevada, it is unsurprising that the Initiative was viewed as manipulative and therefore failed to win great popular support. However, two points make a final analysis of the ethics of the Nevada Initiative less negative than one might first suspect:

1) A main thrust of the campaign was the creation of "scientific truth squads". As much as the opposition hated to admit it, credentials of the scientists chosen for these assignments were impeccable.

2) Perhaps the only people arguing for benefits for Nevada were Kent Oram, Ed Allison and those within the Nevada Initiative. The nuclear industry could easily have walked away from supporting negotiations and compensation given the recalcitrance of Nevada's politicians and the noisiness of the environmental lobby. Indeed, Bennett Johnston and his colleagues in the Senate and House did give up on the idea of compensating Nevada and were only persuaded to continue discussions with Nevada's politicians because of the efforts of those within the Nevada Initiative led by Kent Oram and Ed Allison.

# 51. OIZ Advertising

In the early 1990s, Oram, Ingram and Zurawski Advertising (OIZ) at times became more the focus of the debate over Yucca Mountain than the technical issues themselves. Long known in Nevada as a political consulting firm, in 1991 the American Nuclear Energy Council joined with OIZ and lobbyist Ed Allison to create a public relations campaign in Nevada known as the Nevada Initiative.

For better or worse, OIZ's handling of the ANEC advertising campaign profoundly influenced the perception of the nuclear industry in Nevada for the next two decades. The major problem that OIZ confronted was its image as a "hired gun" of the American Nuclear Energy Council. Kent Oram, the principle partner in OIZ, had originally opposed Yucca Mountain and had even been instrumental in creating a city ordinance stating Las Vegas' opposition to the site. All of a sudden, he'd switched sides.

Naturally, Oram was considered a traitor of the worst kind when he took on the ANEC account. After all, OIZ had helped in the campaigns of many of the state's politicians whose careers depended on defeat of the Yucca Mountain; Governor Miller being the most prominent. Nevada's politicians rightly feared Oram because they knew he had the savvy to hurt them politically if they tried to muscle him too hard. It was bad enough that Oram had sold out the repository opposition in favor of the nuclear industry, but even more threatening was that he was powerful enough to inflict political damage by expose the hypocrisy of anti-Yucca Mountain politicians.

But what were OIZ's true feelings about Yucca Mountain? Had they "sold out" as most pundits and state politicians charged? The problem with that viewpoint, pushed to the extreme by Oram's new-found political enemies, was that it is nearly impossible to fight hard for a cause you don't believe in. However, instead of just picking up their paycheck, OIZ threw themselves into their work with a vengeance. Researching the nuclear issue to its core, interviewing key researchers and visiting nuclear facilities, the staff became convinced that nuclear energy was indeed a valuable and necessary technology, dependent on the building of the Yucca Mountain repository to complete the nuclear fuel cycle.

## DRAWING THE BATTLE LINES

If at first OIZ thought its job was simply to run an innocuous nuclear information campaign supporting the repository, those thoughts were soon put to rest. The political situation surrounding Yucca Mountain was by 1991 so overgrown with political posturing that even minor press releases and nuances in the OIZ ad campaign became bloody slugfests. For example, a hailstorm of criticism erupted over an ad that featured a simulated spent fuel pellet being held by announcer Ron Vitto. The critics claimed this was an attempt to show the pellets were so safe you could hold them in your hand. Actually, the ad had had the modest goal of demonstrating that the pellets were solid, because polls by the survey firm Penn Shoen of New York indicated the public thought the waste could be in liquid or gaseous form. Liquid and gaseous radioactive materials are much harder to safely transport and store, so the whole point was to show the pellets are solid ceramics and not to imply they were safe to hold. Focus groups did not pinpoint any perception that viewers thought Vitto was speaking to the safety of holding pellets themselves.

The fallout from the pellet ad came in the form of vicious attacks on the integrity of on-air spokesman Ron Vitto. Mayor Jones opined that it was all part of a disinformation campaign meant to trick Nevadans. Johnson and Toftee, the KKLZ disk jockeys, sang biting ditty's calling Ron Vitto a sellout. Citizen Alert drew cartoons of Vitto dissolving into a skeleton. Loux and the NWPO staff attacked the ads. The fact of the matter was that the pellets would have taken a number of hours to give a lethal dose, but this was of course a red herring since it is unlikely anyone would knowingly touch a radioactive pellet, even if given the chance. The idea that Ron Vitto would suggest that holding a nuclear waste pellet in one's hand is safe is ludicrous, but this was a war of innuendo, not logic.

Another set of OIZ advertisements which came under attack were those which used file footage to show the robustness of transportation casks subjected to crashes with locomotives, flames and other hazards. Lindsay Audin and Marvin Resnikoff, co-authors of The Next Nuclear Gamble written in 1983, spearheaded the state's attack on the integrity of these ads, claiming the films had been rigged. Of course, Resnikoff is the professional anti-nuclear scientist who started the Sierra Club Radioactive Waste Campaign and with Audin had claimed rigged testing before in England. Consequently, the attacks on OIZ's transportation ads had been preceded by similar campaigns and it is unsurprising that the commercials took a shellacking in the press aided by professional anti-nuclear provocateurs.

Would OIZ purposefully have lied on the pellet ad, the cask testing ads, or for that matter on any other spot they produced, as the state and media accused them of doing? This seems extremely unlikely given Kent Oram's understanding of the dynamics of political warfare. In his words:

**"If you do a political hit, your information has to be a hundred and ten percent correct. If there is even the slightest misinformation in an advertisement, the opposition is going to use it to nail you to the wall!"** [Personal communication May 92]

The attacks against what had been thought to be non-controversial informational ads soon led OIZ to conclude that they weren't confronting disorganized grass roots protesters. Instead they had stirred up a hornet's nest of well-organized professional political opposition complete with its own propaganda mills. OIZ's natural response would have been to go to political war mode and crush the resistance. The OIZ staff had built a sizable collection of information about NWPO and the anti-nuclear environmental movement and certainly had the ammunition to carry out such a war. Indeed, OIZ had been chosen by ANEC in part because they were the big political guns in Nevada and had a proven ability to fight down-and-dirty political battles. Unfortunately, the industry lacked the courage to pull the trigger.

## THE WAR THAT WAS NEVER FOUGHT

As originally outlined in the Nevada Initiative, the war for Nevada's heart and soul was to be a three year, $10 million blitz which not only took the high road of educating Nevadans about the Yucca Mountain repository, but also vigorously countered any misinformation distributed by anti-nuclear forces. Behind the scenes lobbying would also help bring on board the help of prominent citizens.

What actually happened was that the nuclear industry came up with about $2.5 million spread over various stop and start strategies that piddled away their advantage. After the first series of ads run in 1991, a new less-controversial series was to begin in mid 1992 during the election cycle. This was supposed to be a strictly informational advertising campaign in which Nevadans were introduced to both the many ways they benefited from nuclear technology and to the true physics of Yucca Mountain in regard to rising groundwater, volcanoes, etc. Ron Vitto, the announcer for the first series, was to be sidelined because he was viewed as a magnet for criticism.

Hindering OIZ was the micromanagement of ANEC officials who had to approve nearly every step of the ad production process and sit in on what became directionless meetings. Because some of the money supporting the Nevada Initiative was coming from utilities that utilize coal fired plants, it wasn't possible to develop ads that contrasted the belching smokestacks of coal-fired utilities to the clean and pastoral surrounds of nuclear plants. A number of these ads were written and some even produced, but the campaign was halted by a reluctant industry as the 1992 presidential campaign came to a close.

With the election of Bill Clinton, the nuclear industry thought they were about to be destroyed by a Clinton nomination of environmentalist former senator Timothy Wirth to be Secretary of Energy. Industry officials were so relieved at the choice of pro-nuclear lawyer Hazel O'Leary that they felt they could not risk rocking the boat even the slightest with the new Secretary.

OIZ in the meantime developed a new informational campaign inviting Nevadans to the Yucca Mountain site on the well received tours conducted by Science Applications International Corp. Visitor responses to the tours had been highly favorable, with even those opposing the site coming back with an enhanced trust that characterization was being conducted in a careful manner. After extensive filming of a tour group at Yucca Mountain and after equally extensive focus sessions, OIZ thought they had produced about as non-controversial a campaign as could be imagined. However, Secretary O'Leary, also wary of bad publicity and on edge in the Democratic administration, decided Department of Energy scientists could not be seen being interviewed or even in background footage of an industry sponsored ad campaign. This stopped the early 1993 campaign in its tracks because it was impossible to create a convincing commercial without showing testimony from DOE scientists.

Trying to salvage something from this disastrous course, the industry decided to attempt to push a resolution through the 1993 legislature that expressed interest in at least looking at what benefits might be possible from the Yucca Mountain study. The industry pursued this tactic despite advice by both Kent Oram and Ed Allison that the groundwork hadn't been prepared through public relations efforts in the preceding months. Although a last minute ad series was run alongside intense lobbying, the Senate resolution was defeated as one of the last acts of the 1993 Nevada legislature.

## NATIONAL POLITICAL INFLUENCE

An aspect of OIZ's work hidden from most of Nevada was Kent Oram's lobbying influence at both the state and national level. For example, although the ad campaign was going nowhere, Oram was in the background shaping the energy bill of 1992 and its provisions for nuclear energy. Despite public denials, Oram also was in close contact with Governor Miller and the governor's aid, Scott Craigie. In the course of deal making, Oram gained contact with Bennett Johnston of Louisiana and John Dingell from Michigan.

Senator Bryan had done his best to infuriate the powerful Bennett Johnston in the Senate. Senator Johnston with John Dingell in the House and Oram held a common distrust of Bryan's motives. Bryan's CAFE fuel emissions bill sent Dingell to the moon because of its potential disastrous effects on the automobile industry in his native Michigan. Bennett Johnston was made livid by Bryan's refusal to negotiate on the Yucca Mountain issue, no matter what pains were taken to insure the welfare of Nevada. Oram was so incensed with what he considers political opportunism by Richard Bryan on the Yucca Mountain issue, that a plane trip which inadvertently placed Bryan and Oram one row apart nearly led to a nasty confrontation.

**"I talked the entire flight about how sleazy the anti-repository operation was in the State of Nevada, loud enough so Bryan could hear me. He never once acknowledged my presence. When we got off the plane, I called Bryan a wimp, but he wouldn't even turn around." [personal communication, June, 1993]**

Obviously, Oram had become personally involved in the nuclear issue and overreacted by goading Bryan. On the other hand, Bryan had never been shy about baiting the pro-nuclear factions in his many press releases, calling his opposition liars. It's hard to tell who had spoken the first fighting words.

## NUKE LITE 1994

A final attempt to run a professional political campaign was proposed to ANEC in November of 1993. The $1.3 million dollar campaign was to act as a counterweight to Senator Bryan and Governor Miller in the 1994 elections. By pre-producing ads and buying air time far in advance, this would give the nuclear industry the option of either running high-road informational ads if Bryan and Miller were nice, or attacking down-and-dirty in the middle of the elections. Although the campaign was developed and ready to go, it was first delayed by the restructuring of ANEC, USCEA and the other nuclear trade groups into the Nuclear Energy Institute in early 1994. Later, to avoid a political war during the 1994 elections, some agreement was made between the nuclear industry and either Senator Bryan or Governor Miller to not touch the Yucca Mountain issue. Ted Garrish, ANEC's legal representative, referred to this campaign as "Nuke Lite".

Kent Oram's motives for involving OIZ Advertising in what he knew would be a bitter fight-to-the-death over nuclear energy in Nevada are more complex than portrayed in the media. While money played some part, Oram was not exactly poor or hard pressed for accounts. In fact, he and his entire staff became convinced on a personal level that the nuclear cause was worth fighting for, more so than perhaps even the industry executives they worked for. Although Oram thought that the Yucca Mountain account was a challenge, he also thought it a battle that could be won and he had visions of taking the campaign nationally to reinvigorate the building of new nuclear reactors.

Unfortunately, what hindered OIZ's campaign most was not NWPO or the various environmental organizations, but the indecision of pro-nuclear organizations. Nearly every advertising move was nit-picked by industry representatives until commercials became stale and too old to air. The industry was capable of making a principled fight in Nevada, but chose backroom Beltway politics instead and hamstrung any efforts by OIZ to run timely and hard hitting counterattacks in Nevada. Thus, evaluating OIZ's effectiveness in swaying public opinion towards accepting the Yucca Mountain repository is almost impossible. The nuclear industry had never implemented the Nevada Initiative as outlined by OIZ, replacing an aggressive political campaign with dithering and indecision.

# 52. Altamira, Bullfrog County Times

Altamira Communications, a political public relations firm, played the bad cop to the OIZ Advertising good cop in the nuclear industry's repository promotion in the early 1990s. It was Altamira's job to be out front and take the political heat, battling the anti-nuclear forces toe-to-toe. This turned into a bitter public relations war. Though Bob Loux and NWPO denied it, they had their own public relations shock troops, having hired Kamer-Singer Public Relations and a group called Cygnus Scientific to fill the opposing role.

Don Williams headed up the Altamira operation. One of the original founders of Altamira, Williams was known for his ability to attack and prick political opposition. Many people in the community consequently saw Altamira as bullies in the Yucca Mountain battle. When Altamira attacked sympathetic figures like Judy Treichel, they were condemned as heartless. Yet, the anti-nuclear forces in Nevada were hardly helpless despite a cultivated image to that effect.

NWPO regularly spent funds on lobbying and public relations efforts. Citizen Alert, the powerful Safe Energy Communication Council and a dozen or more well heeled environmental groups, some with multimillion dollar budgets, were able to mount coordinated attacks. What black-hat Altamira and white-hat OIZ faced was a well funded and organized professional political advocacy coalition that was ready to fight a long and bitter war over Yucca Mountain. Someone needed to keep the Greens honest and Altamira took this rather thankless task on as its own.

Don Williams attacked NWPO head-on a number of issues that NWPO tried to finesse. When Judy Treichel began sending "informational" material to the high-schools, Williams made an issue out of the bias it contained. It was William's group that hammered Loux on the question of where NWPO money was being funneled, running a letter writing exchange trying to get answers to these money questions out of the Loux fiefdom.

To compensate for media perception already biased against Yucca Mountain, two local reporters were brought on to the Altamira staff to act as liaison with the press and to blunt the attacks of the environmentalists. Brian Gresh and George Knapp, reporters for Channel 8, the local CBS affiliate were hired in November of 1991 for sums claimed to be in the six figures. Unfortunately, the perception that Gresh and Knapp were mercenary hired guns was not lost on either the media or the environmentalists who proceeded to use this to their advantage in press releases. Of course, the fourth estate conveniently overlooked the paychecks being collected by NWPO' staff.

One of the duties of George Knapp and Brian Gresh was to produce a fax attack tabloid called the Bullfrog County Times. Before the age of email, fax machines had become a quick and direct way to reach influential members of the community because they are a precisely targeted information channel. Knapp and Gresh's satiric efforts in the Bullfrog were not without effect.

Bullfrog County refers to an aborted attempt in 1987 by then governor Bryan to carve out a separate county around Yucca Mountain from the already existing Nye County. The purpose of this move was to wring the most concessions from the federal government by creating a special tax district. Naming the fax attack the *Bullfrog County Times* was a none too subtle satire on Senator Bryan's and other state politicians opposition to Yucca Mountain.

Some of the Bullfrog's statements bordered on schoolyard level tauntings of the anti-nuclear cadres. Especially hard hit were Judy Treichel of the Nevada Nuclear Waste Task Force and Bob Loux of the Nuclear Waste Project Office. For example:

**JU-DEE! JU-DEE! JU-DEE!**
**The Treichel Issue**
**Fans of Bob Loux (and who isn't) may be disappointed with the latest edition of the Nevada Nuclear Waste News. According to our official "Loux Watch" tally, the boyish anti-nukester is quoted a mere eight times in the June issue of NNWN. This not only falls far short of Loux's record setting 25 quotes in an earlier issue, it's well below his twelve-quote average.**

The Times is confident that NNWN will make it up to Loux somehow. After all, NNWN publisher Judy Treichel is Loux's main P.R. squeeze. If these two were in high school, Judy would probably be wearing Bob's letter sweater. For those who don't know, Loux is the guy who has awarded nearly a million dollars in contracts to Treichel, including her $337,000 doozie. These fat contracts probably represent the most stupid expenditure in state history. [Bullfrog Times, Vol 1 number 3, June 18, 1992]

The Sawyer Commission also took a lambasting:

Halloween is still two months away but our favorite frightmasters are already gathering together to conjure up their all-too-familiar tall tales of boogeyman, goblins, and other assorted make believe menaces. Yes, the Sawyer Commission is meeting again. These high priests of P.R. hoodoo and scientific voodoo really should consider holding their meetings at night, in the woods, while sitting around a campfire, because their endless stream of "scary" findings are just about as believable as the ghost stories one hears at a summer camp bivouac. And instead of having people like Don Schlesinger or Michon Mastadon serve on the commission, maybe they should appoint Freddy Krueger and the movie psycho who wears the hockey mask. [Bullfrog County Times, Volume 1 number 14]

If the Bullfrog took some liberties with the language, it also asked some hard hitting questions about what was going on in the state, especially probing Loux's control of the Nuclear Waste Project Office:

Loux has always pursued his own agenda, even when he's been in conflict with (Miller) administration policies. You see, Loux is a holdover from the Dick Bryan days, a Bryan mole in the Miller camp. Miller may be the boss, but Loux's loyalties will always be with Bryan. That's because Bobby and his senatorial patron are the baton twirlers in Nevada's never-ending anti-nuke parade. . .

In light of these political realities, Miller has sometimes had to look the other way while Loux uses his office and his $5 million budget to buttress the bank accounts and political activities of his anti-nuke friends, including crusading flower children like Judy Treichel and Citizen Alert. . . .

. . . A few days ago, (Loux) was caught red-handed and red-faced. A report written by his office concerning the nuke rocket project was submitted to the Air Force by Citizen Alert. Word for Word. Pretty nifty, huh? State employees, with salaries paid from public funds, do all the work and then hand over their report to Citizen Alert. Why not make Citizen Alert the official spokespersons for Nevada on all environmental issues? Cut out the middlemen. [Bullfrog County Times, Vol 1 Number 18]

The Bullfrog also came under attack from anonymous sources:

A mysterious plot aimed at muzzling the Bullfrog County Times has been thwarted, and some people are wondering what role state officials may have played in the squalid little drama. If you believe that state officials and state contractors should not use public funds to mail out poison pen letters, then read on.

Someone has mailed dozens of anonymous letters to nuclear industry executives and utility commissioners all over the country. The letters, which were mailed from Oakland, California, complaining about the Times' lambasting of various anti-nuclear crusaders. The phantom author apparently hoped to undercut support for the Times. It didn't work.[Bullfrog County Times, Vol 1 Number 16]

The anonymous letters were likely sent at the request of NWPO staff. The Oakland mail origin fueled speculation that the San Francisco P.R. firm Kamer-Singer under NWPO contract may have had something to do with this (though this is unconfirmed). NWPO was obviously conducting more than science at Yucca Mountain, it was in fact conducting a public relations campaign complete with dirty tricks, something not included in its mandate. This issue was derided in the Bullfrog to seemingly deaf ears:

Loux must have been stifling his giggles when he told his audience (on NWPO's KDWN radio show) that Nevada doesn't spend any of its oversight money on public relations, media, or advertising. Lucky for him the remark was made on radio. Otherwise we could have seen that he was biting his hand to keep from laughing. Nevada isn't spending any money on public relations, huh? What about the $205,000 contract given to a Las Vegas P.R. firm to handle "media relations" What about the other two P.R. firms which were given contracts by Loux? What about the $200 he spends to pay for each of his radio shows? What a jokester he is. He should move to Paris and let the French decide whether he's a greater comic genius than Jerry Lewis. The fact that he would pay for a talk show and then use that same show to claim he isn't spending any money on media is a classic moment in comedy.

Altamira and the Bullfrog County Times played rough in the public relations wars being fought over Yucca Mountain. Yet, we've shown NWPO conducted its own breed of rough politics, with Bob Loux playing the cracker-barrel populist. Altamira's efforts thus may have actually helped balance the public relations wars with the state agency, especially given the flamboyant nature of Nevada politics.

The area where Altamira's effectiveness suffered most was because of its involvement in its reporting on UFOs and "Site 51" (the not-so-secret air force base claimed to house an alien space ship). Discussions with Don Williams of Altamira suggest that this involvement was at least moderately driven by the profit motive, perhaps answering why Altamira sent its reporters Knapp and Gresh to the Soviet Union in 1993 to research stories on Russian UFO sightings. Whether Altamira's UFO reporting had much impact on public perceptions of their Yucca Mountain campaign is hard to tell, however, the media did take some interest:

### RJ-OPINION They Were KidKnapped

**There's nothing quite like ratings week for top-notch investigative TV journalism. Take, for instance, Channel 13's decision to unleash George Knapp, a grand mullah in the Church of Stratospheric Proctology, so he could produce a five-part series on people who claim they've been abducted and violated by bald galactic invaders. Knapp, who has won cash awards from UFO groups for his "reporting," presented the subject with all the skepticism of a wild-eyed toddler approaching Santa Claus. Station honchos even let Knapp go so far as to scare parents with a warning that children are particularly susceptible to being beamed up to the orbiting craft. Now, the fact that there are people who think they've been gynecologically examined by big-headed creatures in spaceships is indeed worthy of study (much like epidemics of medieval peasants who swore they were possessed by demons), but giving true-believer George Knapp 25 minutes of precious air time on the topic is akin to asking Shirley MacLaine to investigate channelers. [Las Vegas Review Journal, Sat. February 13, 1993, p10B]**

# 53. Kamer-Singer, Cygnus Satellite & Radio NWPO

Obviously, the pro-nuclear forces weren't the only ones capable of playing dirty tricks in the public relations wars. In fact the State waged a vigorous campaign that took advantage of a number of foot soldiers: Nevada Nuclear Waste Task Force, Joy Hamann Advertising, Kamer-Singer Public Relations, Cygnus Scientific, and a radio show produced on KDWN. We've already examined the contributions of Judy Treichel's Nuclear Waste Task Force to the war. The late Joy Hamann and her advertising company played a less combative role and were not part of the bloodfest. This leaves three very interesting entities, Kamer-Singer Public Relations, Cygnus Scientific and what we will term Radio NWPO.

## KAMER-SINGER PUBLIC RELATIONS

Kamer-Singer Public Relations of San Francisco was brought on by NWPO to provide the professional muscle to fight the public relations wars. The agency is perhaps best known for helping the Culinary Union in Las Vegas sooth ruffled feathers after tourists were attacked by strikers. The fact that the words Public Relations appears in this company's title highlights the fact that NWPO conveniently ignored the Federal Acquisition Regulations prohibiting such activities by DOE grantees.

Originally, Joy Hamann of Joy Hamann Advertising had the public information contract for the state, but as Hamann's health failed in 1992 Sam Singer was brought on as part of that contract. With Joy Hamann's passing, in 1993 Singer was awarded the full contract for $235,000. Judy Treichel's contract was at the same time diminished from $165,000 to $35,000. Interestingly, Bob Loux rejected results of a competitive bidding process to award Kamer-Singer the later contract, apparently the prerogative of the NWPO director.

Singer had worked for Richard Bryan on his 1988 senate campaign and had been involved in nuclear issues then, helping whip up anti-nuclear passions as a campaign tactic. A former journalist, Singer viewed his objective as a consultant to NWPO as keeping nuclear waste out of Nevada rather than to act as a nuclear information officer:

**"We are up against a formidable opponent, very slick. Our job is to show that most scientists, even nuclear power scientists, don't think it's a safe place to put it. It is only a convenient place to put it." [KDWN Radio, "Yucca Mountain: Fact Not Fiction", sponsored by NWPO, Nov. 24, 1992]**

Of course, most nuclear scientists do believe geologic storage is safe, while Singer himself was not an expert on the Yucca Mountain issue. A caller to Radio NWPO reduced Singer to name calling:

**CALLER: I heard a statement that DOE testing showed water might upwell. But Szymanski has been completely discredited.**
**BAUGHMAN (HOST): Jerry has not been discredited!**
**CALLER: Trench 19 shows proof water never upwelled! If you go there yourself or had any scientific inclinations you would see. It is as clear as it could be!**
**SINGER: Trouble is, it is not so clear!**
**CALLER: Yes it is!**
**SINGER: No! It is not!**
**CALLER: How would you know? You've never been there!**
**SINGER: Bunch of hooey! Bunch of hooey!**
**CALLER: What a phony you are! You've never been there!**
**SINGER; What a phony you are! You know what? You don't have to shoot uphill when you know it's bad for you mister!**

CALLER: Give me a break! You're a politician, not a scientist.

BAUGHMAN: Thank you for calling sir.

[KDWN Radio, "Yucca Mountain: Fact Not Fiction", sponsored by NWPO, Nov. 24, 1992]

Obviously, Singer's display showed him quite capable of playing the political hired gun. Ignorance of physical science did not seem to stop spokespeople for the state from giving technical evaluations to the media and on broadcasts such as KDWN.

## CYGNUS SCIENTIFIC

Another example of NWPO's public relations war was a mysterious advertisement which appeared in the Las Vegas Review Journal in October of 1992. Fashioned to mimic the newspaper ads produced for ANEC, this ad was a rather ham-fisted attempt by NWPO to run a political counter-insurgency campaign against ANEC and OIZ Advertising.

William Bennett, chief executive officer of the Circus Circus gaming conglomerate, had been persuaded to oppose Yucca Mountain and in 1992 donated $100,000 to the state's efforts. Since these funds were a contribution to NWPO and not subject to the guidelines of the grants provided by the Department of Energy, they constituted a political slush fund for Bob Loux's agency. In order to use this money, NWPO contracted with Harry Mortensen, a former Test Site employee who then ran Cygnus Satellite, a company selling satellite dishes to bars and sports pubs.

The nuclear industry first heard of Cygnus Scientific when the mysterious advertisement appeared in the Las Vegas Review Journal. The text of the advertisement follows:

**SHOULD YUCCA MOUNTAIN BE IMMEDIATELY DISQUALIFIED BECAUSE OF RADIOACTIVE CARBON DIOXIDE?**

**Experts on the Environmental Protection Agency's Science Advisory Board after many months of study, have concluded that the geologic media of Yucca Mountain will not contain radioactive carbon dioxide, sufficiently, to meet the standards of the Environmental Protection Agency, for a High Level Radioactive Waste Repository. The Science Advisory Board has analyzed studies indicating that other types of geologic media, particularly bedded salt deposits, would have a very high probability of containing radioactive carbon-14 dioxide.**

**Question**

**Do the EPA Science Advisory Board's findings disqualify Yucca Mountain as a repository?**

**ANSWER: No, because the Department of Energy says it can spend over 3 billion extra dollars for very long life canisters to contain the radioactive carbon dioxide.**

**Question: Then Yucca Mountain will qualify for a High Level Waste Repository?**

**Answer: No, because the Nuclear Regulatory Commission has advised the Department of Energy that " . . . engineered barriers cannot constitute a compensating measure for deficiencies in the geologic media . . . " or " . . .engineered barriers shall not be used to compensate for an inadequate site . . ."**

**Question: Then Yucca Mountain will be disqualified for a High Level Waste Repository?**

**Answer: Probably not, because the U.S. Congress is currently considering a bill (the new [1992] Energy Bill) that will, in effect, force the EPA to allow greater quantities of radioactive carbon dioxide to be emitted from the Repository.**

**Question: What will the consequences be if Congress passes the law allowing more radioactive release?**

**Answer: The bottom line is that Congress will allow more people to die than the Environmental Protection Agency felt justifiable. Under the Environmental Protection Agency's current regulations for the repository, estimates place the most probable number of Fatal Cancer Deaths at 1000 over the life of the repository if the maximum allowable releases occur. If the Congress forces the EPA to allow greater releases . . . then a greater number of people will be allowed to die.**

**If you have any questions about the content of this ad, you may write Cygnus Scientific, 6130 W. Flamingo, #123, Las Vegas, NV 89103**

The nuclear industry wasn't quite sure whether this was a joke or a serious attempt to engage the public through the print media. The address turned out to be a mail box in a Postal Annex. Actually, the question raised by this parody, regarding permissible levels of radioactive carbon dioxide emissions, was a valid issue. Emission standards were changed in favor of higher levels in the 1992 Energy Bill, though the new standards are not by any means permissive. It is interesting to note that the Bill passed 86 to 6, despite Senator Bryan's filibuster.

More interesting is the implication that NWPO was able to direct a political "dirty tricks" campaign through Cygnus Scientific using an anonymous mail drop at a Postal Annex outlet. The Cygnus advertisement's macabre comments about deaths should be put in perspective: if radioactive waste only causes one-tenth death per year, this is negligible compared to the thousands of deaths caused by wastes from hydrocarbon fuel cycles.

## RADIO NWPO

As part of its efforts to derail the repository, the Nuclear Waste Project Office ran a radio show called "Yucca Mountain: Fact Not Fiction", airing for a half hour alternate Tuesdays on KDWN radio. While the listening audience for a show on an AM station was not huge, the content of this show was important as a gauge of where the Yucca Mountain debate was headed. Unfortunately, the show was more propagandistic than an attempt to shed light on the repository; out of 30 shows this author reviewed, only a rare neutral guest appeared and no pro-repository interviews seem to have been allowed. The host for the state, Dennis Baughman, was game enough to take unlimited call-ins, but attempted to deflect all criticism of the state's oversight. Some selected quotes should show the flavor of "Fact Not Fiction":

**April 28, 1992**
**HOST: . . What is your position on earthquakes and Yucca Mountain?**
**JUDY TREICHEL: We have little technical expertise. We just tell the truth and give the public information. There is proof that water did slosh at Yucca Mt. As the paper said, there was a rise and fall in the water table.**

**June 9, 1992**
**CALLER: What are the risks of shipping nuclear waste compared to other dangerous substances like chlorine gas? In other words, what standards are applied to nuclear casks compared to other dangerous materials?**
**RESNIKOFF: That is like apples and oranges. Toxic materials are shipped in tankers.**
**CALLER: I didn't ask that! I said what are the risks of the containers?**
**RESNIKOFF: Generally the risks are different. Chlorine gas versus nuclear fuel. Nuclear gives cancer. There are many more shipments of chlorine gas compared to nuclear.**
**CALLER: I feel more at risk with chlorine. It is very dangerous.**
**RESNIKOFF: It is an additional risk. You get risk with both.**

**June 22, 1992**
**CALLER: Why don't we just cut off the rhetoric and get on with our lives? This is 1992!**
**HALSTEAD (NWPO transportation consultant): I am not taking a position on the Yucca Mt. Repository. I just study transportation issues. The Department of Energy is not willing to adopt maximum safety methods.**

**September 29, 1992**
**BAUGHMAN: The nuclear waste ads seem to be failing miserably. They are galvanizing the people against it, and we believe if we keep up our fight that the nuclear industry will persuade Congress to pull out of Nevada and end this charade!"**

**December 22, 1992**
**CALLER: . . . I think the state needs to work with DOE, actively studying the site and tell us what is really there. They should quit telling the public the site will be built regardless of safety. That simply is not true.**
**LOUX: The state has been involved for ten years. Our scientists tell us the site cannot be found safe. The state cannot get the level of funding to do more. I think the Dept. of Energy's track record has shown, they have not been truthful, and I can provide that evidence to anybody who would like to take a look at it, with**

the state of Nevada, and the people of Nevada, or the country at large, about issues associated with nuclear materials.

Jan. 19, 1993
CALLER: There is already a lot of radiation at the test site. How would you compare the Yucca Mountain thing to the Chernobyl thing in Russia?
BAUGHMAN: It's not the technology I distrust, it's the scientists. Scientists are not known to be accurate in what they have told us over the years. The amount of fissionable material between the test site and Yucca Mountain is like apples and oranges. If you look at a worst case scenario, comparing Yucca Mountain to Chernobyl, Yucca Mountain would be much terribly worse. Even if it didn't explode at the site, when you factor in the shipments going to Yucca Mountain, accidents will occur. It's just a matter of when and where. . . .
CALLER: What are you expecting from the new administration?
BAUGHMAN: It couldn't be any worse. Clinton said that he doesn't want any new nuclear power plants. We hope that he and O'Leary give it a new look. We don't expect miracles, but it should be better.

Feb. 3, 1993
CALLER: You think the storage of nuclear waste is fine, as long as it isn't in Nevada. Could you tell us where you do think it should be stored specifically, rather than giving generalities?
BAUGHMAN: Yea. We believe that most other countries are looking to salt formations or granite formations. There are salt formations in the southern parts of this country. There are granite formations in the New England states.
CALLER: Do you have reports that say these places are better?
BAUGHMAN: I don't know the answer to that. I'm sure there are reports out there that say find a site not so seismically active.
CALLER: I'd have to look at a DOE report. Where would that be?
BAUGHMAN: I don't know.

June 8, 1993
CALLER: I am not for or against the proposed Yucca Mt. dump, but I used to live in Colorado Springs, and a NORAD facility there was built to withstand a direct nuclear hit from a hydrogen bomb. Couldn't the same technology be built into Yucca Mountain to withstand an earthquake or whatever?
BAUGHMAN: Yucca Mountain is seismically active. There are 32 known faults there, the Ghost Dance fault goes through the middle. Scientists will debate the suitability for quite awhile. Can it crack? We don't know. My definition of safe means nobody gets hurt. The DOE says that there is an acceptable number that can get hurt and die. That is built into their definition of safe. It is such a dastardly thing. The potential for catastrophe is so high.

June 22, 1993
MODERATOR: If we didn't have nuclear power, how would other technologies pass modern tests and standards?
TREICHEL: They have cleaner coal through scrubbing and all that. The best answer to energy is conservation; save power. That's a lot of power. There's also thermal power, wind power, hydro power - all sorts of things. A Shearson / Lehman report criticized nuclear power as an investment potential.
MODERATOR: But is that practical?
TREICHEL: It is practical in individual applications; your hot water, your pool, so on. Solar is practical in some areas.
MODERATOR: Aren't taxpayers between a rock and a hard place? You still have cleanup even if nuclear power was ended.
TREICHEL: Most intelligent people I talk to, talk of phasing out nuclear power by not renewing licenses and not building new ones. There are about five plants that shut down already. It makes more sense anyway to keep spent fuel on site at the reactors.

# 54. Media Bias

Unfortunately for both sides of the repository debate, the Las Vegas media embarrassed themselves on the Yucca Mountain issue. Even more disturbing than the bias of some of the journalists (prevalently anti-repository) was the fact that they simply didn't do their homework. The DOE has many faults but no lack of watchdogs, ranging from the Nuclear Waste Technical Review Board, to the GAO, to the bloodhounds of the local media. The organization which had no watchdog was the Nevada Nuclear Waste Project Office, which the media allowed to roam free.

Since high-powered politicians like Senator Bryan, Governor Miller and ex-Governor Sawyer turned a blind eye to the activities of NWPO, it should have been the duty of the fourth estate to keep the state's nuclear agency honest (it isn't as though there wasn't enough material to write a book about it!). Not only did the Las Vegas media fail to verify the credentials of many of the flaky sources provided by NWPO, but they neglected to read NWPO's reports, failed to check peer review findings and in general demonstrated a limited understanding of the scientific peer review processes on which objective science depends.

## LOCAL TV NEWS

One of the most egregious examples of media bias was a series of reports done in 1992 by Dan Burns, then news director of KVBC, the local NBC affiliate. With creative editing, the Burns series gave the impression that interviews with some noted scientists from Sandia Laboratories showed their experiments on nuclear waste transportation casks had been failures. Unfortunately, that is not what had transpired at all.

At the July 2, 1992 meeting of the Sawyer Commission, the state presented the KVBC documentary as evidence that cask designs were faulty. Dr. Robert Luna, who had been interviewed in an unfavorable light in the KVBC spots, gave a complete rebuttal of the Burns' series. Among other things, Dr. Luna told Grant Sawyer and the commission:

**Dan Burns (KVBC) put a little spin on my comments by selective editing . . . . I would never agree with his characterization . . . The spent fuel in the cask was never in any danger of release . . .The casks are safe for the kinds of shipment being planned. . and for things being done in the United States for the last 40 years or so. [statement of Dr. Robert Luna to Sawyer Commission, July 2, 1992]**

The misunderstanding of the cask testing films stems from a misconception about how modern science conducts tests of large scale projects like nuclear waste shipping casks. It is a luxury to be able to test large projects to failure, just as no one tested Boeing 747's to failure by flying a test fleet millions of miles until they crashed. Instead, a few prototypes are made which become the basis for computer models of the system.

Dan Burns and KVBC made much of the fact that cask testing shown in advertisements done by OIZ Advertising for the American Nuclear Energy Council weren't conducted to prove the structural integrity of transportation casks, but to calibrate computer simulations. The connection the technically naive reporter missed was that once calibrated, the computer models could then simulate accident modes barely conceivable in the real world and which would require thousands if not millions of full scale tests to statistically validate. Rather than being a deception, the cask test videos showed only a small part of a much larger validation process that the public, and media, were never aware of because software code doesn't generate splashy footage that makes the evening news.

Fortunately, Burns did not let his aversion to Yucca Mountain impede his career as reported by Jeff German of the Las Vegas Sun:

## PRO-NUKERS LURE NEWSMAN TO YUCCA CAUSE
**The pro-nuke raid on the media continues.**

Another veteran newsman has been ensnared in the clutches of the pro-nuclear forces working to make Yucca Mountain the nation's first permanent high-level nuclear waste dump.

Dan Burns, managing editor at KVBC Channel 3, has announced that he's leaving the station after 10 years to hook up with a group of scientists studying the site.

Burns has accepted a job with Science Applications International Corp., the major contractor at Yucca Mountain. . . .[German, Jeff; Las Vegas Sun, Dec 1992]

Burns was by no means the only TV journalist who took liberties with the Yucca Mountain issue. Another video reporter who dealt loosely with the facts was Tanya Ellis, also of Channel 3, KVBC. For example, on the 10/24/91 report at 6pm, Ellis ran disk jockeys Johnson and Tofte radio spots which spoofed pro-Yucca Mountain ads, then interviewed mayor Jan Jones saying simulated pellets displayed in the ads would kill a person in a matter of minutes, and then opined that ANEC announcer Ron Vitto's arm would have to be amputated in one minute if he were holding a real pellet. On 10/29/91 Ellis followed with an interview of Lindsay Audin claiming cask tests were rigged and framed her report at the spaghetti bowl highway interchange in Las Vegas, implying a transportation disaster there was inevitable.


## NATIONAL NEWS

Sloppy journalism extended well beyond the state level. Most poignant was a telephone encounter this writer had with Roger O'Neil of NBC Nightly News. O'Neil had given a report the night before (July 14, 1992) which examined the possible impact of a number of earthquakes that had occurred in the California and Nevada area on the Yucca Mountain repository. As usual, the piece was long on political hype and shortchanged the science.

In the report, O'Neil suggested DOE had a credibility problem and that the pro-nuclear advertising was failing and making the matter worse in the minds of Nevadans. An interview conducted with NWPO director Bob Loux was particularly slanted. In essence, this was a remake of the local KVBC anchor's position and was hardly surprising.

I called Mr. O'Neil to suggest there was more to this story. Because of my research, I felt I could demonstrate Robert Loux was perhaps not as credible a source as O'Neil imagined and that things were not as they seemed at NWPO. O'Neil stridently defended Loux as a representative of the State of Nevada, claiming no interest in any possibility that the state might be hiding some rather large skeletons in its closet. O'Neil suggested we would never see eye to eye on this subject, though what was being discussed was information he'd never heard before and which at the least may have deserved a neutral hearing, if not further investigation on his part. I asked if I could quote him on his position and his reply was "No! You can't quote me on this. You can't quote me on anything!"

## PRINT MEDIA

In the print media, the Las Vegas Sun took a particularly vitriolic editorial stance against Yucca Mountain, led by its owner, Hank Greenspun. After Hank passed away, his son Brian continued the tradition, often proclaiming pride in the fact that his position was locked in cement, no matter what new information might come available.

The ubiquitous Sun reporter, Mary Manning, provided the longest uninterrupted coverage of anti-nuclear events in Nevada from the early 1980's on. If there was a news release by Citizen Alert or a comment on Judy Treichel's involvement in anti-nuclear testing at the Nevada Test Site, Mary Manning was like as not to be responsible for the coverage. Unlike the television media, Mary Manning's articles were written in a recognizable journalistic style and have provided what would otherwise have been a neglected history of these events. Nevertheless, it should be noted that what Manning was reporting on was often the activities of her friends, and lacked some critical technical analysis.

Manning's sources were frequently members of the environmental movement, whose expertise was accepted perhaps too quickly. For example, Manning quoted the Safe Energy Communication Council as a source for various official sounding pronouncements opposing the repository:

### YUCCA FUNDS SLASH SOUGHT

A national coalition of energy, environmental and public-interest groups has called for the Clinton administration to cut $305.7 million from its high-level nuclear waste project until an independent panel reviews the program.

**The Safe Energy Communication Council recommended in its "Sustainable Energy Budget" Wednesday that the Department of Energy stop work and spending for underground tunneling at Yucca Mountain, temporary storage facilities, transportation, quality assurance and other programs. [Manning, Mary; Las Vegas Sun, Nov 18, 1993 p5A]**

The question is why Manning reported the announcements of this out-of-state organization whose members had little if any technical expertise as an authoritative source. Perhaps it is inevitable that on a subject as emotionally charged as Yucca Mountain, even a solid reporter like Mary Manning would slip from being a reporter to an advocate. Nevertheless, Manning, who at one point planned to write a book on the Yucca Mountain issue, provided a much needed look at the human side of the repository campaign.

In contrast, Keith Rogers of the Las Vegas Review Journal brought a drier style to his reporting on Yucca Mountain than Manning. Rogers ongoing approach was neutral and technical and because it was less controversial he has not been quoted as frequently in this book. This should be thought of as a recommendation and not as ignorance of his work.

## EFFECTIVE REPORTING OF SCIENTIFIC ISSUES

The complexity of the technical issues at Yucca Mountain has exposed a major flaw in the ability of our media to act as watchdogs over major technical issues. It appears an effective science reporter needs to have at least some minimal expertise in the subjects they cover, otherwise investigative questions become meaningless jabbering more likely to confuse than enlighten. The two main technical groups at Yucca Mountain, NWPO and DOE, are both governmental bureaucracies whose political structures render them secretive and overly political. Thus, reporters relying on experts from these groups inevitably found themselves reporting on the politics of Yucca Mountain, even when they thought they were reporting on the scientific substance.

Questions a media more in tune with scientific methods might have asked about the Yucca Mountain project are the following:

1) What are the priority dangers scientists worry about in regard to Yucca Mountain? While earthquakes and volcanoes caught the headlines, they are low on the list of scientific concerns compared to climatic changes, criticality, faulting and heat pipe condensation. This is the opposite of NWPO's concerns, and the DOE may subconsciously have preferred the media to chase these answerable red-herring than investigate more important issues.

2) What do peer reviewers say privately about research being conducted? Jerry Szymanski's rising groundwater theories, and the Mountain West socioeconomic studies both received scathing remarks from insiders.

3) A survey of Nevada's professional engineers was never conducted until the UNR survey presented in this book. Instead, housewives were interviewed about their expert opinions on Yucca Mountain while Nevada's professional engineers, and university faculty, were never polled. Local expertise generally carries a wealth of insider insights.

4) Credentials, so important in academia as a means of filtering out trivial research, were overlooked. While PhDs make errors even in their own field, all things considered an English major may well be less knowledgeable about radiation risks than a scientist who works in the field and holds a doctorate in radiation health science.

The media failed to present critical issues at Yucca Mountain, on both pro and con sides of the scientific debate. Large federal science projects (e.g. Yucca Mountain, the space station, the failed super-collider project, etc.) may suffer from inherent difficulties in having their history's told by reporters more tuned to covering their local sheriff's race. Hopefully, this book will serve as an inspiration for other technical writers to tackle future national scientific endeavors to act as independent oversight before billions are wasted on politicized science.

# 55. Ron Vitto vs The Media

The late Ron Vitto was a long time Las Vegas sportscaster brought on by OIZ Advertising in October 1991 to give credence to their pro-Yucca Mountain campaign. The amount of character assassination Mr. Vitto suffered as a result of having sided with the pro-Yucca forces was substantial, so much so that it calls into question the motives of his attackers, both in the media and the political arena.

After all, Vitto was chosen specifically because of his local reputation as a stand-up kind of guy, one who wouldn't steer you wrong. In an amazing transformation, Vitto was demonized by his former media colleagues and by members of the community, accused of being a liar and having sold out to the nuclear industry. Typical was a parody sung sung by disk jockeys Johnson and Toftee on their show on KKLZ in 1991:

**YUCCA MOUNTAIN MAN**
**(sung to the theme, "Secret Agent Man")**
**There's a man who's selling out his neighbors**
**Ronny Vitto's downplaying the dangers**
**He's getting paid you see, by the sleazy DOE**
**Odds are he don't care about tomorrow**
**Yucca Mountain Man! Yucca Mountain Man!**
**He's taking him a paycheck, to look the other way**
**There's a guy you used to look up to**
**Now you got to wonder what Ron's up to**
**I guess he needs the bucks, to trudge through toxic muck**
**Vitto's sinking low into tomorrow**
**Yucca Mountain Man! Yucca Mountain Man!**
**Ron's making him a fast buck, and giving up his good name**
**Don't listen to him babe!**
**He's just getting a paycheck**
**Ron don't mean nothin' no more**
**You sold out man!**
**[Johnson and Toftee, KKLZ, 1991, as found in NWPO release of Yucca Mountain Man series]**

It was Johnson and Toftee who were hired by Jan Jones (Las Vegas Mayor and Sawyer Commission member) for the Fletcher Jones auto dealership to do the two-headed Yucca Mountain Man parodies in their car ads.

Was Vitto's media transformation because he'd actually sold his soul to the devil? More likely it was because character assassination was simpler for the anti-nuclear forces than trying to win the debate on the basis of science. Especially interesting was the media and NWPO's response to a television advertisement in which Vitto had held a simulated nuclear pellet in his hand, attempting to demonstrate that the pellets were solids and not in volatile liquid or gaseous form. NWPO claimed this was an attempt at subterfuge and that Vitto was portraying the radiation from the pellets as harmless. Bob Loux suggested there was enough radiation in a real pellet to destroy the entire population of Nevada and that Vitto would be dead before the commercial was finished.

Curiously, no one in the media bothered to interview Vitto to find out his side of the story. As someone who suffered a year of vehement attacks in all the media outlets and even rated his own bumper sticker (Bury The Waste In Vitto's Back Yard) , one would think Vitto would have been interviewed by at least one intrepid reporter keen on digging up the dirt on why he had defected to the pro-nuclear side. Instead, the lack of professionalism of local reporters on this issue became a story in itself. Instead of doing their homework, reporters accepted NWPO's aggressive public relations effort to portray Vitto and anyone defending nuclear energy, or the study of the repository, as corrupt.

I asked Ron Vitto whether he thought the infamous pellet ads were meant to hide the dangers of radiation:

**"I have been paid to be the announcer on television ads promoting public education on the subject of a nuclear repository. That does not mean I would knowingly say anything I knew to be untrue. I was encouraged at all times to research and question what was being taped for airing. I accepted the position for one reason. I did it because I believe the nuclear industry is the lesser of environmental evils and that Yucca Mountain is the best site on the planet to bury the wastes. [personal communication, May, 1993]**

Honest people can certainly debate whether Ron Vitto's conclusions are correct concerning things nuclear, but the attacks that were made were brutally personal. For example, in a report done by Dan Burns of Channel 3, the local NBC affiliate, Burns referred to the "so-called nuclear expert, Ron Vitto." In fact, Ron Vitto never claimed to be a nuclear expert, instead relying on the on-air statements of Dr. Phillip Klein and other nuclear professionals to provide expertise. Ironically, Dan Burns himself later laid claim to being a nuclear expert with his own reporting series. The fact that Burns' report was poorly researched and distorted some nuclear issues was made painfully clear at the Sawyer Commission meeting of June 4, 1992 when Dr. Bob Luna from Sandia Laboratories thoroughly discounted the series . Ironically, Dan Burns later went to work for Science Applications International Corporation on Yucca Mountain issues.

The fact that journalists attacked Vitto so rabidly even though he was one of their own shows how heightened emotions were able to take over reason at Yucca Mountain. A comment by Ron Vitto puts this in perspective:

**"The fact is that politicians will always be politicians and businessmen will always look at the bottom line. What is extremely disconcerting to me has been the lack of integrity of the print and broadcast media in regard to Yucca Mountain. This includes obfuscation of facts and the injection of opinion for research in the guise of reporting." [personal communication, May, 1993]**

Vitto seemed to be upset at the local Nevada media more because they didn't do their professional homework regarding the repository than because they were biased in support or opposition to Yucca Mountain. In private conversations, Vitto was never shy about expressing his feeling that you have to keep a watchful eye on the technocrats at DOE and in the nuclear industry, a position that is in complete variance to the media picture of him as a sell-out to nuclear bureaucrats. Vitto was capable of seeing both sides of the issue and had the media interviewed Vitto they would have discovered this for themselves. However, equally important in Vitto's mind seems to be the idea that the media should at least look at both sides of the story and do their job as impartial observers, not as self appointed politicians running for office with predefined agendas. Again from Vitto:

**"It may be time for someone to report on the reporters. The truth is, broadcast journalism is no more than a search for the share-point." [personal communication, May, 1993]**

# 56. UNLV and UNR

The University of Nevada, Las Vegas and University of Nevada, Reno, both up-and-coming state universities, have played the part of sacrificial pawns in NWPO's strategy to keep nuclear waste out of the state at any cost. Both universities were purposely kept out of the decision making loop regarding the early design of technical studies and especially socioeconomic impact studies. This allowed NWPO to fish through the entire United States for sympathetic but unaccountable academics who supported their nuclear doomsday theories. The lack of substantial contracts awarded to the Nevada university system amounted to an admission by the state that it could trust its own academicians. It wasn't a lack of technical expertise, as NWPO claimed, that led to this situation, but the fact that this was the only way the state could influence studies in NWPO's favor.

## NEGLECTED NEVADA SOCIOECONOMIC ANALYSIS

The near monopoly given Mountain West Research for socioeconomic studies preempted involvement in the characterization of Yucca Mountain. For example, the Hotel Management School at UNLV is world recognized and well capable of economic impact analysis. Students from around the globe attend the school because they know Las Vegas' fine convention and hotel industry is intimately linked with the abilities of the management school at the university.

Yet amazingly, an analysis of the impact of Yucca Mountain on the Las Vegas convention industry was not done in Nevada, but in Pennsylvania by the Wharton School, a state not particularly known for tourism, gaming, conventioneering or entertainment. This led to bizarre contracts of the following sort:

**NWPO-SE-021-89 The Convention Planning Process: Potential Impact of a High-Level Nuclear Waste Repository in Nevada. Howard Kunreuther, Doug Easterling and Paul Kleindorfer, Center for Risk and Decision Processes, The Wharton School, University of Pennsylvania (September 1988)**

Why would a study on convention planning ever considered for anyone outside the state of Nevada, when Las Vegas is one of the prime convention destinations in the country (COMDEX, ABA, NAB, Consumer Electronics, etc.) and has a fine Hotel Management school eager to do such analysis? After all, Nevadans don't just have an academic interest in convention planning, they live, eat and die based on the results of the studies they do in this area. They also have intimate experience with the effects of nuclear testing on tourism. One suspects the reason the study was done in Pennsylvania is because Howard Kunreuther, Doug Easterling and Paul Kleindorfer were friends of Roger Kasperson and Paul Slovic from the 1979 studies of Three Mile Island and could be expected to reinforce the economic disaster theories being promoted by the State of Nevada. In Pennsylvania, Three Mile Island had been an economic non-event. In Nevada, the Test Site didn't seem to hinder construction of new extravagant hotel complexes. Yet, in the report from Wharton, the ominous nuclear images once again raised their frightened head

**. . . .we can expect that between 3% and 28% of planners who would normally select Las Vegas for their meeting will choose another city if and when the repository begins operating at Yucca Mountain. [NWPO-SE-021-89, abstract]**

The Wharton report had little relationship to the experience of most of the Las Vegas hotels which experienced negligible impact from fears of the Nevada Test Site affecting convention attendance. While it is true that extensive national media coverage of Yucca Mountain would no doubt have some impact on tourist attendance, the most likely source of such reports would be the state of Nevada's NWPO staff, not the casino industry. NWPO did most of its anti-Yucca Mountain propagandizing through the socioeconomic study groups it imported from out of state and we've covered most of these issues in previous chapters. Just as devastating however was what NWPO did to engineering studies within the state, especially at UNLV, UNR and at the Desert Research Institute.

## PREJUDICING THE ENGINEERING COMMUNITY

While most of NWPO's technical studies were legitimate, there were a number of noteworthy exceptions. One of the most obvious subterfuges was NWPO's attempt to stuff the transportation studies with the ubiquitous Marvin Resnikoff, who was brought in to critique RADTRANS, the transportation software package. Fortunately, this attempt to replace transportation specialists with an environmental activist intent on sabotaging the repository siting process was thwarted by the professionalism of the engineers who filled other positions in the study. UNLV created a top notch Transportation section studying routing problems associated with nuclear waste shipment. Another abuse of scientific process was that millions of dollars in hydrology studies were transferred out -of-state by NWPO to TRAC and friends of Jerry Szymanski, as discussed earlier.

More disturbing were NWPO's attempts to prejudice engineering studies against Yucca Mountain. An article in the May/June 1988 edition of the Nevada Monitor published by the Nevada Nuclear Waste Study Committee shows how NWPO was inclined to distort the findings of in-state technical experts:

## UNIVERSITY SCIENTIST SAYS STATE MAY ONLY WANT BIASED STUDIES

**A memorandum from a Desert Research Institute (DRI) scientist has suggested that the state's Nuclear Waste Project Office (NWPO) director, Bob Loux, may only support funding research that will support the state's opposition to the proposed high level repository at Yucca Mountain.**

**The memorandum was written by Dr. John Fordham, who heads DRI research on Yucca Mountain now being conducted for the state. It was included in a letter sent by Dr. James Taranik, who heads the University Research Council, to University of Nevada Chancellor Mark Dawson and the presidents of both UNR and UNLV. The Research Council was created to oversee various research projects related to the repository throughout the entire university system, which includes DRI.**

**Fordham's memorandum followed a meeting of the council and representatives of the NWPO. It said in part that "it seems there is an agenda on the part of the state office (NWPO) which isn't being put on the table . . . . that he (Loux) believes the University should buy into the present party line of being total adversary of the proposed high level waste site." The memo also indicated that Fordham believes Loux was not abiding by a 1984 contract which specifically provided for DRI's repository program independence to provide unbiased scientific judgment and analysis in its research.**

**The general comments from Dr. Fordham's memorandum are printed below:**

**"In general, it seems that there is an agenda on the part of the State Office which isn't being put on the table. My interpretation of Bob's (Loux) comments is that he believes the University should buy into the present party line of being a total adversary of the proposed high-level waste site. His comments related to conflict of interest and future testimony by University personnel indicate that they presuppose our findings will be counter to those of DOE and its contractors. This falls directly in line with an earlier comment by an attorney from the NWPO that their purpose was to "cast doubt, not determine the truth."**

**"The entire concept of prejudging the results of our research findings runs counter to the concepts upon which we attempt to operate. In our original contract with the State Office (1984), we went to great pains to be explicit about our independence in providing unbiased scientific judgment and analysis. Anything which indicates that we are simply to be a mouthpiece for the current adversarial party line flies in the face of that original agreement." [Nevada Monitor, May/ June 1988]**

If anything, Dr. Fordham's statement understated the seriousness of Bob Loux's and NWPO's attempts to coerce adversarial science from the University system. If a scientist at the University were caught trying to dictate the outcome of a colleagues science, he would be kicked out of the University in disgrace and be lucky not to be reprimanded by the scientific community. When Bob Loux tried to coerce scientific results, he was instead commended by the state's politicians as a defender of truth.

The scientific community guards its independence from political bias with a justified fierceness. Once scientific positions can be bought or coerced by state agencies run as personal fiefdoms, science ceases to exist and is replaced by demagoguery. It is to the credit of the Nevada University system that they avoided being completely co-opted by Bob Loux's peculiar brand of science in which results are preordained. The price they paid was to be cut off from approximately $20 million in funding grants.

Dr. Fordham was by no means a lone dissident; there was a chorus of dissent which was soon stifled by state politicians and key university administrators who also operate on the political plane. Concerns voiced by Dr. William Culbreth from the UNLV Civil Engineering Department pointed to a number of questions regarding

NWPO, Loux and the state's conduct in regard to Yucca Mountain. In a letter to William Wells, Dean of the engineering department, dated December 16, 1987, Dr. Culbreth wrote:

. . . . . **The fundamental problem is the total lack of credentials of the director of the NWPO, Mr. Robert Loux. The nuclear waste repository is a very complex program which the Department of Energy (DOE) has been carrying out for some time and which will continue in the site selection and construction phases into the 21st century. If the program proceeds as planned, waste will not even be accepted until at least 2003 or thereabouts. There are a vast number of issues of vital concern to the State of Nevada. In order to grapple with any one of these issues, a wide variety of expert personnel would be required. Normally, a position of this type would be filled by creating a statement of qualifications for the position (i.e. a description of the qualification that an individual must possess in order to be considered a suitable candidate). Then a national search would be conducted on the basis of this statement of qualifications so as to ensure that the position is filled by the most highly qualified person possible. Yet, supervising this entire oversight process for the state is a man of no apparent qualifications. Mr. Loux has a bachelor's degree in education from the University of Nevada - Reno (UNR). He has NO record of management of large complex projects, management of projects with engineering and scientific components, professional employment in the nuclear industry or any other relevant industry, high level managerial experience in federal government, or any other obvious qualification commensurate with the complex nature and great responsibilities of this position. In fact, the exact nature of Mr. Loux's qualifications and how he came to possess this position is somewhat of a mystery.**

It's clear from this statement that dissatisfaction with Mr. Loux ran deep within the academic community as early as 1987. Dr. Culbreth goes on to describe an even more important problem, the objectivity of the science conducted by the state and how NWPO systematically went about undermining the independence of its independent investigators:

. . . . **Mountain West acts as general contractor (for socioeconomic studies), and subcontracts the bulk of the work mainly to professors at various universities around the U.S. Most of these professors contract with Mountain West privately as opposed to contracting through their respective universities. This has important implications for the control of the research products. These implications will be discussed below.**

**The overall quality of the work is quite poor. The research reports are not turned in on time and have little substance. The research reports are written in jargon and are practically unreadable by a thinking human being. It would be interesting, as a measure of overall quality, to determine the number of journal publications that have resulted from the socioeconomic studies sponsored by NWPO. The various researchers have little knowledge of the state. They have made a variety of blatantly erroneous statements in the various reports. For example, one researcher made the statement that there are no large religious groups within the state that exert political influence. Apparently this researcher had never heard of the Mormon Church or of the Roman Catholic Church. Examples of this sort are legion. The review of these works is haphazard and ill-organized as anyone who has spent time at the annual review sessions can attest. The overall direction of the research is also a matter of concern, i.e. who sets the overall direction of research, how are the research needs chosen, and the responsiveness of the research direction to the needs of local governments.**

**Two important issues are that of the ownership of the data that has been collected and access to the reports that have been generated. Mountain West has, at least by their own assertion, collected a large amount of socioeconomic data. Clark County personnel have asked for this data but were told by Mountain West that it was not in accessible form. The implication being that Mountain West owns this information and that more money would be required to put this information into a viable form. This is information developed with what is ultimately federal money and therefore is in the public not private domain. These same personnel have also requested a complete set of socioeconomic studies which they have never received. Perhaps these are not available in any organized form. In spite of all these problems, Mountain West will receive a contract for $1.5 million for the 1988-89 period. This is the spending limit imposed by congress on socioeconomic studies. Congress imposed limits on these studies and overall spending limits in order to prevent further boondoggle, and not out of spite or to prevent the state from carrying out competent programs as has been asserted by various politicians.**

The geological assessments are handled primarily by UNR and DRI and also by Mifflin and Associates, a private consulting firm located in Las Vegas. The exact level of funding by Mifflin and Associates is difficult to ascertain, but it certainly amounts to several million dollars per year. It would be interesting to determine exactly how this firm obtained this contract and the exact value thereof. Mifflin and Associates does a lot of subcontracting to professionals and to university professors at DRI and UNR and throughout the U.S. again as private consultants. The same concerns about research quality and research direction expressed about the socioeconomic studies would also apply to the geologic studies.

One of the reasons for the use of private consultants is that this gives the NWPO more control over the research results. The NWPO views its relationship with the DOE as adversarial. Any entity that works or wishes to work for NWPO must accept the NWPO party line. Any entity that does work for the DOE is the enemy. This, and not competence, seems to be the selected criterion. A memo to Dr. James Taranik, President of the DRI, that was leaked to the press discussed this problem in some detail. In essence, this memo stated that NWPO is not interested in determining truth but in casting doubt upon the work of DOE, and that NWPO wants only work that accomplishes this goal. In other words, the NWPO wants biased research. This is a very serious allegation which casts doubt upon the NWPO and also upon any organization, particularly academic organizations, which have involvement with the NWPO.

These allegations are borne out by the standard NWPO contract which prohibits publication of research results without prior consent of the NWPO. Neither the method for obtaining this consent nor the basis upon which this consent is granted is specified in the contract. No university would accept such a prohibition, which is why the NWPO prefers to use private contractors. The private contractors then subcontract with university personnel on a private basis, and this prohibition is passed on in the subcontract. The NWPO also wants to prohibit any organization within the University of Nevada system which does work for the NWPO from any research involvement with DOE. This would mean, for example, that if a professor within a particular college had a contract with the NWPO, all of the other professors within that college would be prohibited from doing any contract work for DOE. Again, no university anywhere would or could accept such a prohibition. Any university that accepted such stipulations would lose its reputation for academic integrity. The NWPO is currently holding hostage research funding to the UNLV Department of Geosciences and also $1.25 million to the College of Engineering for transportation studies. Apparently the NWPO is also attempting to force the DRI and UNR to agree to the same stipulations by withholding research funding from these organizations. These organizations have so far declined to do so.

There are a variety of other issues that could be examined. Among these are the lack of oversight by the governor and by the legislative branch, the fear of making facts and disagreements public due to the power of the governor, the Treichel debacle and the employment by the NWPO of long time antinuclear activists who lack credibility, the reasons why Clark County has sought its own funding from the DOE, and the NWPO budget and the general potential for all sorts of misuse of funds.

Obviously, as far back as 1987 NWPO was using strong-arm tactics to muscle the state's Yucca Mountain studies into a preordained antinuclear outcome. The question Dr. Culbreth's letter raises now is whether any of the research conducted by NWPO over the last ten years is free from taint. It seems that the Nevada University system may have avoided having its integrity poisoned by reason of its early protests, but the UNLV and UNR budgets account for at most 25% of the funds that passed through NWPO's hands. Others seem to not have minded restrictions on their academic freedom and were thus funded.

While the scientists and engineers of the university system were for the better part able to resist having the results of their science coerced, the same independence cannot be said of the UNLV administration which under President Maxson caved in to the State's anti-Yucca Mountain party-line. A case in point is what happened to Dr. Richard Wyman, whose outspoken comments in favor of Yucca Mountain were later used to deny him a promotion. In a memorable outburst by university President Maxson, who seemed on the verge of apoplexy so vehemently did he try to silence Dr. Wyman's candid comments. An article from the Nevada Monitor from 1985 gives some background:

## WYMAN TESTIFIES ON SITE SAFETY

A joint hearing of the (Nevada) Senate Human Resources Committee and the Assembly Natural Resources, Agriculture and Mining Committee was held at Carson City on March 15 to consider testimony on the nuclear waste repository.

In testimony before the joint committees, Richard V. Wyman expressed his support for establishment of the site at Yucca Mountain.

Dr. Wyman holds a Ph.D. in Geological Engineering from the University of Arizona, and is Professor of Engineering and Chairman of Civil and Mechanical Engineering at UNLV, as well as a member of the Nuclear Waste Management Committee of the Association of Engineering Geologists.

Wyman has also served as a Peer Reviewer of the geologic, hydrologic and engineering studies involved in the selection of the Yucca Mountain site.

"In my opinion," Wyman stated, "Nevada is the best site for a nuclear waste repository and I favor the selection of Yucca Mountain."

Wyman's testimony focused on an issue of prime concern to many, the possibility of radioactive contamination of the areas groundwater system.

"Ages of the water (from the site) show it takes thousands of years to end up in Death Valley. This is a non-problem," Wyman continued. "We live with higher levels of radiation from coal ash, brick houses, or simply from flying in an airplane at 37,000 feet."

In closing comments, Wyman stated, "Any possible ground water contamination at Yucca Mountain due to fantastically hypothetical situations ignores the fact that over 500 underground nuclear tests have been set off at NTS without any ill effects from radioactive material in the groundwater system at the test site." [ Nevada Monitor, Fall 1985]

After similar public statements made in 1990 to the press, Wyman became the focus of a mini-tornado at UNLV because President Maxson considered Wyman's comments to be out of line with the official University position, which seemed to be opposition to Yucca Mountain no matter what the science said was true. The fact that the University could have a politically correct position on a scientific matter in the first place and that Wyman's credentials as an experienced mining engineer could be questioned should have raised concerns in both the media and the academic community. Sadly, the State had so polarized the issue that they no longer believed in their own university scientists and engineers, instead trusting their fate to ill trained sociologists, environmentalists and politicians.

The unfortunate outcome of this political fratricide in the University of Nevada system is that UNLV and UNR missed a chance to become first rate engineering programs in nuclear engineering, hazardous waste handling and environmental studies. UNLV did succeed in obtaining a sophisticated super computer center as a result of the Yucca Mountain project. Incredibly, the National Science Center for Energy and the Environment came despite opposition by Senator Bryan and former governor Grant Sawyer, head of the Sawyer Commission.

The political opposition to Yucca Mountain effectively drove a wedge between the Nevada university research system and the state's Nuclear Waste Project Office. Tens of millions of dollars in research and benefits were denied the universities in the attempt to scuttle positive results regarding Yucca Mountain. While the loss of these funds should disturb Nevadans, there is an even greater hidden loss. The Nevada university system, because of its proximity to both the Test Site and Yucca Mountain, should have been developed into a first class nuclear research facility benefiting America and the world.

# 57. Opinions of Nevada Engineers

Amazingly enough, no one thought to poll the state's professional engineers whether Yucca Mountain would pose a threat to the health and well being of the citizens until 1994. The professional engineers are the one set of people who might have an educated opinion about the risks a repository might represent. Instead, while the opinions of journalists were ubiquitous, and even casino workers on the street were asked their technical interpretation of earthquakes at the site, the technical community was silent and/or silenced.

Nevada's engineers, especially those in private practice, in non-DOE businesses and in the University system, are one group whose expertise and independence might allow them to think about the interests of their state in a rational way. Opinions of professional engineers have the added weight that comes with licensing, which means a professional engineer's signature on a document certifies his accountability as an experienced professional. The threat of personal attacks by politicians and journalists within Nevada evidently caused many of these engineers to remain silent on the Yucca Mountain issue.

A case in point was the way Dr. Richard Wyman of the UNLV School of engineering was treated when in the spring of 1991 he came forward to say the studies of Yucca Mountain should be allowed to continue. Attacks by Bob Loux from NWPO and from Robert Maxson, then president of UNLV, proved devastating to Wyman's career. Of course, part of the responsibility for Wyman's treatment should go to the engineering community itself which failed to speak up in Wyman's defense. In a discussion with a high official of a Yucca Mountain subcontractor the question of why the engineers of the state didn't aggressively fight back against such distortions was raised. "Do you think anyone would really listen even if we gave our credentials?" was the response.

In fact, the reluctance of the scientific and engineering community to band together and proactively counter the radical environmental opposition left lone engineering professionals at the mercy of uncredentialed critics. The media rarely certified the scientific backgrounds of their sources, allowing outside environmental activists imported by NWPO to speak without review by experienced local engineers and scientists.

In May of 1994 a poll titled Nevada Engineering and Science Survey was to be conducted by the Nevada Transportation Technology Transfer Center, College of Engineering at the University of Nevada Reno, in conjunction with the UNR Department of Political Science. Since the poll had already been conducted and the results only required tabulation (a simple process given present computerized survey techniques), no difficulty was expected in including the results in these pages. Unfortunately, no such results were forthcoming.

The simple explanation is that the UNR engineering department simply lost interest in the project. Unfortunately, it is more likely the poll results were held up until after the 1994 political campaign season as a result of the Clinton administration and nuclear industry's combined "Nuke Lite" strategy. It is apparent the holdup of this survey was in itself an issue, whether or not there was any political conspiracy to delay its release. First, why was NWPO, who it seemed even polled the pet dogs and cats of Nevadans on the Yucca Mountain issue, unwilling to do this critical poll years before? Why did the state's Professional Engineers themselves never clear the air on the repository issue as part of their obligation to Nevada's citizens?

In the hope that these questions are asked in the future, we've included some of the representative poll questions because they demonstrate the importance of the UNR poll:

**3. Evaluate the role Nevada's engineering and scientific community has played to date in site characterization work for the repository.**

**5. Currently the state performs evaluation studies, but does not place engineers and scientists at the Yucca Mountain site. Do you think the state would be better able to evaluate the work being performed at Yucca Mountain if state engineers and scientists were actually on-site with the engineers and scientists currently conducting the evaluation of the proposed Yucca Mountain nuclear waste repository?**

**11. Rate the overall quality of DOE personnel with whom you have worked on past projects.**

15. What is your overall assessment of the quality of engineering and scientific work performed to date by the DOE in assessing the proposed Yucca Mountain nuclear waste repository?

16. What is your overall assessment of the quality of engineering and scientific work performed to date by the State of Nevada in assessing the proposed Yucca Mountain nuclear waste repository?

21. [Indicate how strong your agreement is with the following statements]

    a. Nevada government should do everything possible to prevent building the proposed Yucca Mountain nuclear waste repository.

    b. The nuclear waste repository at Yucca Mountain is inevitable - it will be built whether the state of Nevada opposes it or not.

    d. Nevada is the best place for the repository because the nuclear weapons test site is already here.

    e. The process by which Nevada was chosen as a potential site for the repository was fair.

    i. A repository at Yucca Mountain would hinder Nevada's image as a place to visit.

23. Would you vote in favor of locating the repository at Yucca Mountain if the state of Nevada were provided an unrestricted grant of $100 million per year for the next 20 years?

# PART VII:  REPOSITORY SCIENCE

# 58. Study Design & Oversight

Oversight of Yucca Mountain has been driven by a number of competing goals. The technical goal is to provide long term isolation of nuclear waste from the environment until it is rendered harmless. The legal goal is to satisfy the commitment made by the U.S. Congress to the nuclear industry by 1998. The socioeconomic goal is to mitigate the impacts of the site on residents of Nevada while keeping the nuclear industry viable nationally. The political goal has been to distribute the stakeholder risks equitably, or to provide compensation. These and many subsidiary concerns are the basis for the ongoing study and evaluation of Yucca Mountain as a possible repository.

## TECHNICAL REQUIREMENTS

The driving technical requirements for the repository are the following:
- 10,000 year confinement - This is based on the assumption that after 10,000 years the radioactive elements will have decayed into a form no more radioactive than naturally occurring uranium ore. Releases of radioactivity to the environment are limited over that 10,000 year period.
- 300 to 1000 year package containment - The repository must remain a working barrier for the full 10,000 years, but waste containers are required to provide substantially complete containment for only a lesser period.
- Multiple barriers - "Defense in depth" is required from 1) the waste package, 2) the repository itself and 3) the host rock.
- Emissions levels - The Environmental Protection Agency is still defining acceptable radiation levels, no easy task given the complex escape mechanisms and the uncertainties in health effects of low level exposure. The goal is to maintain a rate of fewer than 1/10th death per year per total population (not just Nevadans).
- Retrievability - At least fifty years of retrievability of the spent fuel is required, though monitoring and retrieval for 100 years is being explored.

To address the scientific and engineering criteria, a by no means all inclusive list of studies have included the following broad topics:

- **GEOLOGY** - Determining the load bearing ability, porosity, temperature stability and chemistry of the volcanic tuff at Yucca Mountain.
- **HYDROLOGY**- Determining the groundwater flowfield, origins, depth, and historical height due to climate and tectonic influences. Determining the effects of heat-pipe condensation flows of water in fractured rocks.
- **CHEMISTRY** - Determining the absorptive ability of zeolyte minerals for radioactive isotopes, determining the corrosive conditions of the rocks and canisters, chemistry of the isotopes.
- **TRANSPORTATION** - Design of canisters and overpack, route designations, emergency response, materials handling, canister metallurgy.
- **MINING TECHNOLOGY** - Tunnel boring, tunnel reinforcement, the dynamic reaction of tunnels to seismic events.

The Nuclear Waste Policy Act also defines a number of stakeholders including Indian tribes, the State of Nevada, affected units of government, nuclear ratepayers and others. To satisfy their needs, socioeconomic studies also are a part of the research being conducted and are integral to the implementation of the technical results.

Sites studied in the past include salt domes, granite structures and sub-groundwater disposal. Yucca Mountain holds an advantage over other studied sites in dryness, depth of water table, presence of zeolytes, mechanical properties of the rock, lack of valuable minerals, and a number of other attributes.

## STUDY EXECUTION

The study is directed by the Department of Energy Yucca Mountain Project Office under the Office of Civilian Radioactive Waste Management. Influencing the project office's direction are a multitude of governmental and private entities. Technical input from government institutions includes (but is not limited to) the following:

- United States Geologic Survey
- Environmental Protection Agency
- National Academy of Science
- Nuclear Waste Technical Review Board
- National Laboratories (Sandia, Livermore, etc.)

On the private side there are two prime contractors and a variety of subcontractors. Originally, Science Applications International Corporation (SAIC) ran both characterization and support services (such as public information). In 1991, DOE hired a team of companies, led by TRW, as the management and operating contractor for the nation's Civilian Radioactive Waste Management Program. SAIC took up support duties and has continued to maintain a presence in large part because of its extensive expertise at Yucca Mountain and its knowledge of the political climate. The consortium led by TRW includes:

- TRW - Prime contractor, systems engineering, design, development, technical direction, management and operation.
- BABCOCK AND WILCOX - Engineered barrier design and development.
- DUKE ENGINEERING - Licensing, outreach, MRS design, quality assurance.
- E.R. JOHNSON ASSOCIATES - Strategic Planning and Policy Analysis
- INTERA - Performance assessment.
- FLUOR DANIEL - Surface facility design and development.
- INTERA - Performance Assessment.
- J.K. RESEARCH ASSOCIATES - Strategic planning and policy analysis.
- LOGICON RDA - Systems engineering and integration support
- MORRISON KNUDSON - Underground facility design.
- WOODWARD-CLYDE - Site characterization.

The Office of Civilian Radioactive Waste Management and the Yucca Mountain Project Office in Las Vegas have jointly run the study of Yucca Mountain, a job that at times has been an organizational nightmare.

## OVERSIGHT

Adding to the bureaucratic complexity of designing a nuclear waste repository has been the creation of multiple layers of oversight institutions. There is more oversight of nuclear waste disposal than of any other project in U.S. history. Among the most prominent oversight entities are:

- NUCLEAR WASTE TECHNICAL REVIEW BOARD
- NATIONAL RESEARCH COUNCIL
- OF THE NATIONAL ACADEMY OF SCIENCES,
- NUCLEAR REGULATORY COMMISSION
- ENVIRONMENTAL PROTECTION AGENCY
- UNITED STATES GEOLOGIC SURVEY
- GENERAL ACCOUNTING OFFICE
- OFFICE OF MANAGEMENT AND BUDGET
- SENATE ENERGY AND
- NATURAL RESOURCES COMMITTEE
- HOUSE ENERGY AND COMMERCE COMMITTEE'S
- SUBCOMMITTEE ON ENERGY AND POWER
- NEVADA AGENCY FOR NUCLEAR PROJECTS /NUCLEAR WASTE PROJECT OFFICE

Consequently, if there is one criticism of the study of Yucca Mountain, it may be that there is too much oversight of the science and engineering being conducted. This can have the unintended result of causing the project to bog down in the minutia of the study rather than pursuing a workable solution. Nevertheless, because

Congress recognized that many of the actors on the stage had vested interests in certain outcomes for the repository study, one necessary level of oversight was the creation of an impartial Nuclear Waste Technical Review Board:

**PUBLIC LAW 100-203**
**NUCLEAR WASTE TECHNICAL REVIEW BOARD**
   **(B) The National Academy of Sciences shall nominate 22 persons for appointment to the Board who meet the qualifications described in subparagraph (C).**
   **(C)**
         **(i) Each person nominated for appointment to the Board shall be ---**
         **(I) Eminent in a field of science or engineering, including environmental sciences; and**
         **(II) Selected solely on the basis of established records of distinguished service.**
         **(ii) The membership of the Board shall be representative of the broad range of scientific and engineering disciplines related to activities under this title.**
         **(iii) No person shall be nominated for appointment to the Board who is an employee of---**
         **(I) the Department of Energy;**
         **(II) a national laboratory under contract with the Department of Energy; or**
         **(III) an entity performing high level radioactive waste or spent nuclear fuel activities under contract with the Department of Energy.**

The Nuclear Waste Technical Review Board has proven to be an irritant to DOE on a number of occasions, and it appears to be fulfilling its oversight responsibilities in good faith. The issues of concern at Yucca Mountain stretch across a wide panorama of sciences. The main technical issues broadly divide into the areas of: on-site versus geologic storage, transportation, groundwater, earthquakes, volcanism and radiation emissions. The next chapters will address each of these issues in turn.

# 59. On-Site, MRS, Geologic Storage

Spent nuclear fuel is presently stored on site at 65 reactor facilities around the country. Held in large pools of water, these storage facilities are substantially full and need to either be expanded at substantial cost, or an alternative storage plan needs to be implemented. The three candidate solutions to this problem are on-site storage, an MRS facility, or geologic storage at the Yucca Mountain repository. It is likely that a combination of these approaches will eventually make up the nation's nuclear waste disposal program.

No matter how the battle over Yucca Mountain turns out, the nation will still be faced with the question of how and where to store 70,000 tons of nuclear waste already created and with more in the pipeline. Even the critics of Yucca Mountain have never seriously questioned the need for geologic storage of the waste, they have only contended that the timing is inappropriate and that present technology is inadequate. Thus, on-site storage is suggested only as a temporary solution, although it is often presented to the public as a long term remedy.

## ON-SITE STORAGE

Because neither a geologic repository nor an interim Monitored Retrievable Storage site will be available by the deadlines imposed by the Nuclear Waste Policy Act, the Nuclear Regulatory Commission looked at the question of whether spent fuel could be kept on site for periods up to 100 years. The NRC's finding that this would be safe, although not necessarily advisable, left an opening for opponents of Yucca Mountain to argue there is no rush to build a geologic repository.

Environmentalists, coordinated by groups like Nader's Safe Energy Communication Council, long proposed that nuclear spent fuel assemblies could be left on site at various reactors around the nation in large "dry cask" units until a better solution was derived. The environmentalists hope that in thirty or forty years, science will have advanced to the point where our knowledge of how to build underground geologic storage will be vastly superior, at which point we could presumably proceed with building a repository. However, this approach does not address whether the expertise now available could be orchestrated forty or fifty years into the future to dispose of the waste.

The Nevada Nuclear Waste Project Office also adopted this on-site storage philosophy, often citing a study by researchers at the Institute of Safety and Systems Management at the University of Southern California.

**An alternative that looks like it potentially possesses many desirable characteristics is to establish an endowment at the current time to provide the financial resources 100 years from now to manage the spent nuclear fuel that would be stored between now and then in dry storage facilities at the sites of nuclear power plants. The base case analysis suggests that the on-site facility with a $6 billion fund for the future is better by the equivalent of at least $20 billion than the monitored retrievable storage facility and $30 billion better than a repository. With on-site storage and a $3 billion fund, the equivalent costs are at least $10 billion better than the monitored retrievable storage facility and $21 billion better than the repository. [Keeney, Ralph; von Winterfeldt, Detlof; "An Analysis of Strategies to Manage Nuclear Waste from Power Plants", Institute of Safety and Systems Management, USC, Report 93-1, p82]**

Whether or not the Keeney and von Winterfeldt economic analysis was correct, there should be some concern that it is possible to keep a fund with $3 to $20 billion in funds inaccessible to the political establishment. A similar fund supported by highway taxes was recently usurped by Congress.

Interestingly, the anti-nuclear lobby has in effect admitted that its position favoring on-site storage is a sham meant to diminish support for geologic storage at Yucca Mountain rather than offer a viable solution. Attempts to store waste in 17 on-site casks at the Prairie Island nuclear powerplant near Red Wing Minnesota were also opposed by environmental activists:

Storage Of Nuclear Wastes Fuels A Debate

In what is seen as a test case for other states, Minnesota is giving a new lease on life to a nuclear power plant that is running out of storage space for its radioactive waste. . . .

Without the temporary storage, the Red Wing, Minn., plant would have been forced to close one reactor next May and the other in early 1996 - eliminating 500 jobs, raising electricity rates and straining demand on other utilities.

It was the first time a plant threatened to close because it lacked temporary nuclear waste storage capacity.

I'm acutely aware of the problem," says Energy Secretary Hazel O'Leary, a former Northern States Power executive. "Even if a plant shuts down, you have to do something with the waste."

By the year 2000, she says, 32 plants in 21 states will need similar above-ground storage. By 2010 - the soonest a federal site may be open - almost half of the nation's 108 plants will have run out of space. . . .

Among those opposed: Prairie Island Mdewakanton Dakota Indians, whose reservation is next door. The tribe fears the casks will become permanent, creating health risks.

Though Prairie Island has a good safety record, it is "putting off the inevitable" by using temporary storage, says Michael Mariotte, Nuclear Information and Resource Service. USA Today, Monday, May 9, 1994 p 7A.

Both the National Resource and Information Service and the Sierra Club actively opposed the Prairie Island on-site storage site as active members of the Safe Energy Communication Council. Consequently, these groups favor on-site storage when speaking to Nevadans, but oppose it when speaking to Minnesotans.

Fortunately, the Mdewakanton Indians recognized the problem with this strategy, as have other stakeholders in nuclear states. Without a permanent storage site, such as Yucca Mountain, states could face indefinite storage of nuclear waste on-site, at best in storage casks and at worst in spent-fuel pools. The long term radiation risks from such scenarios are substantial.

## MONITORED RETRIEVABLE STORAGE

An option intermediate to on-site storage and geologic storage is to build one or more aboveground temporary storage and handling facilities as part of a Monitored Retrievable Storage (MRS) system. These sites would act as collection and processing centers for nuclear waste, perhaps doing the final preparation of the casks for final disposal below ground.

One possible advantage of an MRS facility is that spent fuel cools with time (an advantage in a cold repository design). Another advantage is that a centralized MRS facility would allow standardization of storage and handling technology. Nevertheless, the GAO concluded on the basis of both safety and costs that the MRS solution does not offer significant advantages over a system without an MRS intermediary.

One obvious choice for an MRS facility would be at Yucca Flats at the entrance to Yucca Mountain. The Nuclear Waste Policy Act now prohibits an MRS from being sited in the same state as the repository, however, the act may be amended in 1995.

## GEOLOGIC STORAGE

The concept of geologic repositories goes back to the mid 1950s after the National Academy of Science recommended deep storage. Field studies of salt formations as a possible disposal medium began in the 1960s and with studies in basaltic rocks being added in the 1970s. Studies of the salt beds near Lyons, Kansas in the mid 70s led to embarrassment when it was found that nearby commercial drilling for brine had contaminated the area. Drilling in basalt near Hanford, Washington, would have faced severe technological hurdles from severe water flows. A site in Deaf Smith County, Texas, also was below the Ogallala aquifer.

The singular thing about Yucca Mountain and the Nevada Test Site is the great depth from the ground surface to the water table. Consequently, while many consider the choice of Yucca Mountain as a purely political outcome, Yucca Mountain was the only site under study which offered disposal of nuclear wastes in unsaturated rock.

## COMPARING THE OPTIONS

**TRANSPORTATION:** There is more immediate transportation of nuclear waste using the Yucca Mountain repository instead of on-site storage, but in the long term the waste will still need to be moved. One problem with waiting is the possible decomposition of fuel rod cladding over time, making repackaging more difficult and an accident more likely to spread contamination (pointed out by Marvin Resnikoff).

**COOL DOWN PERIOD:** While it is true that nuclear spent fuel becomes exponentially safer to handle with time, the waste has already become substantially cooler (both radioactively and thermally) after the five years required before shipment. Moreover, most waste will experience more than ten years wait before being stored at Yucca Mountain given present timetables.

**TERRORIST ATTACK:** The odds of being able to protect 65 reactor sites from terrorist attack are certainly less promising than protecting a deep geologic disposal site. Dry-casks are extremely robust, however the 911 terrorist attack on the World Trade Center indicates that attack is possible and that powerful and sophisticated conventional explosives are available.

**SURVEILLANCE:** On-site storage requires a significantly larger surveillance effort than associated with a centralized repository. These costs can be prohibitive when a number of sites are factored into the equation. Surveillance would be needed during transportation for both scenarios, although delayed by a number of decades for on-site storage.

**INSTITUTIONAL STABILITY:** Given present world instability, assuming America will be politically stable 50 to 100 years from now is in itself a risky prediction. Delaying Yucca Mountain increases the risks that a geological repository will never be properly constructed, if it is even built.

**HAZARDS TO NEVADANS:** While it is true that siting Yucca Mountain in Nevada will expose Nevadans to slightly higher radiation risk (approximately 1 millirem per year for someone within three miles of the site), it is also true that radiation pollution of the American continent could well be higher utilizing the on-site solution. While a Nevadan might be marginally affected long term by Yucca Mountain, their health and safety might be immediately threatened from on-site nuclear waste accidents during their trips to other cities.

**TECHNOLOGICAL SOPHISTICATION:** On-site, MRS and geologic storage canisters are likely to be patterned after the Multi-Purpose Canister system. Consequently, in a system that utilizes on-site, MRS and geologic storage, the technology for each leg is likely to converge.

**RESPONSIBILITY TO FUTURE GENERATIONS:** On-site storage and MRS facilities both risk the possibility that nuclear waste will be left aboveground indefinitely. While one can argue that interim on-site storage or the MRS option allow us to take advantage of advances in nuclear waste disposal technology, Yucca Mountain is likely to remain open for 100 years. Consequently, future reconfiguring of the waste is not ruled out even for deep geologic storage.

After analysis, the supposedly superior benefits of dry-cask storage promoted by anti-nuclear advocates become at best equivalent to the Yucca Mountain option. In a rational world, the anti-nuclear forces would be ambivalent about whether on-site or immediate geologic storage was chosen. Instead, their nearly hysterical opposition to geologic storage suggests their motives may be more ideological than measured by technological arguments.

# 60. Geology & Chemistry

Yucca Mountain was formed twelve to fifteen million years ago by multiple eruptions of a composite volcano type known as a caldera. The events that formed Yucca Mountain created four major layers of ash-flow tuff interspersed with minor ash-fall layers. These events produced more than 600 cubic miles of silic magma alternating in layers of relatively porous and nonporous rock. The volcanic tuff at Yucca Mountain is at least 6,000 feet thick (see following Figure).

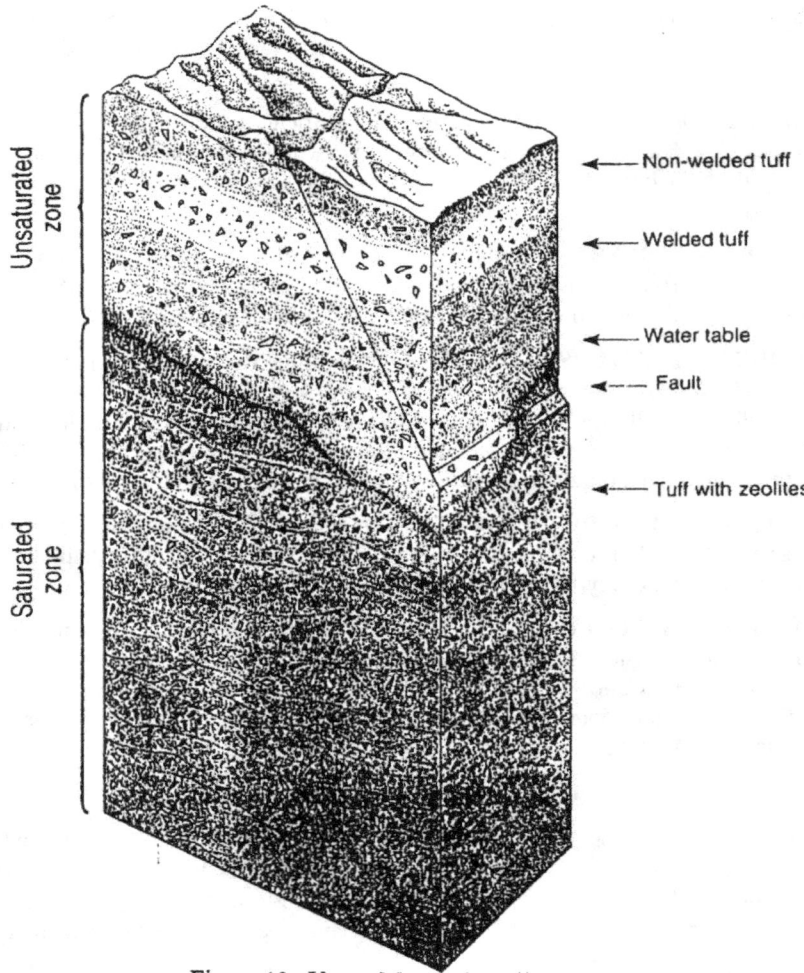

It is the unique properties of this welded tuff combined with deep dispersal of groundwater and lack of rainfall that make Yucca Mountain a promising site for a nuclear waste repository. Among these critical reasons are:

- Deep groundwater, 1300 feet below the top of Yucca Mountain and 800 feet below the repository level.
- The absence of mineral resources. Although there are gold and silver mines in nearby mountains, the volcanic tuff of the caldera is not commercially interesting.
- The rock has a high yield strength - 25,000 psi - important for carrying the load of heavy nuclear waste containers.
- The presence of zeolytes, a class of minerals that retards the diffusion of radioactive substances through the rock.

- A closed drainage system. Groundwater from Yucca Mountain drains eventually into Death Valley, below sea level, and not into aquifers which drain towards the ocean.

About 1,000 feet below the surface of Yucca Mountain is a densely welded rock formation known as the Topopah Spring welded tuff. This ancient layer is being studied as the possible site for the nuclear waste repository and the research is by no means trivial.

## TEMPERATURE GRADIENTS

One factor affecting both the geology and chemistry of the repository is the temperature at which the site is designed. At full capacity, there will be an energy load of approximately 50 to 100 megawatts on the site which can have some environmental impact at the surface by raising the ground temperature by approximately a degree Centigrade.

A critical question is whether to design a "hot" or "cold" repository, where hot is defined as temperatures high enough to drive away water from the area as vapor. Unfortunately, even a hot repository may not ensure dryness because vapor driven from the site will tend to condense at a distance and may flow back towards the drifts through fractures.

High temperatures will also have other impacts. The mechanical structure of the rock can be degraded by long term exposure to heat, perhaps necessitating rock-bolting of tunnels and backfilling of drifts to prevent cave-ins. Corrosive chemical reactions are also accelerated by higher temperatures.

Finally, the higher the repository temperature, the more hostile the working environment for human beings. Air conditioning the drifts may be problematic because of the volumes of air to be moved and the need to control humidity, velocity, local temperature and other factors in an efficient manner.

## MINERALS

The Nevada Nuclear Waste Project Office claims a number of geologic features disqualify the repository site, including a fault through the repository (the Sundance fault), the presence of earthquakes and the nearness of volcanoes. They at times suggested granite would be a better repository material, though they were reluctant to discuss an alternative site.

One NWPO claim was that there are valuable minerals on the site that would invite exploration and intrusion by future generations, as expressed in the following citation.

**Yucca Mountain is situated within a world-class precious metal mining district. Millions of dollars of gold and silver may be located in the area. ["Why Nevada Opposes Yucca Mtn.", NWPO, June 1993]**

This is a half-truth. Yucca Mountain is composed of volcanic tuff left over from massive volcanic activity from millions of years ago. The area around these calderas is loaded with precious metals, but the tuff from which Yucca Mountain is itself formed is mineralogically worthless. The nearest bedrock that might conceivably contain precious minerals is six thousand feet below the surface of Yucca Mountain and five to ten kilometers away on the surface.

## LONG TERM RADIOLOGIC QUALITY OF WASTE
One question that arises is what the spent fuel will look like after various periods of time and when when residual dangers will become trivial.

The first thing to note is that the canisters holding the waste are likely to break down after as little as three hundred years even under the best of conditions. This will be caused by corrosive chemistry present even in a dry environment and from the effects of radiation bombardment of the container materials.

Long term, it was expected that the spent fuel would tend to degrade towards a form in some manner equivalent in radiologic risk (though not chemical or isotopic composition) to naturally occurring uranium ore. Recent long term criticality studies by Dr. Bill Culbreth of the University of Nevada Las Vegas and Doctoral candidate Paige Zelinsky suggest the Actinide series elements may decay more towards uranium 235, a radiologically active isotope. The Culbreth studies also suggest possible complications from the use of boron as a moderator because of its solubility.

## ZEOLYTES

Zeolytes are a mineral that composes much of the welded tuff that underlies Yucca Mountain. Zeolytes are a family of hydrated silicate minerals which are found in sedimentary and volcanic rocks. The most common zeolyte minerals at Yucca Mountain are Clinoptilolite and Mordenite.

The interesting thing about zeolytes is their ability to absorb the metallic compounds that radioactive substances represent. What this means is that radioactive substances being transported by water within the rock will travel at a substantially slower speed than the water. This occurs because the radioactive particles are alternately bound and unbound to the rock minerals, retarding their movement. While this is an added protective safeguard at Yucca Mountain, it is not totally foolproof. Technetium and some other isotopes are able to move through zeolytes nearly unimpeded.

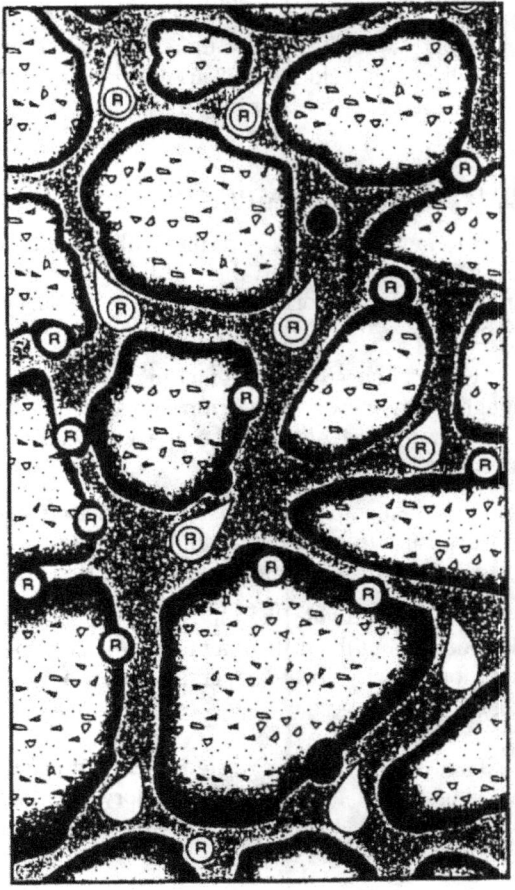

 *Moisture carrying radionuclides*

 *Zeolites*

 *Radionuclides sticking to zeolites*

 *Tuff*

Artist's concept showing the interaction of zeolites and radionuclides. In this drawing, moisture is shown moving in a cross-section of tuff. Some of the drops of moisture carry radionuclides (radioactive particles). As the radionuclides come in contact with the zeolites, they would be attracted to the zeolites, sticking to them, and, with less intensity, to the tuff.

# 61. Volcanoes, Earthquakes & Faults

Three issues that the State of Nevada argues make the Yucca Mountain area unsuitable for a nuclear waste repository are active earthquake activity, volcanism and faulting. While both earthquakes and volcanoes affect Yucca Mountain, from a scientific standpoint they are a relatively insignificant threat to the site. Faulting may pose more problems, not as much from shearing of the repository block as from presenting a path for water flow through the repository.

In fact, tectonic and volcanic activity created Yucca Mountain. Unfortunately, the emotional volatility of the geologic catastrophe theories proposed by the state became imprinted in the minds of Nevadans in the ongoing propaganda wars. Massive earth upheavals and flowing magma are visions quite able of stirring mass hysteria but have little to do with the safety of the nuclear repository. Nevertheless, there are real questions still unresolved regarding geologic dynamism at the site which still deserve study. Some background is in order.

## VOLCANOES

Yucca Mountain was formed millions of years ago by a series of volcanic eruptions that deposited ash and material that compressed together to create layers of rock called tuff (see Figure 14). The explosive type of volcano that formed Yucca Mountain is extinct. There are, however, seven small and dormant volcanoes scientists are studying in the Yucca Mountain area to determine whether one might erupt in the 10,000 years the repository is designed to last, and whether they could affect the site.

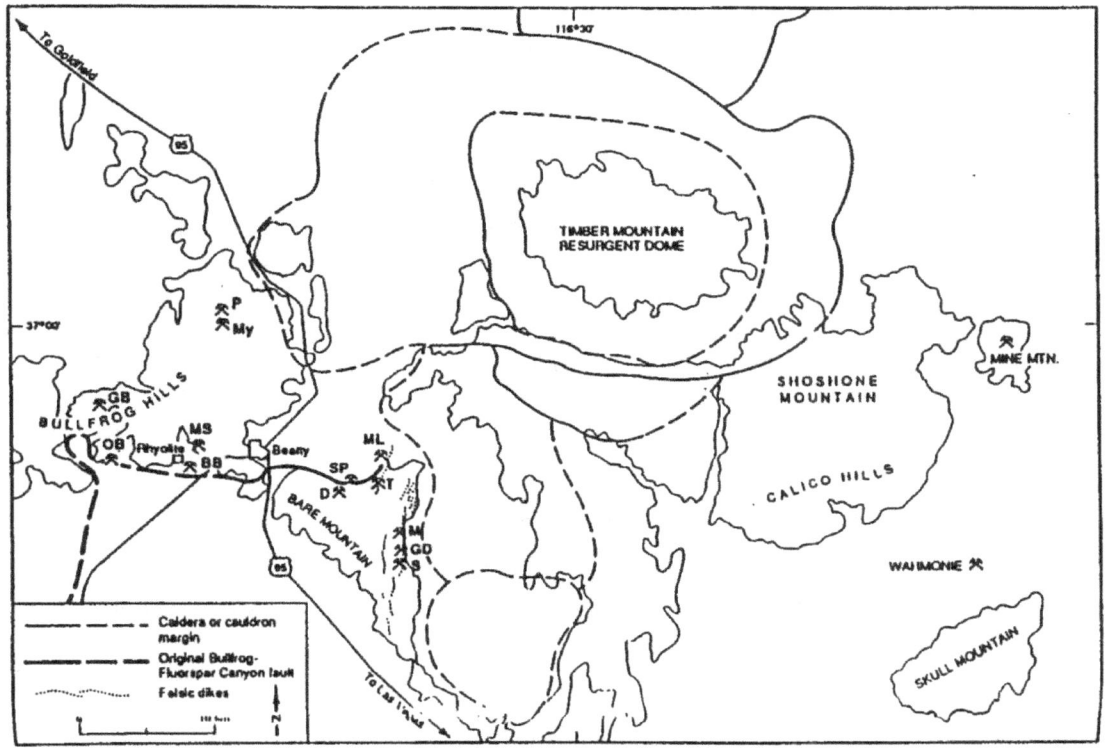

**Figure 1  Volcanic caldera system responsible for Yucca Mountain's creation.**

There are three main types of volcanoes worldwide: composite, shield and cinder cone. Composite volcanoes have explosive eruptions, such as Mount St. Helens in Washington State. Shield volcanoes have less explosive eruptions and slow moving lava, such as the Hawaiian volcanoes. Cinder cones generally are the smallest volcanoes. The seven dormant volcanoes near Yucca Mountain are cinder cones.

The seven cinder cone volcanoes located near Yucca Mountain are among the most common types of volcanoes on earth. Two cones located 12 and 27 miles away may have been active within the last 10,000 years. The other five, located 8 to 27 miles away, had their eruptions from 300,000 years to 1.2 million years ago. All seven cones consist of less than .06 cubic miles of material, a relatively small amount.

The presence of volcanoes near Yucca Mountain leads to questions: of how likely are volcanic intrusions s near the site and what effect would such an eruption have. After intensive studies of the soil and rock to determine the age and type of volcanoes in the Yucca Mountain area, the conclusion most scientists have come to is that the chances of a volcano affecting the repository are remote. An extensive review of these issues conducted by the National Research Council in its efforts to understand local groundwater issues led to the following statement:

**In assessing what processes are likely to cause a perturbation of the water table, the panel considered the long and complex Tertiary volcanic history of the region. A possible recurrence of the earlier highly explosive silicic volcanism that produced the ash flow tuffs, which are the predominant bedrock of the Yucca Mountain area, was dismissed because the subduction zone that caused it is now extinct in the Great Basin region. Concurrent and subsequent basaltic basaltic volcanism, related to the change in extensional tectonic that created the Basin and Range structure, has experienced a progressive decline in volume, as expressed in the low-volume volcanic eruptions of Crater Flat which bounds Yucca Mountain on the west, and the latest and lowest in volume, Lathrop Wells cone, a short distance to the south.**

**Thus the geologic record of waning basaltic volcanism indicates that the only likely style of intrusion into the Yucca Mountain area during the lifetime of the repository is a low-volume basaltic dyke. . . . Thus, dike intrusion appears to be inadequate to cause the rise of the water table level of more than 10-20 meters.**

**The calculated probability of occurrence of a dike intrusion that would affect the proposed repository is a very small number, on the order of $10^{-8}$ per year.**

**["Groundwater At Yucca Mountain: How High Can It Rise?", Report of the Panel on Coupled Hydrologic/Tectonic/Hydrothermal Systems at Yucca Mountain, National Academy Press, April 1992, p143]**

The activity line of area volcanoes seems to have been moving northward over the last few hundred thousand years and the size of eruptions has been diminishing with time. Non-explosive basaltic eruptions seem to be occurring near Yucca Mountain only about every 200,000 years. Moreover, the fact that Yucca Mountain is composed of many layers of volcanic tuff that haven't been disrupted would seem to indicate that the chance of the site being disrupted is virtually nil.

## EARTHQUAKES

The very real possibility of earthquakes occurring near Yucca Mountain turns out to be a less pressing issue than is often portrayed by opponents of Yucca Mountain repository. The two main threats earthquakes pose are through effects on the water table and disruption of tunnels and support structures. However, historically earthquakes seem to cause only minor fluctuations in water tables (less than 10 meters) and ground tremors do not affect underground structures the same as above surface facilities.

When an earthquake occurs, the layers of rock take on a movement that is in some ways related to the movement of jello. If one visualizes one cherry buried in the jello and one on the top, it is clear that the cherry on the free surface is much more likely to be affected by movement. Similarly, as one nears the surface of the earth, earthquake movement is accentuated because the surface has greater freedom to move. This is because the stiffness of the rock increases with depth and the degrees of freedom with which material can move is limited below ground compared to surface structures. Consequently, mining tunnels buried in a mountain act like cherries buried in the jello and are relatively unaffected by earthquake movement. Indeed, tunnels at Skull Mountain near Yucca Mountain were little affected by a series of earthquakes in 1992.

Over the last hundred years many engineered ways to protect buildings have been developed that allow us to build skyscrapers capable of withstanding 7.0 Richter scale readings, a rather violent earthquake. Designing the repository to withstand such stresses actually requires less finesse. Unlike a skyscraper, the repository will be embedded in rock and therefore not free to move and sway the way surface buildings do.

A surface receiving facility for the nuclear waste won't have the advantage of being situated a thousand feet below the surface in rock, however, the surface waste handling facility will be built just sub surface to maximize earthquake resistance. With five foot thick walls used as part of the radiation shielding of the receiving building, its vulnerability to earthquakes will be inconsequential.

Of course, the possibility that Yucca Mountain is not very vulnerable to earthquakes has been ignored by the State of Nevada in its attempts to raise fear levels. Judy Treichel, the state's one time nuclear information expert was heard to exclaim, "Earthquakes, push earthquakes!" at the June 1992 Sawyer Commission meeting, shortly after a number of earthquakes had hit both California and Nevada, obviously aware of the sensationalism of those words.

Carl Johnson, then State geologist for NWPO, correctly admitted to this author at that same Commission meeting that buildings could be designed for foreseeable earthquake threats and admitted that surface structures were much more vulnerable than subsurface structures like the repository itself. The defense Johnson made for the State's contention that earthquakes disqualified Yucca Mountain as a site was that there wasn't enough data from buildings monitored during earthquakes to make an informed decision about the vulnerabilities of the site. It appears that Johnson was avoiding the empirical evidence and fact that many of our cities have been subjected to earthquakes and building designs are regularly validated by this real world testing.

Another issue sometimes raised is the possibility of complex interactions between volcanoes and earthquakes, but these effects if present seem secondary.

## REPOSITORY FAULTING

More interesting than volcanoes or earthquakes in terms of their possible effects at Yucca Mountain is the existence of faults within the repository block. Intensive mapping of these faults, notably the Ghost Dance and Sundance fault systems, have been a focus of research.

The existence of deep faults crossing the repository does not in itself disqualify the site but it does raise a number of difficult questions. The obvious concern, though not necessarily the most problematic, is the question of future movement along these faults causing the fracturing of storage canisters. The more important concern is that the faults present an access channel for groundwater, concentrating the flux through the repository to vulnerable hot-spots. Resolving this question is not straightforward because the accretion of various calcites within water channels can lead to the cementing of water channels like faults. This makes it conceivable that the faults may in many circumstances be less permeable to water than the normal tuff matrix. According to Science News:

**In its preliminary designs for Yucca Mountain, the Energy Department has indicated it will avoid placing waste directly within the Ghost Dance fault zone, which is believed to reach the depth of the repository 300 m below the surface. If the Sundance fault zone also extends to this depth, DOE will have to decide whether to work around these faults as well.**

**At present, geologists do not know whether the Sundance or Ghost Dance faults have generated earthquakes within the last several million years. But even if these structures are not active, they may still threaten the storage facility. Because fractured rocks fill these faults, they could provide a path for groundwater to reach the repository, potentially speeding up the rate at which radionuclides leak into the environment. Faults could have the opposite effect, however, if they contain natural mineral cement that inhibits water flow. [Science News, 5/14/94]**

# 62. Groundwater

A prime reason Yucca Mountain has been thought acceptable as a site to store radioactive waste is because of the dryness and deep depth of groundwater at the site. Yucca Mountain was one of the few sites where the repository could be placed substantially above the water table.

Some of the groundwater issues at Yucca Mountain were already discussed in an earlier chapter on geologist Jerry Szymanski. However, Szymanski's theories tend to bias the discussion of groundwater issues toward rather extreme hypotheses. There are a number of traditional hydrological reasons the site is considered unique and worthy of study:
- Historically dry climate.
- Extremely low water table.
- The presence of minerals called zeolytes that would slow transport of radionuclides through the rock.
- Isolation from the groundwater of the rest of Nevada.
- The fact that the surface tends to shed water rather than absorb it.
- The absence of a thermal source near the earth's surface (magma) to cause thermal upwelling.

Some concerns still exist over surface and groundwater issues, but they are unrelated to the Szymanski theories. The Szymanski theory that thermal upwelling of deep waters could cause the inundation of the site has generally been dismissed by his peers as poorly conceived. More important concerns include:
- The effect of increased rainwater over the next ten thousand years
- The possibility of focused recharge (focused water infiltration).
- The heat pipe effect causing thermal vaporization / condensation cycles.

## HISTORICAL RECORD

The historical fact of many millennia of relatively dry climate has been established in a number of ways. One is the paleobiology of the area, i.e., the traces of ancient plant and animal life in the ground that are indicative of dry or wet climates. Rat middens, the nests left by rats over many thousands of years, have proven a valuable insight from the fossilized plant remains, crystallized urine and droppings left there.

Another method of dating the underlying groundwater is by analysis of the isotopes of various elements including strontium, oxygen-13, oxygen-18, carbon-13, etc. Also, by testing the ratio of strontium-87 to strontium-86, it is possible to compare the originating sources of calcium deposits. The importance of this has been that sensitive areas, such as Ash Meadows (where resides the endangered pupfish) have been checked to determine whether water from the Yucca Mountain drainage system, invades this aquifer. Fortunately, it does not.

Other relative rather than absolute means of aging groundwater process stem from the ordering and style of layering of various calcite deposits. The cumulative data from all the above data presently does not support the theory that Yucca Mountain has within recent millennia experienced a significantly wetter climate capable of raising the local water table, though certainly somewhat wetter periods have existed.

## POSSIBLE GROUNDWATER INTRUSION MECHANISMS

Yucca Mountain is made of many layers of volcanic ash of various consistencies. The repository itself is to be sited in one of these layers called the Topopah Spring tuff. This layer ranges from about 800 to 1300 above the water table and up to 1500 feet below the top of the mountain.

There are a number of conceivable (though unlikely) mechanisms for water to rise to the level of the repository: earthquakes, volcanic dikes and increased rainfall. The National Research Council, in its report "Groundwater at Yucca Mountain: How High Can It Rise" considered a number of these and other scenarios and offered some observations:

**VOLCANIC INTRUSION** **In considering the long and complex history of the region, the possibility of a recurrence of the highly explosive volcanism of the Tertiary was dismissed because the subduction zone**

origin of the activity has been replaced by extensional tectonics that has resulted in the basaltic volcanism of more recent geologic time. The progressive decline in volume of these eruptions convinces the panel that the only likely volcanic intrusion in the region during the lifetime of a repository is a low-volume basaltic dike. . . . .

Calculating the probabilities of a dike intrusion close to Yucca Mountain results in a very small number, 10-8 per year. Although there may be considerable uncertainty in the probability values, the panel considers that the small effect a basaltic dike intrusion would have on the water table and the low probability of a dike forming close to Yucca Mountain mean that volcanic intrusions can be discounted as potentially disruptive events with respect to water table stability. ["Ground Water At Yucca Mountain: How High Can It Rise", National Academy Press, 1992, p6]

The National Research Council also commented on the earthquake problem in regard to groundwater:

EARTHQUAKES The panel concludes, given the experience from historic earthquakes, the small modeled response of the water table to earthquakes consistent with historic experience, the low strain rates and low seismicity both in magnitude and frequency of occurrence of the Yucca Mountain area, that significant water table excursions to the design level of the repository are unlikely. ["Ground Water At Yucca Mountain: How High Can It Rise", National Academy Press, 1992, p6]

Further backing the NRC's conclusions on the minimal effects of earthquakes on the Yucca Mountain groundwater is the report, Earthquake-Induced Water-level Fluctuations at Yucca Mountain, Nevada, April 1992 issued by the U.S. Geologic Survey. This presents the results of measurements taken at wells near Yucca Mountain during a series of earthquakes that rattled the California and Nevada area in March and April of 1992:

ABSTRACT

This report presents earthquake-induced water-level and fluid-pressure data for well USW H-5 during April 1992. Well USW H-5 is located in the Yucca Mountain area, Nevada. On April 22, 1992 a 6.2-magnitude earthquake occurred in southern California which caused a maximum fluid-pressure change of approximately 50 centimeters (approximately 1.6 feet) in well USW H-5. Within 18 hours on April 25-26, 1992, three major earthquakes occurred in northern California. The water-level responses to these earthquakes were detected in well USW H-5. The maximum water-level fluctuation from the northern California earthquakes was in excess of 52.5 cm. [Earthquake-Induced Water-level Fluctuations at Yucca Mountain, Nevada, U.S. Geologic Survey, Open File Report 92-137, April 1992, p1]

Water fluctuations due to earthquakes, hydroseisms, are generally small. The USGS report continues:

Hydroseisms, or water-level fluctuations in response to earthquakes, are relatively common phenomena observed in wells penetrating confined aquifers. For example, the Anchorage, Alaska earthquake of 1964, the largest North American earthquake thus far in the 20th century, caused water-level fluctuations throughout the world; the largest peak-to-trough range was about 7.0 meters (22 feet) in a well in South Dakota. Hydroseisms are more commonly in the centimeters to meter range and typically are observed for minutes to tens of minutes. [U.S. Geologic Survey, Open File Report 92-137, April 1992, p1]

Although the State of Nevada makes much of the earthquake and volcanic activity at Yucca Mountain affecting the level of the water table, in reality these are peripheral issues. The most difficult problem now challenging the scientists at Yucca Mountain is verifying that long term changes in rainfall cannot sufficiently affect the levels of the groundwater, at least not enough to endanger the repository.

Rainwater in the Yucca Mountain area now ranges around 6 inches per year. There is evidence in the area and the Vegas valley that the climate was wetter in the past. The presence of sediment along the Vegas valley floor and the bones of horses, camels and even mastodons argue for a more lush climate. How wet the climate has to be to cause an increase in height of the water table is still a matter of debate, however:

. . . Analysis of the paleoecological and paleoclimatic information of the area suggests that even at the last glacial maximum during the Pleistocene Wisconsin 18 ka (thousand years before present) the Yucca Mountain area experienced no more than a 40 percent increase in rainfall over the present. ["Ground Water At Yucca Mountain: How High Can It Rise", National Academy Press, 1992, p141]

One of the important considerations is where groundwater drains from the Yucca Mountain area. Unlike areas surrounding the site, this area drains into the hydrological dead-end of the Amargosa River and Death Valley rather than into the river systems that flow towards the Colorado River and drinking supplies. One of the purposes of the isotopic studies being conducted at the Mountain is to verify that all the water in that area does indeed flow towards Death Valley and not into other drainage systems that enter the ocean.

A common misconception among the public is that contamination of the groundwater at Yucca Mountain would somehow quickly contaminate the Las Vegas water supply. There are two reasons this is absurd, the first being the already cited observation that Las Vegas is part of the Colorado River drainage system and Yucca Mountain of the Amargosa River, an underground river which empties into Death Valley. The second is that the movement of radionuclides through the rock, even in water saturated rock, is slow. It would require thousands of years for radioactive waste to migrate towards the surface through groundwater flow.

An environmental concern often expressed is that the radioactive materials threaten the habitat of the endangered pupfish that reside in isolated pools in an area called Ash Meadows. The pupfish are goldfish-sized creatures that are vestiges of a lake that existed in the valley west of Yucca Mountain about 40,000 years ago. The evaporation of that lake left the pupfish isolated in a few deep pools balanced precariously at the mercy of stable groundwater levels. Isotopic analysis of the water in Ash Meadows shows, however, that it derives from the nearby Spring Mountains and is not from the Yucca Mountain area. In any event, water levels are being monitored in these pools and the dangers to the pupfish from Yucca Mountain seem less than from intrusions from overanxious biologists.

## APOCALYPTIC GROUNDWATER THEORIES

Nevada NWPO raised over the years a number of groundwater issues regarding Yucca Mountain. Unfortunately, because of their concentration on anti-repository sensationalism, some of these theories have been extremist and not in agreement with empirical evidence.

For example, at the heart of geologist Jerry Szymanski's theory that water had risen to the Yucca Mountain surface was the presence of calcite veins in Trench 14, a trench dug into the side of the mountain to expose its history (see Figure 11). When Szymanski viewed this trench, only part of the calcite veins were exposed and they certainly appeared to point to the possible presence of springs in the area. Further trenching at the site however, showed a different story. The calcite veins tapered as they reached further below the surface showing this to be a surface water phenomena rather than the result of massive rises in the water table.

Another argument expressed by the state was that fractures within the rock could not be adequately modeled and that water could seep quickly along these fractures. Kristin Shrader-Frechette, the science philosopher whose ethical theories we analyzed in an earlier chapter, claims expertise in hydrology, among other fields. In a work published in the Dutch philosophy journal, Synthese, she dismissed the entire basis of the field of hydrology, using a one dimensional form of D'Arcy's Law [Shrader-Frechette, Kristin; "Idealized Laws, Anti-Realism, and Applied Science: A Case In Hydrogeology", Synthese, vol. 81, 1989, p329]. Shrader-Frechette seems to believe thousands of experimental and analytical hydrologists have not already considered the problems she associates with the use of D'Arcy's Law. As someone with a background in fluid dynamics, this author's experience is that rather than ignoring these effects as Shrader-Frechette claims, hydrologists and fluid dynamics professionals are in fact consumed with these issues.

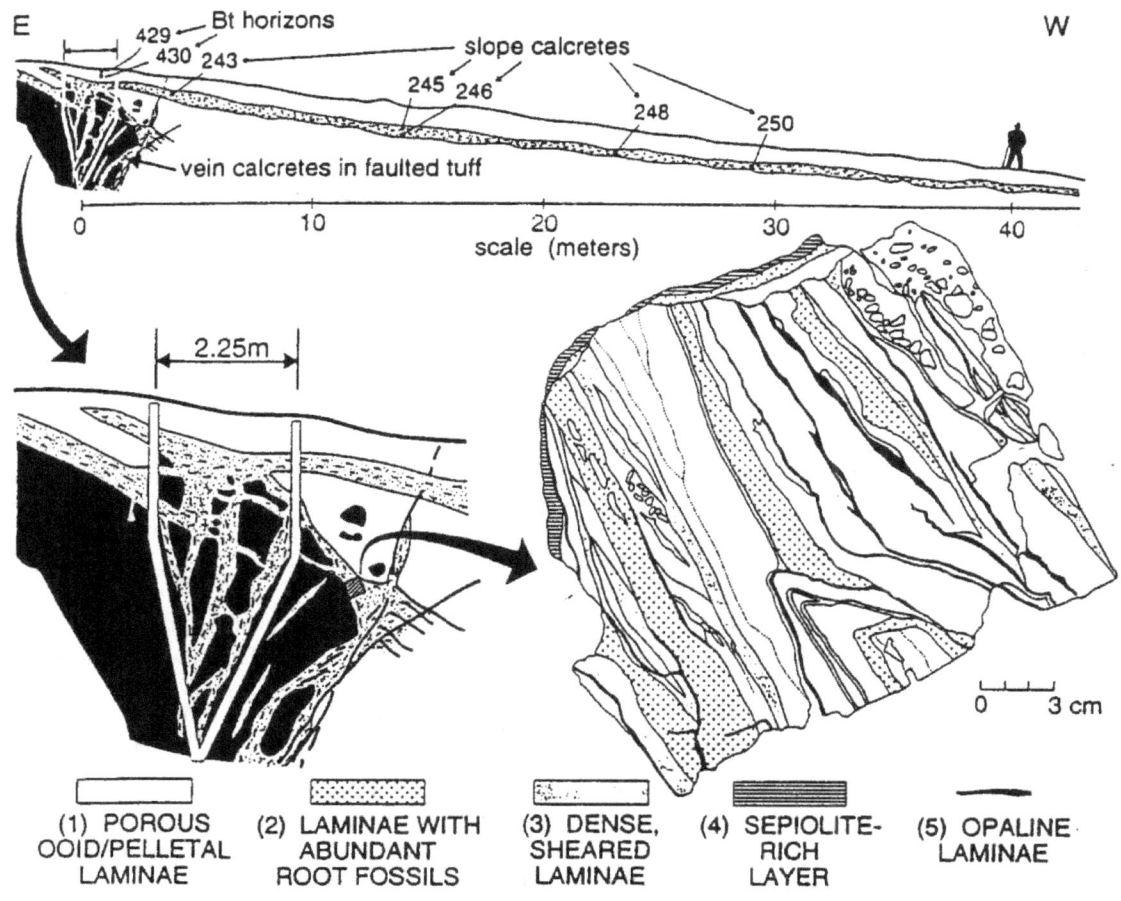

E
429
430 243
Bt horizons
slope calcretes
245 246
248
250
W
vein calcretes in faulted tuff

0          10          20          30          40
scale (meters)

2.25m

0    3 cm

(1) POROUS OOID/PELLETAL LAMINAE     (2) LAMINAE WITH ABUNDANT ROOT FOSSILS     (3) DENSE, SHEARED LAMINAE     (4) SEPIOLITE-RICH LAYER     (5) OPALINE LAMINAE

ABSTRACT: Siliceous calcretes in faults near Yucca Mountain, Nevada, have been interpreted by some observers as evidence of seismically triggered eruptions of deep water.

# 63. Transportation

The major risk avenue for release of radionuclides may not be at Yucca Mountain through geologic mechanisms, but during transportation and placement of nuclear spent fuel at the site. Fortunately, the disaster scenarios envisioned by NWPO consultants appear to be overstated. Nevertheless, there are a number of legitimate concerns that have been raised regarding the transportation of nuclear spent fuel:

1) Cask design, durability and testing methods.
2) Transportation routes to be utilized.
3) Perceived risks along transportation routes.
4) Acceptable exposure levels of bystanders during transport.
5) Emergency response to worst case disaster scenarios.

Originally, the plan was to transport spent fuel rods in shipping casks capable of holding one to three assemblies, both on truck beds and on railcars. If a design option called Multi Purpose Canisters (or similar alternatives) is chosen, almost all shipments would be by rail in large casks weighing from 75 to 125 tons and carrying up to 24 spent fuel assemblies.

## CASK DESIGN AND VALIDATION

From the outside, nuclear shipping casks appear to be simple shiny stainless steel cylinders. In reality, the casks are sophisticated devices meant to provide both radiation shielding and isolation from the environment, even in severe accident conditions. Spent fuel has been shipped safely in casks for decades and a number of designs have evolved. Present spent fuel transport in the United States is mostly an attempt to shift current waste among various storage pools, however these shipments have also provided a record of safe nuclear waste transportation.

A typical state-of-the-art nuclear waste shipping container has a number of layers which function synergistically as structural, shielding, heat transfer and moderating elements. An example is shown in the accompanying figure, a GA-4 transportation cask showing layers which function in the following manner:

- **DEPLETED URANIUM** - A good shield for gamma rays.
- **POLYETHYLENE** - The hydrogen and carbon of the plastic are good at absorbing neutrons.
- **STAINLESS STEEL SHELL**- For structural rigidity, crash resistance and added radiation shielding.

Among the purposes of this layered approach is to reduce the exposure of anyone approaching the casks to low levels, prevent leakage of radioactive material to the environment, provide structural support and to act as armor against physical and chemical assault.

The present effort is to move to 75 to 125 ton Multi-Purpose Canister (MPC) canisters capable of being used in on-site, transportation and storage modes depending on the overpack associated with a base container (see Figure for MPC design concept) . There are a number of advantages inherent in the large MPC cask design:

1) Existing on-site storage casks are not licensed for transportation or disposal. This would require them to return to pools for unloading and reloading into transportation casks. Large MPCs would minimize handling at all stages.
2) MPC technology is standardized, diminishing compatibility and safety concerns.
3) Large rail delivered casks diminishes total shipments perhaps by a factor of five or more over present designs.
4) This concept allow for at-reactor storage, allowing shutdown reactors to proceed with the decommissioning of spent fuels.
5) Total system cost may be $500 million less than concepts using Monitored Retrievable Storage.

# GA-4 LEGAL WEIGHT TRUCK SHIPPING CASK

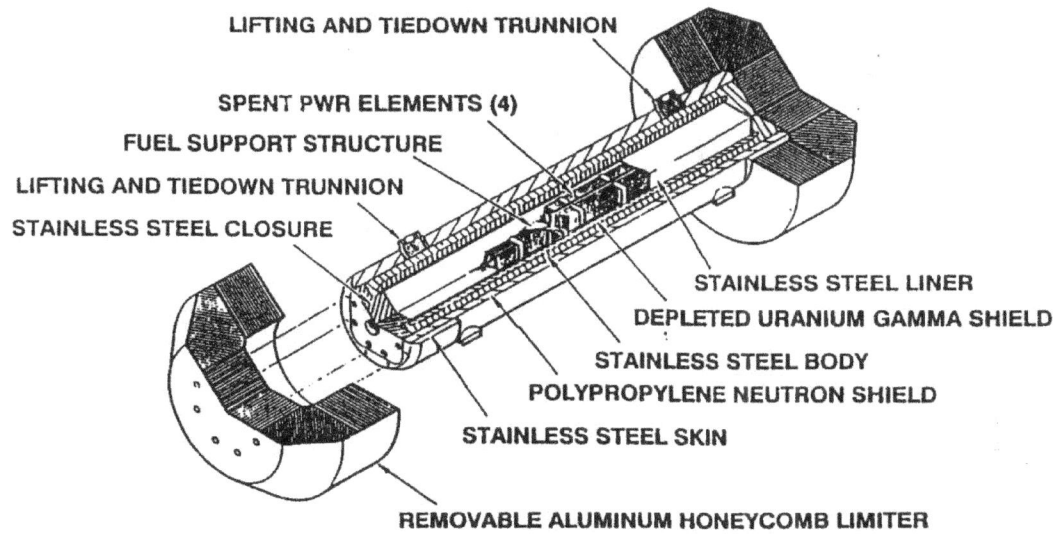

**Figure 2 SHIPPING CASK: While many designs exist, this shows a typical shielding arrangement and how spent fuel is carried. The GA-9 is a proposed shipping cask capable of carrying nine fuel assemblies simultaneously.**

Despite the promise of the MPC concept, there are a number of technical issues include:

1) Burnup credit, a technical issue regarding how much internal shielding is required to avoid criticality.

2) Canister retrievability. The MPC concept is designed around permanent closure of the units, but Yucca Mountain may remain open for one hundred years as retrievability technology is improves.

3) Criticality issues arise with a submersed repository and leaching of moderators from breached canisters. Since the likelihood of this scenario is small, this may be a moot issue.

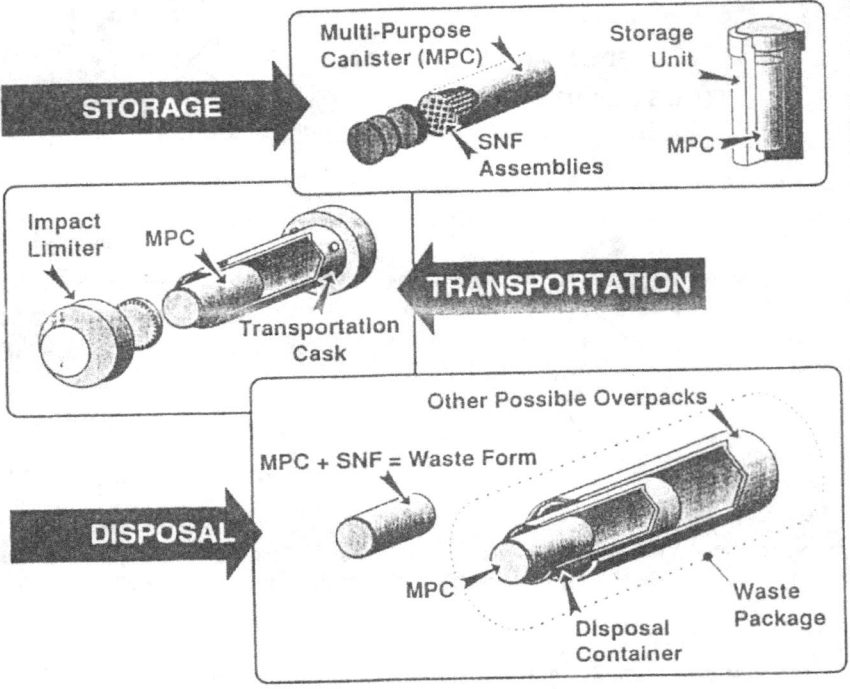

## Multi-Purpose Canister (MPC) System

Figure 18: MULTI PURPOSE CANISTER: Or MPC, a system of standardized storage canisters which can be enclosed in outer packs to allow for on-site storage, transportation or final burial.

**TRANSPORTATION ROUTES**

The challenge in choosing transportation routes for nuclear spent fuel has been to minimize radiation and accident risks without breaking the budget for infrastructure. Taking advantage of existing roads and rail lines minimizes the cost but assures the waste will be shipped through population centers. Building dedicated routes, however, could be so expensive it would make the project impossible. This is especially a problem in regard to building a railroad spur to service both Yucca Mountain and the Nevada Test Site, a project projected to cost multiple billions. Within Nevada, a number of routes have been studied over the years, but now appears to narrow to a few as shown by the accompanying figures. The preferred transportation method appears to be by rail if large Multi Purpose Canister are used. The large casks would lower the number of shipments and other problems.

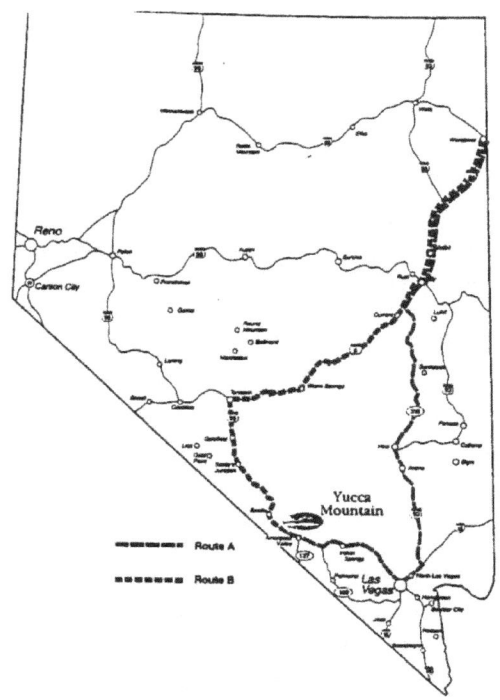

Polls conducted by Mountain West in 1987 showed that one of the issues the public feared most about the proposed nuclear waste repository was the risk of transportation accidents involving spent fuel being shipped to the Yucca Mountain site. The scenario the state focused media attention on was the danger at the "Spaghetti Bowl" highway interchange. The very center of Las Vegas sports a traffic nightmare, an interchange complex that threads highways I-15 and I-95 together very near Fremont Street and the downtown casinos. Because the interchange was designed in days when Las Vegas was never expected to be a booming metropolis, and even after substantial upgrades the Spaghetti Bowl is often a traffic nightmare. The state suggested the risks of a nuclear waste shipping container being overturned in the middle the Spaghetti Bowl during rush hour traffic were non-negligible, possibly spilling radiation over the entire city.-

Studies have considered the Las Vegas route because it takes advantage of already existing roadways, but the downtown Las Vegas overpass is just one route among many and not very likely due to political considerations. Shipment of nuclear waste through Las Vegas is unlikely because the Governor of the State of Nevada has the right to reject any such route by the Code of Federal Regulations:

**49 CFR Ch.I section 177.825 Routing and Training requirements for Class 7 (radioactive) Materials . . .**
**(b) . . . a carrier or any person operating a motor vehicle containing a highway route controlled quantity of Class 7 (radioactive) materials . . . shall operate the vehicle only over controlled routes. An Interstate System bypass or Interstate System beltway around a city, when available, shall be used in place of a preferred route through a city, unless a State routing agency has designated an alternate route.**

**(1) A preferred route is either or both an Interstate Highway System for which an alternative route is not designated by a State routing agency as provided in this section in accordance with the following conditions:**

**(i) The State routing agency shall select routes to minimize radiologic risk under "Guidelines for Selecting Preferred Highway Routes for Highway Route Controlled Quantity Shipments of Radioactive Material", or an equivalent routing analysis which adequately considers overall risk to the public. . . . .**

**(ii) State routing agencies may designate preferred routes as an alternative to, or in addition to, one or more Interstate System highways, including an Interstate System bypass or an Interstate System Beltway.**

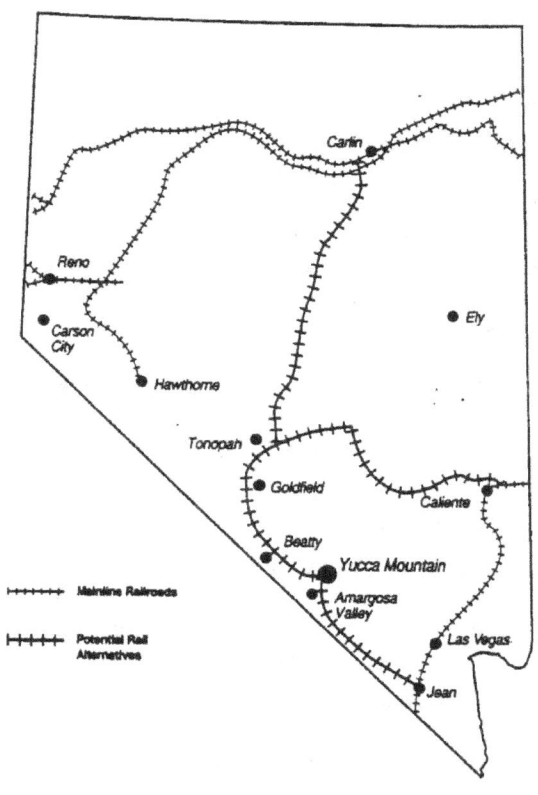

While nuclear waste shipments could theoretically travel through the Spaghetti Bowl, neither the Governor nor NWPO will ever sanction this route, making the issue moot. Consequently, the words "Spaghetti Bowl" in relation to Yucca Mountain are something of a political football.

## PERCEIVED RISKS

One of the prime reasons the perceived risks of transportation safety and routes has been debated so strongly by the state is because of the New Mexico court ruling that awarded a couple $337,815 for the loss of property value due to the perceived fears of neighbors. On November 14, 1988, the City of Santa Fe condemned 43.431 acres of a 673.77 acre parcel owned by John and Lemonia Komis. The property was condemned to permit the construction of a bypass around the City to be used for transportation of hazardous nuclear waste from Los Alamos to the Waste Isolation Pilot Project (WIPP) in Carlsbad, New Mexico. At the time of the taking the highest and best use of the parcel was speculative investment for subdivision into rural homesites or for recreational purposes.

A jury made the following awards to the Komises:
1. $489,582 for the 43.431 acres condemned.
2. $60,794 for severance damages to the buffer zone.
3. $337,815 for severance damages for perceived loss due to public perception.

The New Mexico Supreme Court held that property owners are entitled to compensation for the loss in value to land which is caused by public fear, whether or not that fear is reasonable. [City of Santa Fe, 845 P.2d at 756-757].

Consequently, discussions of the transportation safety issue must now include not only a rigorous analysis of the cask hardware and crash testing, but perhaps more importantly an analysis of the public's perceived risks. Unfortunately, this means that if welding the fins from a '57 Cadillac to the canisters improved the risk perception of the casks, though not diminishing the real risk, such a design might be well preferred.

The fact that Nevada sees the transportation perceived risk issue as the Achilles heel of the repository is evidenced by the efforts of Bob Loux and NWPO transportation consultant Bob Halstead to spread fear of nuclear waste transportation nationwide:

### STATE SENDS ANTI-NUKE PITCH ON ROAD
. . . With words like "Every American is a Nevadan when it comes to the disposal of radioactive waste" and transporting it "would unleash a Pandora's box of health and safety risks," Loux and his staff set out to reinforce the fears that the nuclear power industry is trying to erase.

Loux's cross-country, anti-dump crusade began in Omaha Neb., on a Monday in April.

It ended 11 days later when he and four staff members returned to Reno on a commercial jet. They had spent $25,689 and contacted 72 media outlets in Cleveland, St Louis, Chicago, Nashville, Washington D.C., Kansas City and Omaha.

The money for the trip came from a special fund the Legislature appropriated to the Nuclear Projects Commission for informational efforts of this type, Loux said.

[Rogers, Keith; Las Vegas Review Journal, August 1, 1993, p1B]

## EXPOSURE LEVELS

A mitigating factor in transportation risks is that most of the spent fuel now in storage pools has already been decaying radioactively for more than ten years. Since radioactive decay is exponential, we are actually faced with much less of a radiation exposure problem than what the casks were originally designed for. Estimates publicized by DOE handouts suggested casual radiation exposure to bystanders would be minimal:

Evaluations show that the additional exposure to people situated 30 meters (100 feet) from the route of a vehicle carrying spent fuel or high-level radioactive waste moving 24 kilometers per hour (15 mph) is one-thousandth to one millionth of a millirem per shipment. [Transportation: Overview of the Issues, February, 1993]

These figures reflect exposures utilizing older style casks with limited capacity. The situation changes for Multi Purpose Canisters which while potentially carrying 125 tons of waste, are unlikely to increase bystander exposure because of their shipment by rail. The large MPC units also have some self-shielding qualities because of their bulk, so surface emissions would not be substantially changed even for the greater mass carried. There has been some argument that pregnant women stuck in traffic at the I-15 spaghetti bowl in mid Las Vegas next to a waste canister might receive unacceptable doses, but this seems to be a rather contrived scenario given present designs and routing.

## EMERGENCY RESPONSE

Bob Halstead, who served as a $125,000 per year NWPO transportation consultant and replaced Bob Loux as agency head, was not a transportation engineer but something of a science historian by training. While not a technician, he attempted to earn his keep by traveling and photographing nearly every inch of the possible truck and rail routes through Nevada, producing a complete slideshow of potential problems along the highway and rail routes leading to Yucca Mountain. Halstead originally worked in Wisconsin as part of a state nuclear advisory board and was an active opponent of nuclear energy since the early eighties, in part perhaps because of a belief that a daughter has a disease condition caused by atmospheric testing.

We mention this history not to criticize Halstead's tenacious study of the issues regarding nuclear waste transportation in Nevada, because he has proven to be a diligent worker. However, when it comes to designing emergency response measures to possible accidents along a nuclear waste shipping route, we are left with the uneasy feeling that political appointees may not be able to produce convincing technical arguments. Indeed, that is why the studies at UNLV's Transportation Research Center utilized Civil Engineers with PhDs as consultants, at much smaller wages.

NWPO transportation risk assessment tends to depend on the analysis of Marvin Resnikoff, who projected in a report for the state that a nuclear waste transportation accident could cause thousands of deaths and cost $9.5 million. According to Resnikoff, drastic measures would be required in the 1.5 square mile contaminated area. Bulldozers would have to scrape up all the earth in the area and then be dismantled and buried. Workers and residents would have to undergo medical tests for years after the accident.

In contrast, estimates from national labs tend to suggest an accident which released 1 percent of the contents of a canister in the presence of a petroleum tanker explosion would cause perhaps 22 cancer deaths over 50 years within a fifty mile radius. Such a disaster, already improbable, could be nearly eliminated by rules governing the proximity of fuels and other canisters in multi-car trains.

Yet, even given the improbability of major disasters resulting from the shipment of spent nuclear fuel, it is likely that the occasion might arise when workers would have to handle a derailment type accident in which there was little or no breach of containment but a sizable mess. One problem with the Multi Purpose Canister is that finding cranes to lift 125 ton steel casks is not that easy, especially in rough terrain. Safety crews themselves would also need to be trained to deal with worst case disasters and respond to each derailing as if it were such a case. Among the equipment crews reporting to such an accident might need are:
- Sensitive monitoring devices to determine whether radiation is leaking from a breached cask.
- Radiation shielding clothing that covers the bottom of the feet and the top of the head.
- Electronic equipment that allows safety personnel to communicate with each other and officials at a base station.
- Decontamination equipment to wash radioactive dust and particles from safety workers clothes and equipment.
- Lead shielding on trucks and other equipment that carry public safety workers.
- A lead shielded mobile command post containing a computerized link with a national data bank that provides details on how to combat a spill.
- Giant vacuums to suck up contaminated materials.
- Radiation absorbent substances that encapsulate contaminated materials.

Thus, even if the most extreme disaster scenarios are discounted, nuclear waste transportation safety response is likely to be expensive. It may well cost over $1 million to equip each five man response team, and such teams might be spaced every fifty to two hundred miles along the route.

# 64. Chernobyl, Three Mile Island, Fukushima

One of the distortions used in the past to try to convince Nevadans of the dangers of nuclear waste has been to compare Yucca Mountain to the Three Mile Island reactor accident in 1979 or to the Chernobyl reactor disaster that occurred in Russia in 1988. For example, NWPO consultant Dr. Shrader-Frechette in a quoted speech to students from the UNLV environmental studies program stated:

**"Chernobyl was not a worst case disaster."** [Shrader-Frechette, Kristin; speech to UNLV **Environmental Studies sponsored by NWPO, 3/15/93]**

This was meant to imply Yucca Mountain could even be more dangerous. In fact, Chernobyl was exceedingly close to being a worst case nuclear reactor disaster. Three Mile Island (TMI) was not a disaster at all and is classed as an accident because there were no casualties. Yucca Mountain in contrast is not a nuclear reactor, but a nuclear waste repository, an exceedingly different technology from nuclear reactors.

Now with the Fukushima reactor disasters caused by the earthquake and tsunami of 2011, there will be renewed effort to claim a restarting of the Yucca Mountain project is impossible. However, unlike all three of the reactor centered disasters, a geologic repository is designed to minimize the energy levels of the nuclear materials it encloses by keeping away moderators (like water) that enhance nuclear reactions and by using neutron poisons (like Boron) to stop chain reactions. Chernobyl, TMI, Fukushima and other nuclear reactors are on the other hand designed to produce tremendous amounts of energy by promoting nuclear reactions. Power creating engines like reactors are by nature subject to catastrophic failure, while a repository is by nature designed to minimize energetic processes.

Commercial nuclear waste in the U.S. is mostly in the form of inert reactor pellets and some glassified material. Chemically, nuclear spent fuel is a non-explosive ceramic, related to an inert ashtray. It takes either a runaway nuclear reaction, a very hot furnace, or an intense explosion to create the conditions necessary to disperse these ceramic materials on the wind.

Chernobyl created exactly those conditions, becoming an incredibly hot, explosive burning graphite nuclear furnace. This chain of events is extremely unlikely to occur in nuclear reactors as implemented in the United States, evidenced by the lack of casualties and impact at Three Mile Island where there was no massive dispersal of radioactive material. Even at the disaster at Fukushima, a much worse situation than TMI complete with hydrogen explosions, air born dispersal of radioactive materials was limited. Instead, the Chernobyl disaster has everything to do with the criminally sloppy design of Russian power plants. In contrast to Chernobyl, Fukushima and TMI, nuclear waste to be stored at Yucca Mountain will lack the reaction mechanisms necessary to turn it into a massively destructive release of nuclear material. Reviewing the Chernobyl disaster will highlight the huge differences in failure modes separating a nuclear waste repository and a nuclear reactor.

## DESIGN FACTORS LEADING TO CHERNOBYL DISASTER

The first nuclear reactor that was built was done under the stadium at the University of Chicago in 1942 The reactor was composed of a "pile" of blocks of graphite (carbon) on the order of six inches square, and from this we derive the term "reactor pile". Selected blocks of the graphite contained a slug of uranium fit into a cylindrical cavity within the block so that it was possible to position uranium precisely within the graphite pile. In other words, the nuclear reactor was a pile of carbon bricks in which uranium was dispersed.

The reason for this design is that the graphite blocks slow radioactively emitted neutrons down to the right speed needed to keep a nuclear chain reaction going. In the game of pool, a billiard ball that goes too fast will jump off the table and not hit anything; the same holds true for neutrons within a reactor which must travel at the right speed to split uranium atoms and cause energy to be emitted. The graphite blocks not only slow neutrons down, but they also keep the uranium in the pile in a fixed geometry and provide a heat conduction medium to draw away the energy that is generated. This is the same general design as used at the Chernobyl plant forty years later.

Most U. S. nuclear reactors (including Three Mile Island) evolved away from the graphite pile design to water enclosed systems (Pressurized Water Reactors (PWR) or Boiling Water Reactors (BWR)) for a number of reasons both historical and involving safety. Our power reactors were originally modeled after the reactors used in the Navy's nuclear submarine program, developed during Admiral Hyman Rickover's long and memorable career. A high-performance submarine requires compactness, high power density and water cooling and for better or worse our land based reactors owe much of their design emphasis to Admiral Rickover's nuclear submarine fleet.

A number of safety factors inherent in water moderated reactors decrease the risk of nuclear accidents by American nuclear power plants in comparison to graphite pile reactors of the type Chernobyl represents. When a pressurized water reactor, or PWR for short, loses its coolant the reactor automatically shuts down. This is because the water itself is used as a moderator in place of graphite to slow neutrons to the proper speed to continue chain reactions in fissionable uranium atoms. Added measures of safety are also added to American reactors in the form of various safety systems, the most obvious being the reinforced concrete containment building which houses the reactor core. These safety features all played a role in diminishing the impact at TMI.

The reason some graphite core piles still exist in the U.S. and the reason the Russians used this design in predominance is that the nuclear reactions these reactors generate produce as a byproduct large quantities of weapons grade plutonium. It's also cheaper to build such a pile and the energy efficiency is greater, though we need to stress that efficiency comes at the expense of safety.

The Russian reactors are continuously "milked" of their plutonium on a regular basis by removing spent fuel, while U.S. commercial reactors requires a yearly shutdown; another difference between American and Russian reactor designs. The Russians omitted reinforced concrete containment shell in their reactor designs because the use of overhead cranes to pull fuel assemblies out of the core made a bulky containment structure impractical. In other words, the Russian graphite moderated reactors walk a safety tightrope without the failsafe net a containment structure provides.

Consequently, you cannot design a much more efficient way to disperse radioactive contamination over the countryside than Chernobyl. Graphite only burns at high flame temperatures, around 1800 degrees Fahrenheit. Once Chernobyl reached a meltdown state and its huge stack of graphite ignited, blowing the top off the reactor building, there was no stopping such an inferno, or even slowing it down. Yucca Mountain has zero in common with such a nuclear incinerator.

Many accounts of the Chernobyl disaster blame the problems on operator errors as they conducted a test on the energy producing capacity of the turbines after a power shutdown. While there is no doubt operator errors initiated the meltdown chain of events, according to V.M. Chernousenko, a Director for the Task Force for the Rectification of the Consequences of the Accident at Chernobyl, the power plant was designed to fail. In Chernousenko's book Chernobyl: Insight From The Inside, he writes:

**It was the very emergency system which is supposed to shut the reactor down reliably and quickly, whatever its condition, that caused it to run away. The explosion occurred 5 seconds after the emergency button was pressed. Three seconds after the button was pressed, when the power level was at 520 MW, the emergency alarms were activated by the power rise from the preceding level of 200 MW and by the sharpness of the rise.**

**Thus the accident prevention system failed to trip the reactor, not just in the normal situation, but even in this extreme situation. . . .**

**The 211 control rods could be inserted into the core at a rate of about 40 cm per second. The height of the core is 7 meters. Thus the time required for full insertion of the rods into the core averages 18-20 seconds. . .**

**Thus, Professor B.G. Dubovskii states:**

**"An emergency system which takes 18-20 seconds to operate is not a protection system at all - it is a parody of such a system. Normal emergency systems, as used in reactors all over the world, come into operation in just a few seconds (up to 5 seconds at most). At least, that is true of the rapidly acting subsystems. [V.M. Chernousenko; Chernobyl: Insight from the Inside, Springer Verlag, 1991, p. 76]**

Chernobyl was from the beginning a failure prone reactor in which most normal safety measures were compromised or ignored. At Three Mile Island when a failure of primary cooling pumps occurred, backup systems responded quickly in time to prevent a disaster. The unsafe elements of the Russian design are completely absent from the technology being envisioned for Yucca Mountain.

## THE ACCIDENT AT THREE MILE ISLAND

In contrast to the Chernobyl disaster, the accident at Three Mile Island led to no fatalities because automatic shutdown features and backup systems prevented the runaway situation in the Russian design. Because of the massive containment structure at Three Mile Island, the amounts of radioactive material the local population was exposed to was trivial, mostly in the form of gaseous bleed off of radioactive iodine. According to Bernard Cohen, the differences between the TMI accident and Chernobyl are substantial:

**. . . containment provides a broad range of protection to (a) reactor against external forces, such as a tornado hurling an automobile, a tree, or a house against it, an airplane flying into it, or a large charge of chemical explosive detonated against it. In a meltdown accident, however, the function of the containment is to hold the radioactive material inside. Actually, it need only do this for several hours, because there are systems inside the containment for removing radioactivity from the atmosphere. One type blows the air through filters in an operation similar in principle to that of household vacuum cleaners. In another, water sprinklers remove the dust from the air. Thus, if the containment holds even for several hours, the health consequences of a meltdown would be greatly mitigated. In the Three Mile accident, there was no threat to the containment. The investigations have therefore concluded that even if there had been a complete meltdown and the molten fuel had escaped from the reactor, the containment would very probably have prevented the escape of any large amount of radioactivity. In other words, even if the Three Mile Island accident was a "near miss" to a complete meltdown (a highly debatable point), it was definitely not a near miss to a health disaster.**

**The Chernobyl reactor did not have a containment anything like those used in U.S. reactors. Analyses have shown, that if it had used one, virtually no radioactivity would have escaped, there would have been no threat to human health, and the world probably would have never heard about it. [Cohen, Bernard; The Nuclear Energy Option, Plenum, 1990, p77]**

## Accident at Fukushima-Daiichi

The March 12, 2011 earthquake and tsunami that destroyed the Fukushima-Daiichi nuclear power complex in Japan may be considered the worst case possible set of circumstances for nuclear disaster. While it is unclear what the total long term casualty numbers will be, it is certain they were not anywhere near the magnitude of Chernobyl, even though these were 40 year old boiling water reactors based on fifty year old designs. These reactors had already weathered 7.0 Richter scale earthquakes before the thousand year 9.1 Richter earthquake and 40 foot tsunami caused chains of events leading to partial meltdowns.

The designers of the reactors at Fukushima-Daiichi had anticipated situations where pressure would be rising in the reactor cores. So long as power was available, pumps would circulate hot fluid from the reactor to the wetwell where it would be condensed. Heat removal could continue indefinitely in this way, but it all relied on a power source, and power had been lost due to the tsunami's destruction of the diesel generators.

Water in reactors is susceptible to damage from radiation, causing it to split into its components, hydrogen and oxygen. Normally, circulation would channel the hydrogen and oxygen to a recombiner where they would be restored back to water, but in the hours after the Fukushima reactors were shut down, hydrogen was accumulating and separating in the wetwell and reached a point where it was vented into the sparse steel-frame structure at the top of the reactor building. It was only a matter of time before the hydrogen reached a level where it would detonate, and one after another, the first unit, then the third unit, and finally the second unit, suffered hydrogen explosions that blew off the steel panels and left the top of the reactor building exposed. The reactor vessels remained intact as did the reinforced concrete containment buildings, but each reactor building lost its hat due to the hydrogen explosions.

Initially there was hope of saving the reactors to generate power again after the crisis had passed. But as that hope faded and the need to remove the decay heat remained, operators at Fukushima-Daiichi took measures to cool the reactors that would ruin them for future operation, such as the decision to try to cool the reactors with seawater. It would be necessary for some time to actively cool the reactors while the decay heat continued to decrease, but within a few months it will be possible to depressurize the reactors and assess their internal states.

This is a situation very different than Chernobyl or Three Mile Island. There was no operator error involved at Fukushima-Daiichi, and each reactor was successfully shut down within moments of detecting the quake. The situation evolved in a manner that was not anticipated by designers who had not assumed that electrical power to run emergency pumps would be unavailable for days after the shutdown. They built an impressive array of redundant pumps and power generating equipment to preclude against this problem. Unfortunately, the tsunami destroyed it.

There are some characteristics of a nuclear fission reactor that will be common to every nuclear fission reactor. They will always have to contend with decay heat. They will always have to produce heat at high temperatures to generate electricity. But they do not have to use coolant fluids like water that must operate at high pressures in order to achieve high temperatures. Other fluids like fluoride salts can operate at high temperatures yet at the same pressures as the outside. Fluoride salts are impervious to radiation damage, unlike water, and don't evolve hydrogen gas which can lead to an explosion. Solid nuclear fuel like that used at Fukushima-Daiichi can melt and release radioactive materials if not cooled consistently during shutdown. Fluoride salts can carry fuel in chemically-stable forms that can be passively cooled without pumps driven by emergency power generation. There are solutions to the extreme situation that was encountered at Fukushima-Daiichi, and it may be in our best interest to pursue them.

## THE YUCCA MOUNTAIN REPOSITORY

In contrast to Chernobyl, Three Mile Island and Fukushima reactors is Yucca Mountain, a repository specifically designed to:

1)  Operate in a non-critical state (by excluding groundwater that would increase the decay rate of radioactive materials).
2)  Have no combustible materials (no graphite) which could disperse radioactive materials.
3)  Be under an 800 feet of rock containment (versus in essence zero containment at Chernobyl).
4)  "Poison" the radioactive material with neutron absorbing materials like boron (to further diminish nuclear reactions).
5)  Make radioactive material escape pathways as long as possible, through hundreds of feet of rock vertically and miles horizontally.
6)  Slow radioactive migration through natural absorbents like zeolyte mineral.
7)  Be impervious to earthquake damage (which little effects underground tunnel structures).

Combined with other safety features, the Yucca Mountain nuclear waste repository would be substantially safer than nuclear power reactors and their storage pool. Of all the potential disaster mechanism remotely possible, local intrusions of water into the repository seems to be the greatest cause of concern.

Unfortunately, because of somewhat artificial language in the Nuclear Waste Policy Act and other regulations, for licensing purposes safety mechanisms are now limited to "non-engineered" barriers. Of course, engineered solutions might well enhance safety beyond the natural barriers to radiation transport already present at Yucca Mountain. One such mechanism would be a drainage system for the repository. Another would be continuous monitoring for ten thousand years. Many other engineered solutions could also enhance long term safety.

Even so, Yucca Mountain has an inherent degree of safety because it is by design the opposite technology from Chernobyl, Three Mile Island, Fukushima and other reactor systems – a passive rather than reactive system. The repository lacks the escalation factors through its absence of moderators that might create a criticality situation. It also lacks the incinerator-like characteristics (graphite blocks) that made Chernobyl a verifiable disaster. Consequently, reactor disasters are a poor substitute for evaluating the safety of the mountain.

# 65. Alternative Nuclear Technologies

Most nuclear reactors now in use in the United States are water moderated reactors whose designs are based on designs found in our nuclear submarine fleet since the 1950s. This is an historical anomaly, many other nuclear technologies are now available that promise a new era of safety, efficiency, and best of all, lowered creation of high-level nuclear waste. Moreover, a number of technologies on the horizon would either reprocess nuclear waste or render it significantly less threatening.

If we can survive the current onslaught against nuclear energy and high-technology in general, we can look forward to seeing improved technologies such as High-Temperature-Gas-Reactors, thorium cycle reactors, breeder reactors, transmutation by linear accelerators, fuel reprocessing, fusion energy and a host of other nuclear technologies. However, all these advances still require some disposal or storage of high-level nuclear waste, such as at a site like Yucca Mountain. These alternative technologies impinge on the study of Yucca Mountain because they increase the value of reconditioning nuclear waste. The more valuable the waste is for potential reuse, the more likely Yucca Mountain will be utilized as an interim solution to storage of nuclear spent fuel rather than a final resting place.

Ninety percent of the energy that is available from uranium nuclear fuel is still contained in the waste. In fact, the main reason spent fuel is removed from reactors is not because of a lack of usable energy, but because of the degradation of the cladding that keeps the fuel intact. A number of processes have been advanced with mixed success that attempt to recover the leftover energy available in spent fuel. Other processes seek to recover some of this energy while at the same time decreasing the radiologic hazards of the remaining waste. We'll touch briefly on a number of these technologies.

## BREEDER REACTORS

By tuning nuclear processes within what is called a breeder reactor, much of the energy left in spent fuel can be recovered by transforming portions of the waste into usable radioactive isotopes. Unfortunately, breeder reactors generate plutonium, which besides being a fissionable reactor material that can fuel power plants, is also a bomb making material. Because of President Carter's concern over the proliferation of nuclear weapons, America stopped its development of the Clinch River Breeder Reactor demonstration plant in 1980. Generating a bomb from the fissionable uranium found within reactor spent fuel requires high-tech separators and sophisticated processing. In contrast, plutonium is readily separated from nuclear waste (though not in someone's garage) and is more easily constructed into a crude bomb, hence the fear of weapons proliferation and terrorism. This is likely a moot point at this time because the fall of the former Soviet Union and the existence of other stockpiles of fissionable materials makes it relatively easy for nations which wish to obtain nuclear status to obtain the seed materials from black-market warhead stockpiles.

President Reagan overturned Carter's edict against breeder reactors, but escalating costs made restarting the Clinch River program unfeasible.

## REPROCESSING

Even without a breeder reactor, reprocessing spent fuel by separating out usable fissionable material was long thought a viable option for recycling at least part of our nuclear waste. Spent fuel rods actually contain almost 4% $U235$, but are removed from service because of degradation of their mechanical properties rather than because of a decrease in their power generating capability. Separation and reprocessing of usable fuel is already done in France as an integral part of their nuclear energy program at their Phoenix facility. However, the reprocessing process is presently 1.5 to 2 times as expensive as using primary uranium feedstock. Thus the reprocessing option has been killed by economics rather than the viability of the process. Some hope exists, however, that expensive chemical separation may not be necessary and that spent fuel could merely be ground up and reused after minimal repackaging. Such a possibility would make retrievability at Yucca Mountain a key issue.

## TRANSMUTATION PROCESSES

In a breeder reactor, neutrons from a reactor core irradiate a blanket of uranium and transform it into plutonium through the capture of a neutron by the uranium atom. Advanced research on such a breeder reactor has been conducted at the Argonne National Laboratories Integral Fast Reactor, in which neutrons are used not only for the production of new fuel, but also to generate heat to run power turbines. The Integral Fast Breeder Reactor is designed to irradiate its own waste, allowing the reactor to generate its own fuel as well as reduce the amount of high-level radioactive waste all reactors create at some level. With modifications, the Integral Fast Reactor could also irradiate the waste from other plants as well and depending on the isotope either convert elements into useful radioactive substances or to non-radioactive forms.

**Accelerator Transmutation of Waste**

- **Driven sub-critical system**
- **Intense thermal neutron fluxes**
- **Favorable transmutation cross sections**
- **Minimal material inventories**
- **Continuous material feed and separation**

An alternative way to transmute nuclear waste is by irradiating it with neutrons from a particle accelerator (see Figure). Protons are accelerated into a lead target which spews neutrons that transmute the waste and also generate enough heat to produce surplus energy. One advantage of accelerator based transmutation technology is that if problems arise, for example the creation of excessive heat, the accelerator can be simply shut down to stop the process. Another advantage is that substantial amounts of energy can also be recovered from the process. Both the Los Alamos and Brookhaven National Laboratories have been studying accelerator based transmutation technology and private industry is also involved. Once again, cost is the driving factor

## ALTERNATIVE REACTOR TECHNOLOGIES

Part of the disposal equation is also the availability of cleaner and more efficient reactor designs. Among such options are high-temperature gas reactors, modular designs which take advantage of mass production techniques to eliminate production error, fast integral reactors which use fast neutrons to degrade or process waste materials, ad infinitum. Designs already available would create substantially less waste per amount of energy produced. Unfortunately, because of the bottleneck created by Yucca Mountain, it is unlikely any advanced reactors will be built given the possibility that long term storage will not be available at the end of the process. This creates a situation in which not only are safer and more efficient designs kept from production, but older designs are operated towards or past their design lifetimes, further compromising safety.

THORIUM: One such alternative is to move to a thorium based fuel rather than uranium. The thorium fuel cycle uses the naturally abundant isotope of thorium, 232Th, as the fertile material which is transmuted into the fissile artificial uranium isotope 233U which is the nuclear fuel. However, unlike natural uranium, natural thorium contains only trace amounts of fissile material (such as 231Th), and these are insufficient to initiate a nuclear chain reaction. Thus, some fissile material or other neutron source must be supplied to initiate the fuel cycle. In a thorium-fueled reactor, 232Th will absorb neutrons eventually to produce 233U, similar to the process in uranium-fueled reactors whereby fertile 238U absorbs neutrons to form fissile 239Pu. Depending on the design of the reactor and fuel cycle, the 233U generated is either utilized in situ or chemically separated from the used nuclear fuel and used in new nuclear fuel. The thorium fuel cycle claims several potential advantages over a uranium fuel cycle, including greater abundance, superior physical and nuclear properties of fuel, enhanced proliferation resistance, and reduced plutonium and actinide production.

HIGH-TEMPERATURE GAS REACTOR (HTGR): Current Light Water Reactors (LWTRs) Operate at an approximate maximum temperature of 300°C and generally produce just electricity. HTGRs operate at temperatures of above 950°C, allowing them to produce electricity more efficiently, while being hot enough to support a variety of chemical processes. Specifically, HTGRs can provide the heat needed to produce hydrogen and liquid fuels (including diesel and jet fuel) from natural gas, biomass, coal, oil sands or oil shale. All these resources are abundant in North America; therefore, the United United States could improve its energy security by using HTGRs in our energy supply chain.

As a side benefit, HTGR based energy production will greatly reduce greenhouse gas production. Conventional methods of converting feedstocks to liquid fuels require combustion to obtain a high temperatures needed. Replacing combustion with a HTGR produces no $CO_2$, so the increased energy security can be achieved with significant environmental benefit. For example, replacing combustion with a 600 MW HTGR in the process of converting coal to jet fuel, diesel, or gasoline would displace approximately 41,500 barrels per day of imported crude oil while essentially eliminating $CO_2$ emissions. Even if we continue to rely on petroleum refining to produce liquid fuels, integrating a HTGR into the process greatly reduces $CO_2$ emissions. As an example, peaceful refining with a HT GR potentially achieves approximately 75% lower $CO_2$ emissions.

Despite the benefits, there are two major challenges to implementing HTGR's: 1) the high front capital investment required and 2) public perceptions about the safety of nuclear power. However, HTGR's are self-regulating below the temperature at which the fuel will melt, and they are gas cooled, eliminating the danger of superheated liquids exploding.

Other advanced reactor concepts are being evaluated throughout the world for the next generation of nuclear energy. The U.S. Congress initiated the Next Generation Nuclear Plant (NGNP) project in 2005. Based on a systematic evaluation of several next-generation concepts, the Department of Energy (DOE) selected a Modular High Temperature Gas-cooled Reactor (MHR) as the concept for NGNP. A key design feature of the MHR is intrinsic safety. The MHR can survive a complete loss-of-coolant accident, including failure to insert control rods, without reliance on any emergency systems. As the reactor heats up, natural processes will shut it down. Because the reactor core and nuclear fuel are composed entirely of refractory and ceramic materials with capacity to absorb heat at high temperatures without structural degradation, there is no damage to the reactor, i.e., the reactor cannot melt down under any circumstances. No public evacuation is required, even next to the plant's entrance gate. With its high temperature capability and efficient heat utilization, the MHR can generate electricity with high efficiency and displace fossil fuels for a number of petrochemical and industrial applications, including production of hydrogen for future clean fuel utilization. San Diego-based General Atomics (GA) is a pioneer of this technology. With partners from the U.S., Japan, and South Korea, GA is now completing the conceptual design of the NGNP demonstration plant for the DOE.

Whatever technology eventually evolves, it is clear America is at least a generation and perhaps two behind the types of technologies that could be implemented, as shown in the accompanying figure.

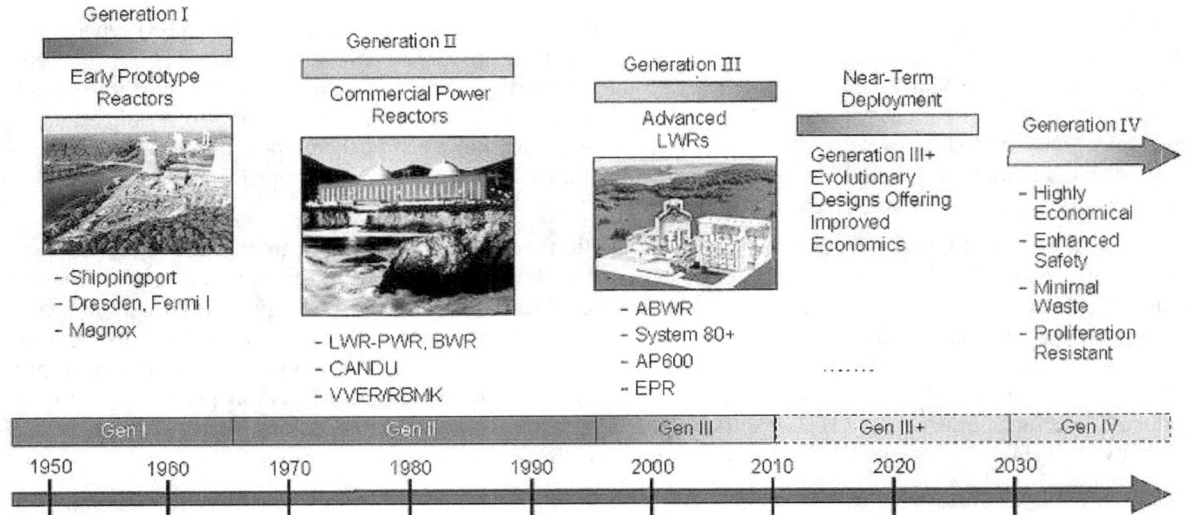

Generation IV: Nuclear Energy Systems Deployable no later than 2030 and offering significant advances in sustainability, safety and reliability, and economics

## FUSION ENERGY

Fusion energy, the energy source of the future, also requires the disposal of high-level nuclear waste. Where nuclear reactors represent the release of energy from controlled atomic bombs, fusion harnesses the even greater energy availability from the processes that drive hydrogen bombs and the sun. When two hydrogen atoms are compressed and heated to extreme densities and temperatures, they fuse to form a helium atom that is slightly less massive than the two hydrogens. This small mass differential is converted through Einstein's famous equation, $E = MC^2$, into large amounts of energy. In practice, two isotopes of hydrogen called deuterium and tritium are used because they are easier to fuse than elemental hydrogen.

Both deuterium and tritium are radioactive and because they combine with oxygen to make water, are easily dispersed through the environment. More importantly as the source of an energetic radiation spectrum ranging from gamma rays to x-rays to atomic nuclei, fusion processes generate energetic neutrons which irradiate the containment walls and transform various elements into long lived high-level nuclear waste material. For example, if 316SS (a stainless steel) is used in the walls of a plasma fusion reactor, the following induced radioactive materials would be created:

**55Fe Iron-55 2.94 year half life**
**58Co Cobalt 58 72 year half-life**
**54Mn Magnesium-54 310 day half-life**
**60Co Cobalt-60 525 year half-life**
**[after Nuclear Fusion, Keishiro Niu, Cambridge University, 1989, p221]**

Although the half lives of these radioactive substances are relatively shorter than those found in fission reactors, their quantities will still be substantial. In any event, the promise of fusion power still appears to be decades away from being fulfilled because the technology is exotic. Even with the technical problems of fusion power solved, it is not clear that it will be economically competitive in the near future, especially for intermediate sized plants.

## FUTURE USES OF NUCLEAR WASTE

Although reprocessing and transmutation are not now cost effective and fusion energy is many years distant, we have taken the time to explain these technologies because of what they imply for the Yucca Mountain project:

- At some time in the future, perhaps less than 100 years, it may become economical to reprocess or transmute spent fuel. This implies that Yucca Mountain may someday become a valuable energy resource repository. The question this raises is who will have deed over this resource and for what period of time will the repository be designed to remain open for retrieval of spent fuel.

- The Third World cannot afford to overlook nuclear power as an energy source if it is to thrive. Even if the materials stored at Yucca Mountain may never have economic benefit to the United States, they may have benefit to other world nations. In that case, the total risk to the environment from nuclear materials may well be advanced if Yucca Mountain becomes an internationally recognized and competent center for nuclear waste disposal, transmutation and reprocessing technology.

- If Nevada became the energy capital to the world, compensation for this service would not presently go to Nevada. Nevada's current political incumbents may be in the process of walking away from an energy bonanza and benefits that would dwarf the Alaska pipeline profits paid to that northernmost state. Consequently, forward thinking politicians might be negotiating now for the rights to these potentially invaluable resources, recognizing that Nevada may at some point become the central nuclear technology center in the world.

- Reprocessing, transmutation, new reactor designs, etc., do not eliminate all high-level nuclear waste and often actually increase the volume of waste. This means foreseeable technologies will not eliminate the need for a long term geologic repository, such as Yucca Mountain.

- Unless we wish to eliminate nuclear power and depend solely on hydrocarbon and solar technologies, which have environmental and technical problems of their own, there will be a continuing need for geologic storage of high-level nuclear waste. The Third World may well remain impoverished and potential enemies if we force them to compete for limited traditional energy resources. Yucca Mountain could conceivably evolve into the energy "bank" of the world.

Shutting down Yucca Mountain may delay the inevitable transformation of our nation to nuclear energy sources, but it does not stop the international movement towards nuclear energy nor the need for our nation to find alternative means for disposing of the waste we have already created. A moratorium on new nuclear technology (imposed de facto by delay of the Yucca Mountain project) thus poses serious risks for both ourselves and the world environment by hobbling America's design of new generations of efficient and safe reactors that would keep us economically competitive. Although these reactors would create less than a quarter as much nuclear waste as present designs, they are unlikely to be pursued if geologic storage is cut off.

# 66. Relative Risks

In a rational world, humans would determine their actions by an objective analysis of risks. In the case of Yucca Mountain, an entire new field of science, i.e. nuclear risk perception analysis, has arisen to try and prove the opposite; that perceived risks should be the determining factor behind technological advance. In the preceding chapters, we have outlined the attempts by the Mountain West socioeconomic teams, funded by $15 million in Nevada Nuclear Waste Project Office grants, to impose an "availability heuristic" and "negative images" theory on Nevada. These theories suggest humans cannot change their perceptions of unknown fears through education and that negative images and fears dictate their every action. It seems appropriate to attempt to present a non-psychological, objective analysis of the risks of Yucca Mountain in the hope that citizens can decide for themselves whether to judge Yucca Mountain on the basis of emotional fears or cold hard statistics.

## YUCCA MOUNTAIN RISKS

Saying the risks from Yucca Mountain are low is not the same as saying they are non-existent; technology and life in general are inherently risky. The risks associated with Yucca Mountain are of the following sort:

**TRANSPORTATION ACCIDENTS** - The breach of a transportation cask within a city boundary would obviously cause damage requiring expensive remediation. Fortunately, even in an accident in which a transportation cask is hit by an airplane or engulfed in flames as in the Caldicott tunnel fire, the integrity of the robust casks is unlikely to be much affected. Even if such a breach occurred under these extreme conditions, cleanup is relatively straightforward because the radiation can be detected with simple apparatus (Geiger counters).

**CRITICALITY** - If water invaded the repository site through some rise in groundwater, there are certain design situations now being evaluated in which containers could reach a critical state (given complete submersion of some fraction of the repository plus the leaching of moderators from the container package).. This is not the same as a "China Syndrome" meltdown, but a situation in which a nuclear reaction would generate heat in the repository. These possibilities are under investigation.

**EARTHQUAKES, VOLCANOES** - As discussed before, geologic movement is perhaps the least likely to cause actual risks to the public through any affect on Yucca Mountain. Unfortunately, NWPO and the media find these to be emotional triggers.

**HEAT PIPE WATER CORROSION** - A hot repository can cause water vapor to accumulate and condense in fractures, allowing water to concentrate and corrode waste casks. Radioactive releases would likely be spread over substantial time. This problem is still being addressed.

**GROUNDWATER CONTAMINATION** - Pathways for contamination of the groundwater outside the Yucca Mountain area are effectively non-existent. The Las Vegas valley groundwater system and that of Yucca Mountain do not interconnect. Contamination of Yucca Flats would be over a timescale of millennia and then at low levels.

**CARBON 14 AND IODINE 138** - Releases of both these substances for the entire Yucca Mountain site would be near those of a single reactor. Release of these gases would likely be the primary radiation exposure mechanism for humans.

These are not the only risk mechanisms, but represent the majority of risk. Present analysis suggests that there are hundreds of risks that Nevadans face each day that are orders of magnitude greater than any danger the Yucca Mountain repository poses. The limiting factor in nuclear waste risk has much to do with the inert chemistry of the spent fuel oxides.

Nuclear spent-fuel waste is composed of uranium dioxide and oxides of other transuranic metals. In other words it is a ceramic, similar to a ceramic pot or an ashtray. This material is relatively inactive: chemically, it won't explode, it melts only at extremely high temperature and if it is spilled, it is difficult to spread beyond the accident site. That isn't to say nuclear waste isn't dangerous, but most kitchens have Clorox and ammonia and other killer substances under the sink which pose significantly greater immediate health risks, though no one panics about these deadly chemicals.

Preliminary estimates are that a person lives within five kilometers of the Yucca Mountain will receive less than one additional millirem per year of total background radiation, about one tenth the radiation of an x-ray. This compares favorably to the U.S. annual average dose of 360 mrem

## SIMILARITIES TO NEVADA TEST SITE EXPERIENCE

NWPO generally argues that estimates of risks prepared by DOE are inaccurate. They have also argued that even one death per ten years, DOE's conservative estimate for total possible worldwide deaths from Yucca Mountain, is too high. Yet, Nevadans already have been exposed for thirty years to a series of nuclear incidents as deadly as the worst case scenario for a transportation accident at Yucca Mountain, all with negligible effect. In the 1950s and early 1960s the government conducted above-ground nuclear explosions at the Test Site that Las Vegans watched from Glitter Gulch and their backyards. Compared to the threat posed by Yucca Mountain, those previous planned accidents spewed tons of radioactive material into the air.

The risk levels Nevadans will suffer from Yucca Mountain are at worst of similar magnitude as those already experienced from Test Site atomic explosions. To address the possibility that DOE has underestimated potential repository related deaths, one can gain some insight by looking at data on so-called downwinders, those who were downwind of the above-ground nuclear tests of the 1950s. Recent studies by University of Utah epidemiologist Richard Kerber and others studied these effects in thyroid tumors and cancers:

**STUDY LINKS CANCERS, FALLOUT**
**RESEARCHERS SAY UTAH'S HIGHER THAN NORMAL RATE OF THYROID CANCER MAY BE TIED TO 1950S NUCLEAR TESTS**
Schoolchildren living in Utah and Nevada during above-ground nuclear tests in the 1950s had a higher than normal incidence of thyroid cancer 30 years later, according to researchers at the University of Utah.
The study of 2,473 people likely to have been exposed to radioactive fallout from the Nevada Test Site shows "an association between thyroid tumors and exposure to radioiodine, researchers said.
The study funded by the National Cancer Institute, was published in this week's Journal of the American Medical Association.
It was the second part of a study that looked at both thyroid diseases and leukemia. The researchers reported in 1990 the results of their leukemia work: that radioactive fallout may have been responsible for about 3 percent to 6 percent of Utah's leukemia deaths from 1952 to 1981.
The thyroid findings are similar, said University of Utah epidemiologist Richard Kerber, one of the study's authors.
**[Green, Jan; Las Vegas Review Journal November 4, 1993 pp 1A]**

Since the levels of radiation exposure due to weapons tests are much larger for the test group than the population could expect from Yucca Mountain accident scenarios, applying these risk levels to Yucca Mountain would be conservative. But what do these studies really say?

For humans under the age of 65, risk of death from leukemia is approximately 1 in 10,000 per year. Consequently, even for an increased risk of 6%, this would suggest that for the exposed population of Utah and Nevada (perhaps 200,000) there would be an extra .6 cancer deaths per year.

Consequently, we can presume the risk factors presented by DOE in regard to Yucca Mountain (1/10th death per year from Yucca Mountain) are no more than an order of magnitude incorrect given the findings of the Utah study. Even a factor of ten, however, is unable to much move Yucca Mountain from the bottom of the relative risk heap in regard to other technologies and other accident mechanisms. For example, the construction of the Luxor hotel in Las Vegas, a $500 million project, saw two outright construction deaths, for a risk level of 2 deaths per year, twenty times that predicted for Yucca Mountain, even based on the whole population of the Las Vegas Valley.

If we make a rough worse case extrapolation of possible deaths from the Yucca Mountain repository, we might estimate that as many as one death per year could be caused long term from excess radiation. For a population of nearly two million in the Yucca Mountain extended area, this corresponds to a risk of 1 in 2 million. In comparison, the risk of death from a falling airplane is 1 in 10 million, from lightening 1 in 10 million, from tornadoes 1 in 455,000, from earthquakes 1 in 588,000 (California), and from being struck by an automobile 1 in 20,000.

## OBJECTIVE RISK ANALYSIS

Many risk assessments have been made of nuclear energy, nuclear waste disposal and transportation. Some of the most important general reviews are contained in the National Academy of Science's BEIR Committee reports (Biological Effects of Ionizing Radiation).

There are many ways of expressing quantified risk, but the one which probably puts the risks associated with Yucca Mountain in best perspective is the loss of life expectancy (LLE); i.e., the average amount by which life is shortened by a given risk. This contrasts with other measures of risk which may estimate, for example, the probability of obtaining a cancer from a radiation exposure, but do not give a feel for how this might affect one's quality of life. Certainly, a hazard that produces an immediate death at a given rate affects our quality of life more negatively than a death that occurs only after eighty years of life, albeit at the same probability.

The physicist Bernard Cohen presents an in depth discussion of lowered life expectancies from various sources in his book, The Nuclear Energy Option [Chapter 8, Plenum, 1992]. According to Cohen's analysis (see accompanying Table), based on numerous supporting references including the environmental community, a full nuclear program in the United States would be expected to increase average exposures by .2 millirem per year, leading to a reduced life expectancy of 37 minutes. Since local radiation exposure due to Yucca Mountain, even for those living exceptionally near the facility, is expected to be less than one millirem per year, the LLE for Yucca Mountain would be approximately 185 minutes or three hours. In comparison, the LLE of naturally occurring radon gas in homes is 35 days and the LLE from motor vehicle accidents is 180 days.

Unfortunately, the opposite question does not seem to have been analyzed thoroughly - i.e., what is the lowered life expectancy due to implementing indefinite on-site storage as an alternative to Yucca Mountain, which now seems to be our present course. Without a definitive study, we can only make educated guesses, but there are some reasons to believe the order of magnitude of risks would be just as great as for implementing Yucca Mountain, even for Nevadans. The first reason this is so is because emissions of gaseous iodine-129 and carbon-14 will be no better isolated in on-site storage casks than in geologic storage. Moreover, since on-site storage is likely to be in the same Multi Purpose Units as proposed for Yucca Mountain, similar levels of release of other solid radioactive materials is likely to result from either scenario in the short term (long term, isolation under 800 feet of rock would make the repository safer).

Interestingly, the lowered life expectancy to Nevadans might be the same or even worse with on-site storage as with storage at Yucca Mountain. First, Nevadans are a notoriously transient population, so residents are likely to have moved from sites near on-site storage facilities and carry much of that risk with them. Secondly, on-site storage could lead to cancellation of the nuclear energy option in the U.S. making us less economically competitive. Since Cohen points out that living in poverty has an LLE of 3500 days, even a marginal negative economic impact from not building Yucca Mountain could have an LLE effect that swamps the radiation risks associated with living near a repository.

# PART VIII:  ENERGY ECOLOGY

# 67. Murphy's Law vs Second Law

*MURPHY'S LAW: Anything that can happen, will happen!*

The fundamental question at Yucca Mountain is whether we can utilize the sophisticated resource called nuclear energy safely or whether we are doomed by some sort of universal Murphy's Law to destroy our planet through the use of this forbidden fruit.

At a June 1992 Sawyer Commission meeting, rural sociologist William Freudenburg and NWPO consultant applied Murphy's Law to risk analysis at Yucca Mountain. Roger Kasperson and his sociologists at CENTED as well as Paul Slovic of Decision Research also used similar lines of reasoning to argue for NWPO that the potential for human error at Yucca Mountain is insurmountable. They argued that it is impossible to accurately detail the risk trees of complex technological ventures and that we therefore cannot guarantee the safety of the nuclear waste repository. Marvin Resnikoff, a consulting physicist for NWPO, has also argued that nuclear disaster is inevitable along transportation routes because human error is insurmountable due to a vague statistical Murphy's Law.

Murphy's Law, however, is not a law! At best, it is a bastardized form of the physical law of entropy, and its use by NWPO scientists and others within the environmental movement has led to a distortion of the risks of nuclear waste transportation and in the construction of the repository. However, it is absolutely false that anything that can happen, will happen. These events *may* happen, often with a probability so close to zero as to be impossible.

The false assumption that Murphy's Law is a scientific verity leads to an apocalyptic theory of impending environmental doom in which every technological invention is impossibly dangerous. For example, an automobile could crash into a nuclear weapons silo, possibly sending a warhead into space, which might collide with a meteorite, possibly deflecting the missile to the moon, which might split the moon in two. Fortunately, instead of Murphy's Law, science uses principles of thermodynamics to study the likelihood of technology proving catastrophic. This provides a much less cataclysmic view of our world than Murphy's Law and fortunately doesn't require humans to return to caves to save the environment.

The two laws of thermodynamics which govern everything at the macroscopic level (i.e. all things above the level of subatomic particles) are the following:

*FIRST LAW: Conservation of Mass/Energy. Mass and energy can neither be created nor destroyed, only transformed.*

**dM + dE = 0**

*SECOND LAW: Entropy. The order of a closed system can at best remain the same, or it can degrade, but it cannot become more ordered without an external mass/energy source.*

**dS >= dQ / T**

For those uncomfortable with physics, the laws say in simple English that:

*FIRST LAW: Energy and matter don't appear out of thin air.*

and:

*SECOND LAW: If you want to increase the organization of your world, you'll have to put some work into it*

These two laws have been known for better than a hundred years to the physical scientists. However, many social scientists and environmental activists have failed to comprehend that the First and Second Laws work in

---

conjunction with each other. Engineers are familiar with the symbiosis of the two laws in a relation called the Gibb's Free Energy equation (although it appears in other forms depending on the scientific discipline):

*GIBB'S FREE ENERGY: A reaction only occurs if it results in a decrease in free energy, G, given by the relation:*

**dG = dH - T dS**

In translation, it takes brains and/or brawn to accomplish useful work. Intuitively, there are multiple ways to get a rock to the top of the hill: (you can push it yourself by hand [brawn] or you can drive a bulldozer [brains and brawn] or you can hire someone else to do it [pure brains]). What is generally not understood is that given a sufficient source of energy, there is little that can't be accomplished. Specifically, if humans can harness enough energy from solar, nuclear or other traditional sources (and use that energy wisely), there is no limit to how far our civilization can progress in an environmentally sound way!

Where environmental philosophers have gotten into trouble is by theorizing exclusively on the basis of the First Law, or the Second Law without reference to the combining Gibb's equation. This has led to endless scenarios of impending environmental doom because the thermodynamic laws taken individually do not describe a complete physical system. There is no way to reverse the effects of pollution within such crippled physics.

In a worldview dominated by an inordinate fear of Second Law entropy processes (bastardized further as Murphy's Law), nuclear radiation becomes particularly frightening. The paranoid physics that results assumes that radioactive substances cannot be kept isolated from the environment and that even minuscule levels of radiation will lead to irremediable damage.

Researchers from the Nuclear Waste Project Office have viewed Yucca Mountain as an impossible technology because we supposedly cannot see to the end of every fault tree. Every risk, no matter how statistically unlikely, thus becomes for NWPO an insurmountable entropy problem. Every chance emission of radiation becomes horrendously frightening. Thermodynamics instead suggests that while we may not know every possible fault path, we can put outer limits on potential disaster scenarios.

For example, we know the nuclear waste will not explode with the force of a nuclear bomb because there are no mechanisms for creating a super critical reaction at the repository. We know nuclear spent fuel pellets are ceramic and highly unlikely to be widely dispersed even after an accident. We know the heat load on Yucca Mountain will be about 56 megawatts. We know that transportation casks will not melt and vaporize unless a well characterized amount of energy is available in an accident. Most importantly, we know that even in the event of a release, radioactive substances can be cleaned up if sufficient effort is expended.

Nature also argues that radioactive substances can be handled effectively, though they must be treated with care. Biological systems already have billions of years of experience successfully dealing with natural radiation ranging from cosmic rays, radiation from the sun, background radiation from soils and rock, radiation from ingested substances like radioactive potassium and radon gas, etc. Humans have the added advantage of being able to "see" radioactive emissions with simple devices like Geiger counters and dosimeters.

Consequently, while everything that can happen may happen, this does not imply they will happen, or that they cannot be reversed. If Murphy's Law were a law, biological organisms would long ago have become extinct due to radiation mechanisms. Fortunately, neither the biosphere nor Yucca Mountain are inherently doomed by some arbitrary law of human error.


## THE ENVIRONMENTAL ENTROPY CRISIS

If the Earth were a closed entropy box, disorder could enter or be created on planet earth but never leave and chaos would soon overwhelm us. Just such a misinformed theory has become embedded in environmental policy, perhaps because it promises ecological catastrophe and buttresses environmental lobbying efforts in Washington. However, misusing the laws of thermodynamics to support popular political ideology can prove dangerous.

Historically, the First Law (i.e., the conservation of Mass/Energy), was used and abused by communist regimes until the eventual collapse of their entire social system. Workers were valued for their efforts in mass or energy production (how many shoes they made, how many tons of energy rich coal they dug), completely ignoring the entropy and information content of their efforts (The Second Law of entropy), which in this case was the order or quality of the products they produced. The managerial, information element of civilization was ignored, resulting in a society in which senseless production was preferred to quality and timeliness. This eventually resulted in the massive degradation of their environment.

Where the Marxists over-emphasized the First Law, the new Greens over-emphasize the consequences of the Second Law, which says that closed systems degenerate into chaos over time. They theorize that the earth is a closed entropy box doomed to self destruction from man's technology (epitomized by Yucca Mountain). Fortunately, we do not live in a closed entropy box. Noted climatologists Jose P. Peixoto and Abraham H. Oort have the following to say about the entropy balance of the earth:

**Thus solar photons are richer in energy than terrestrial photons. In other words, the amount of entropy associated with the incoming solar radiation is much lower than the amount of entropy associated with the emitted terrestrial radiation, and the climate receives high-quality "rich" energy and returns low-quality "impoverished" energy to space. Thus, solar radiation revitalizes the meteorological phenomena, feeds the hydrological cycle, and renovates the biosphere. If earth were an isolated system there would be an unavoidable increase in entropy leading to a death-like uniformity of the planet earth. It is this capacity for permanent renovation that makes all natural phenomena possible in the climate system. [Peixoto, Jose P. and Oort, Abraham H.; Physics of Climate, American Institute of Physics, 1992, p 403]**

Also:

**As we see from the values given . . . the total amount of entropy exported by the climate system to space is -925 mWm-2K-1. This value is 22 times the amount of entropy imported by the incoming solar radiation at the top of the atmosphere (41.3mWm-2K-1). [ Physics of Climate, p409]**

This points out why understanding the thermodynamic laws is so important; literally the difference between the life and death of civilization. As stated by Peixoto, the ratio of entropy influx into the earth is twenty-two times less than the efflux going out, so we are not doomed to increasing randomness and disorder as a civilization, even given any technological advances we might make. Natural energy input from the sun to the earth is approximately $3.5 \times 10^{17}$ watts versus $5 \times 10^{12}$ watts from manmade energy, a difference on the order of 1 in 100,000. In other words, any entropy we produce will be a negligible fraction of the entropy 'current' already flowing through Mother Earth.

Because Earth is not a closed system, if we can find sufficient energy reserves there is no reason we can't effectively export our entropy (i.e., our pollution) to the universe. The trick is to find an appropriate energy source.

Solar and alternative energies have inherent inefficiencies; hydrocarbon resources are finite and polluting. Nuclear energy, if not artificially restricted by political processes, promises a substantial energy reserve capable of stretching thousands of years. This would allow us to avoid entropy death indefinitely and reverse the polluting effects of man on his earth environment. The Yucca Mountain repository obviously plays a critical role in such a scenario; without it nuclear energy ceases to be a positive part of the human entropy equation.

# 68. Fossil Fuels, Global Warming & War

The specter of Kuwaiti oil fields burning and spewing black clouds of pollution across the landscape at the end of our war with Iraq in 1991 brought into focus the entire range of environmental and security problems that surround our dependence on fossil fuels. Fossil fuels promise not only environmental problems, but also entanglement with world politics on far away foreign soils. American men and women will be repeatedly called upon to die for the sake of protecting hydrocarbon lifelines.

One tenet of the opposition to Yucca Mountain has been the notion that by opposing and destroying the nuclear energy option in the United States, we could somehow stop the proliferation of weapons of mass destruction. However, Saddam Hussein, Moamar Khadaffi and other Middle East dictators, funded in large part by oil revenues, were able to cause wars without help from American radioactive materials. In fact, Iran has built its nuclear weapons capabilities on the back of oil proceeds. The stagnation of America's nuclear energy capacity over the past decades, caused by the successful delay tactics of the Green movement, has thus contributed to provoking hostilities throughout the Middle East while not slowing nuclear proliferation. America has no alternative but to respond to threats to its oil supplies while France, which receives 73% of its energy from nuclear power, has been notably smug knowing its energy jugular isn't being severed.

The world economy's increasing electrical energy needs can only be met through one of three alternatives: hydrocarbon, nuclear or solar energy. Unfortunately, solar energy can only supply part of the energy needs of civilization and nuclear energy is in the process of being blocked, specifically at Yucca Mountain. Surprisingly, fossil fuel in conjunction with coercive energy conservation measures, seems the default policy choice being advocated by environmental activists, despite the fact that fossil fuels are by their very nature a polluting and unrenewable source of energy. But just how polluting are fossil fuels compared to, say, nuclear energy? A prime example is coal:

**As an initial perspective, it is interesting to compare nuclear waste with the analogous waste from a single large coal-burning power plant. The largest component of the coal-burning waste is carbon dioxide gas, produced at a rate of 500 pounds every second, 15 tons every minute. It is not a particularly dangerous gas, but it is the principle contributor to the "greenhouse effect." . . . (of the other wastes) first and probably foremost is sulfur dioxide, the principle cause of acid rain and perhaps the main source of air pollution's health effects, released at a rate of a ton every five minutes. Then there are nitrogen oxides, the second leading cause of acid rain and perhaps also air pollution. Nitrogen oxides are best known as the principle pollutant from automobiles and are the reason why cars need expensive pollution control equipment which requires them to use lead-free gasoline; a single large coal-burning plant emits as much nitrogen oxide as 200,000 automobiles. The third major coal burning waste is particulates including smoke, another important culprit in the negative effects of air pollution. Particulates are released at the rate of several pounds per second. And next comes the ash, the solid material produced at a rate of 1,000 pounds per minute, which is left behind to cause serious environmental problems and long-term damage to our health. Coal-burning plants also emit thousands of different organic compounds, many of which are known carcinogens. Each plant releases enough of these compounds to cause two or three cancer deaths per year. And then there are the heavy metals like lead, cadmium and many others that are known or suspected of causing cancer, plus a myriad of other health impacts. Finally, there is uranium, thorium and radium, radioactive wastes released from coal-burning that serve as a source of radon gas. The impact of this radioactive radon gas from coal burning on the public's health far exceeds the effects of all the radioactive waste released from nuclear plants. [Bernard Cohen, The Nuclear Energy Option, Plenum, 1990, p174]**

If fossil fuels are so environmentally destructive, what options are left? The two industrial scale alternative energy sources which have underutilized potential but which are also environmentally sound are nuclear power and solar energy.

Despite the claims of environmentalists about the rosy future of solar power, there are fundamental technical hurdles that will keep solar from being the major source of energy in the foreseeable future. Chief among the problems are solar power's intermittency and low energy density, and the fact that it is resource intensive, requiring large amounts of infrastructure. Other alternative energies have problems too. Hydroelectric power generation is not likely to grow from its present level because the U.S. river system has already been dammed to near capacity. Geothermal is geographically limited. Wind power is capital intensive, is plagued by structural failures, and suffers from the same intermittency problems as other solar alternatives.

Clearly, some part of the energy equation will have to be reformulated and it is difficult to imagine air-conditioners and microwave ovens in our future unless nuclear energy (and therefore Yucca Mountain) plays some role in our energy mix. This may not sound like much of a choice until one considers the form of wastes from hydrocarbon and nuclear power plants. The Second Law of Thermodynamics says that all processes create waste, so no matter what energy generation system we use, it will create a waste problem in some form. The only choice is which type of waste poses the least potential disposal problems.

The big difference between nuclear and hydrocarbon wastes is that nuclear residues are concentrated and eventually end up as solids while the waste from coal and petroleum powered plants is gaseous carbon dioxide, carbon monoxide, nitrous oxides, sulfur oxides, ash and a host of other chemicals that end up dispersed worldwide. True environmentalists might endorse nuclear power plants whose waste is compact and can be buried in one spot at a site like Yucca Mountain rather than being dumped on the wind for all to suffer. It may well be that future generations will view the the most environmentally destructive agents of our age as those who opposed nuclear energy and the construction of a geologic repository, thereby promoting the dispersal of carbon based combustion products into the atmosphere.

Unfortunately, Nevada's politicians and Green activists seem to prefer global pollution to the thought of burying the waste from nuclear energy creation in a drift 800 feet below the surface. Apparently, they would rather risk the threat of massive global warming from fossil fuel waste gases than admit the possibility that nuclear waste can be safely disposed of in a geological repository. This traces to the political agendas of these groups more than to their environmental concerns with radiation. Restructuring society seems to be a higher priority than promoting technologies that actually solve environmental problems.

Particularly disturbing is the possibility that by opposing nuclear power, environmentalists are ensuring we all will be subject to more radiation than before. As mentioned by Cohen above, coal fired plants are in themselves a source of radiation, releasing more radioactive elements into the air than nuclear plants. Perhaps ignorance of the physics and chemistry of nuclear and fossil fuel energy cycles is bliss. In the real world, however, the record of nuclear energy in preserving the environment stands on its own. According to the United States Council for Energy Awareness:

**NUCLEAR POWERS GREEN BENEFITS**
- **Today, 420 nuclear power plants produce about one-sixth of the world's electricity. This is more than the electricity generated from all fuel sources as recently as 1958. As well as enhancing international energy security, nuclear power has reduced greenhouse gases and air pollution in the 26 countries that use the atom to generate electricity. Since the first oil embargo in 1973, nuclear power plants worldwide have:**
    - **reduced carbon dioxide emissions by 13.4 billion tons**
    - **cut sulfur dioxide emissions by 109 million tons**
    - **eliminated 48 million tons of nitrogen oxides**
    **NUCLEAR POWER CONSERVES WORLD ENERGY SOURCES**
    - **Since 1973, nuclear power plants worldwide have cut fossil fuels used to generate electricity by:**
    - **17.6 billion barrels of oil, worth $470 billion**
    - **2.2 billion tons of coal**
    - **26 trillion cubic feet of natural gas**
    - **Since 1973, nuclear power in the United States has saved:**
    - **4.6 billion barrels of oil, worth $135 billion**
    - **1.1 billion tons of coal**
    - **7.9 trillion cubic feet of natural gas**
    **[USCEA, InfoBank, 1993]**

## ENERGY WARS

Environmentalists seem driven by an apocalyptic vision in which nuclear technology eventually pollutes the world through dispersal of radioactive substances, possibly through nuclear war. However, an alternative scenario is that the anti-nuclear solar utopianism of the Greens becomes a nightmare because of its internal contradictions. Some elements of this nightmare are easy to envision:

1)  For lack of American nuclear expertise (both in reactor construction and waste disposal), China fails to clean up its ever widening nuclear contamination.

2)  Iran finally constructs a missile system for its burgeoning nuclear bombmaking abilities. Oil production is halted by Iran as it seeks retribution and power.

3)  America loses 22% of its electrical energy producing capacity as aging reactors are retired. To compete in global markets, it is forced to import ever increasing amounts of petroleum products.

4)  For lack of sophisticated U.S. nuclear reactor designs, the Third World turns to increasingly "dirty" reactors that are environmental time bombs.

5)  Mixed oxide reactors which include plutonium in their reaction mix proliferate as America's involvement in world nuclear matters diminishes. Weapons proliferation is the end result.

The combination of the above factors obviously lead to war scenarios. While some of the above scenarios are more believable than others, it is clear that environmental pacifists need not be the only ones with visions of Armageddon. The anti-nuclear movement portrays itself as both a pacifist and environmental movement, opposed to destruction of Mother Earth through pollution of any form. However, a war over hydrocarbon energy, fought with tactical nuclear warheads, would certainly be the most environmentally destructive event imaginable. Unless we coercively limit third World populations, or industrial nations return to subsistence societies, there is little hope of preventing frictional global competition for hydrocarbon fuels. Consequently, nuclear energy is a critical element in the effort to save the environment, and perhaps save a significant portion of the human race from energy wars.

# 69. Solar Alternatives

One of the fantasies shared by everyone, even those in the nuclear industry, is that it will someday become possible to provide all of our power needs with clean, non-polluting solar power. Standing in the way of this dream are solar energy's inherent thermodynamic limitations which when taken into account suggest we will never achieve energy sufficiency without a mix of established as well as alternative sources. Bluntly, solar energy has three problems that cannot be wished away:

1) Solar is a low density energy source. That means there may be a lot of energy packed in natural energy systems, but it is dispersed over large geographical areas, requiring an extensive concrete and steel infrastructure to collect the power.

2) Solar systems are intermittent (the sun sets at night, the wind sometimes fails to blow, the tides are cyclical). This requires massive storage capacity to even-out the fluctuations in power

3) As a consequence of the first two problems, solar energy is resource intensive. Because of their low power density and need for load shifting storage, solar systems require massive amounts of infrastructure in the form of solar collectors, windmills, transmission systems, maintenance facilities, storage reservoirs, etc.

The problems associated with large scale natural energy systems are thus formidable. That does not mean they cannot be overcome to some degree, only that solar power must be competitive in the economic arena with other energy systems to justify its use. Solar is competitive in certain energy niches, but is unlikely to be competitive in all energy domains and this is where a mixed energy economy and nuclear energy come into play.

Of course, a rational energy policy is not what's being debated at Yucca Mountain, but a social and political policy. Solar is often promoted by people like Ken Bossong of Ralph Nader's Public Citizen's Critical Mass Energy Project for reasons other than its efficiency or safety:

**It is growing increasingly more apparent that achieving an economy based on solar technologies in no way assures the realization of any political goals. It is now clear that the U.S. can move from a nuclear/fossil fuel economy with little change in the status quo. Without a concerted effort to promote solar technologies in the context of social change, we face continuation of the problems of maldistribution of wealth, of environmental degradation, of energy waste, of health threats to workers and individual citizens, and of continued centralization of power and decision-making. [Bossong, Ken; A Solar Critique, Citizen's Energy Project, 1980, p1)**

Solar as an energy solution is quite a different animal from solar as a social policy: the former can be judged on technical merits while the later is merely a theory of Washington's environmental policy activists. The keyword from Bossong quote is centralization (supposedly the crime nuclear energy and industrialized solar energy are guilty of) as opposed to decentralization, which we've seen is a codeword for radical political advocacy. American's may not realize that a simple thing like a solar water heater is in some circles considered a revolutionary device.

Criticism of extremist proponents of solar energy representing groups like the SECC should not be confused with criticism of solar energy itself. In fact, the much maligned Department of Energy is one of the main researchers and boosters of solar energy, though at the same time they are accused by the anti-nuclear lobby of being totally inept in their conduct at Yucca Mountain. In an ironic twist, many of the DOE's solar studies have been used as the core of the repository opposition's Utopian fantasies in which the world transitions to a solar civilization.

Sandia National Laboratories conducted research for the Department of Energy on one of the most promising technologies called central receiver solar plants. Early experiments at the ten megawatt Solar One demonstration plant just outside Nevada in Barstow California led to calls for a second demonstration plant called Solar Two. Solar Two is a more advanced design that used elements from the Solar One site but employed molten nitrate salts as an energy storage medium. Unfortunately, even though Solar Two isn't even operating and in any event would generate at most 50 megawatts (compared to the 500 to 1200 megawatts of commercial nuclear plants) many environmentalists already want us to foreclose our nuclear future in favor of this just-emerging technology.

But what does Sandia National Laboratories, which has hands-on experience actually operating solar plants rather than the wishful theorizing of armchair engineers in the Green movement, have to say about the potential of central receivers?

**IT'S RELIABLE, CLEAN -- Under development for more than 15 years; solar central receiver power plants can play a large part in supplying the world's increasing needs for electric energy. They can produce electricity more cheaply than can any other utility scaled solar power plant. In addition, they will always meet even the most stringent environmental regulations.**

**IT'S UNIQUE -- These are the only solar power plants that can be designed with a capacity factor ranging from 25 to more than 60 percent -- meaning a central receiver power plant can operate at capacity for up to 60 percent of the year without using fossil fuel as a back-up, thus delivering power during most peak demands. And central receivers have the ability to store energy very cheaply. Because of this, they can deliver electricity on demand, even at night. This ability is known as load shifting. ["A Solar Electric Power Plant", Sandia National Laboratories SAND960235]**

Note first that Sandia claims for its plant only the ability to produce electricity more cheaply than other utility scaled solar plants, a far cry from actually beating the cost of nuclear. Secondly, they only claim a capacity factor of "from 25 to more than 60 percent" compared to nuclear's 65 to 85 percent. Even more importantly, the solar capacity factor comes at the wrong time of the year (summer) to take care of the major home electrical heating load in winter. Nighttime loads, though serviceable, are not guaranteed.

In other words, even as sophisticated as Sandia's solar central receiver is, it still needs back-up fossil fuel plants (or perhaps nuclear!) to make it year round viable. Sandia goes on in their brochure:

**It will be economical -- Using the most modern central receiver technology, a 200-megawatt plant could supply electricity at a cost competitive with power from fossil fuel. All the advanced generation technology needed for such a plant is proven. The cost of electricity from this solar plant is only 1 to 2 cents per kilowatt hour higher than from a coal plant of similar size. Many believe this is a small price to pay for such a clean and environmentally sound source of energy.**

**IF -- The initial investment, however, is several hundred million dollars, and this size plant has never been built before. An investment in such a venture -- in spite of the evidence to the contrary -- seems risky to many. A smaller demonstration plant could be the first step toward commercialization of the technology, thereby assuring a clean, inexhaustible, and secure domestic supply for power plants of the future. ["A Solar Electric Power Plant", Sandia National Laboratories SAND960235]**

Note here that 1 to 2 cents higher per kilowatt hour than coal plants of 200 megawatt size may translate into up to three times the cost of nuclear power, hardly a bargain here. This is without allowing for the introduction of next generation nuclear plants which have much higher efficiencies, greater safety and lower costs than present models.

Also, the startup costs of solar are large even compared to the startup costs of nuclear (which are also front end loaded), so any savings from solar technology would not be seen until far into the future even if it were employed today. And if 1 to 2 cents a kilowatt hour is a small price to pay for such a clean and environmentally sound source of energy (which, by the way, nuclear is also) why hasn't anyone in the Green lobby made themselves rich by putting a consortium together to build solar receivers in such anti-nuclear, energy starved states like New York?

This is not to heap abuse on solar; the people at Sandia have made huge strides in solar energy research and should be encouraged to bring these technologies on line whenever and wherever economically feasible. Instead, this strikes at the heart of the political motivations of some in the solar community and in the anti-nuclear opposition at Yucca Mountain. Before Nevada and the United States head down the path of rosy solar scenarios, we should ask whether there is compelling evidence that solar is our best option.

Most disturbing about the Sandia brochure is not its contents, but that Judy Treichel of the Nevada Nuclear Waste Task Force gives out this brochure as evidence that a solar future is just around the corner. While Treichel is quick to accept a solar sales brochure from the DOE (represented by Sandia), she has been unwilling to accept studies from this same agency about the safety of Yucca Mountain and about new developments in the nuclear industry (such as High Temperature Gas Reactors). A dispassionate engineering analysis at best rates solar as a potential large scale element in a mixed energy economy and it is typical of the technological naiveté of the state's representatives that they feel solar is a magic bullet for our nation's energy needs.

Another area that Nevada's anti-nuclear/pro-solar activists have sidestepped is the adverse environmental and economic costs of solar energy. The Second Law of Thermodynamics declares that no machine can run without waste or some form of pollution. In the early 1980s, Ken Bossong (later at Public Citizens) worked with Scott Denman (later SECC executive director) as part of the Citizens Energy Project. Bossong was candid about solar energy problems in an article titled Hazards of Solar Energy:

**Solar is being touted by its growing cadre of supporters as a technology that is environmentally benign -- one that promises to be a panacea for problems ranging from pollution to social and economic injustice to national security. While dispersed, small-scale solar technologies can offer many advantages over their fossil fuel and nuclear power competitors, they fall short of the qualities they are now being credited with . . . . . Moreover, there could be a backlash against solar technologies from the general public when it realizes solar is not all it stacks up to be now. . . . . projections are that over two million homes by 1985 will be employing one or more solar technologies and that will present an immense pollution control problem. [Bossong, Ken; "Hazards of Solar Energy", Solar Compendium, v2., A Solar Critique, 1980, p65]**

So even in solar utopia, there are problems. At least in 1980, the Citizens Energy Project (which becomes the Safe Energy Communications Council) wasn't afraid to reveal solar shortcomings. Bossong goes on:

**There is a price to be paid with every energy technology. The only way to lessen the costs associated with an energy-intensive society is to make it less energy intensive. Simplifying lifestyles, developing an ethic of conservation, and being mindful that anything we do has an impact are really the only ways to curb problems of pollution, poverty, etc. Solar is not a total solution; it will help but only if used intelligently. ["Hazards of Solar Energy", p66]**

Is a less energy-intensive society in mankind's best interest? Only if there is no alternative and solar is pushed coercively as the only option. But even a nationwide transition to solar energy would not necessarily solve every environmental problem and Bossong is a wealth of information on the possible dangers:

**PASSIVE SOLAR ENERGY --**
**While probably the least environmentally offensive of all the possible solar technologies, there are still a number of drawbacks associated with passive solar systems. A primary environmental concern is potential degradation of interior air quality as measured by temperature, humidity, and air circulation patterns (e.g. stuffiness, high humidity, and mold or fungus accumulations) . . . This problem could be increased if indoor pollutants are retained in the building's interior . . .**

**The Federation of American Scientists suggested that radon in buildings accounts for 20,000 cases of lung cancer per year; even if this estimate is a gross over-estimate, it can be seen that there is a possibility of increased danger from radon build-up in tightly sealed buildings. . . .**

**A second problem posed by passive solar systems is the possibility that passively designed homes will incorporate air circulation passages that may compromise the fire integrity of a building structure -- that is, make it easier for a fire to spread through a building. ["Hazards of Solar Energy", p66]**

Thus, even with passive solar heating there are dangers. Ironically we might be exposed to more radiation from trapped radon in our passive solar homes than if we lived next door to a nuclear plant. To be fair, Bossong does suggest solutions to these problems, although Bossong et.al. have never accepted the nuclear industry's solutions to their problems. Active solar systems pose similar questions:

**ACTIVE SOLAR ENERGY**
**. . . . A large demand for active systems could also pose a serious resource depletion problem for such materials as copper and aluminum (keep in mind that the U.S. already imports a portion of its copper; an increased demand for copper tubing for collectors could increase imports so that oil imports will decrease while copper imports will rise).**

**In addition, workers would be exposed to a range of chemical substances used in collector manufacturing such as the materials used in selective coatings for collectors or any plastics employed.**

**The most immediate problem posed by active systems once installed is that they might adversely impact upon the structural safety of a home; . . . . the glass employed in an active system is subject to breakage due**

to vandalism or other causes such as hail or hurricane with consequent injury to a building's occupants. . . . Decomposition of the selective coatings of the collectors could likewise release toxic gases; a related hazard is "out gassing" -- i.e. some solar systems use insulation materials that may discharge toxic or corrosive fumes when the collector overheats.

Certain working fluids . . .(which could also include such other commonly used chemicals as nitrates, nitrites, chromates, sulfites and sulfates) degrade over time . . . . If not properly disposed, these working fluids could degrade water supplies and affect aquatic life. . . .

A related problem is potential contamination of potable water by the solar working fluid. ["Hazards of Solar Energy", p67]

Reading Bossong, one feels that the promise of solar utopia is like Icarus crashing from the sky at the hands of one of its own proponents. Moreover, we haven't included every conceivable possible solar disaster as usually done when nuclear systems are critiqued. Still other problems arise:

## PHOTOVOLTAICS

While posing fewer problems once installed and operating than active solar collector systems, photovoltaics pose potentially more serious problems in the earlier stages of manufacture -- including mining and refining of the materials used and the subsequent production of the (solar) cells themselves.

The manufacturing and refining operations entail a range of environmental pollution problems. In the refining process for silicon, gaseous carbon monoxide and submicron-size particulates of silicon oxide are discharged . . . . silicon oxide can become a respiratory irritant. For every metric ton of silicon processed, 28 kilograms of solid soluble metal chloride and undetermined amounts of gaseous hydrochloric acid can be produced at the workplace. . . . .

Similar and additional problems may be posed by cells that are produced using highly toxic cadmium sulfide or gallium arsenide. Extraction of gallium from zinc and aluminum ores yields mercuric and acidic effluents as well as large volumes of alumina sludge which pose disposal problems. The use of arsenic in the production of gallium arsenide cells poses worker health problems. The cadmium used in the production of cadmium sulfide cells is a highly toxic substance whose dust can be a cause of kidney disease, emphysema, and pulmonary edema and is suspected of being a cause of hypertension.

Further, the Argonne National Laboratory has reported that: "The relative risk of workers involved in cell-related production activities is among the highest occupational risks in the U.S. In addition, environmental effluents emitted during cell production contain potentially toxic substances. Large scale development of terrestrial photovoltaic systems could result in significant release of these toxic substances with substantial public health risk.

There is also a possibility of localized climatic changes in areas where large numbers of photovoltaic arrays are located; "heat islands" could be created . . . ." ["Hazards of Solar Energy", p69]

Photovoltaics are obviously not an energy panacea, recognized as far back as 1980.. The pollution problem from photovoltaics is real and Bossong neglects to mention that the energy needed to create solar cells is substantial, meaning it takes a long time before the system breaks even as far as energy consumption. If photovoltaics have problems, what about wind systems?

## SMALL WIND SYSTEMS

A starting concern with small wind systems is the possibility that the blades could pop loose and be thrown resulting in possible injury to persons nearby . Structural collapse of the system is also a possibility.

Possibly a more significant health problem is that of burns, shock and electrocution from improper handling or poorly maintained equipment. . . . Personal injury could also result from falls from towers (not that far-out a possibility when one considers the frequency of serious injuries in cases of falls from roofs by persons repairing them).

A wind machine can result in decreased wind speed, increased soil moisture, temperature changes, increased relative humidity, and other impacts downwind of the structure. Likewise there could be other minor impacts on local ecosystems. . . . There is also the possibility of birdkills . . . And the noise produced by the whirring of the blades may be audible to nearby residents . . . ["Hazards of Solar Energy", p71]

These seem like a lot of concerns, yet we have culled the list presented by Bossong. Another area where many problems will occur in any large scale solar application is energy storage. According to Bossong:

## SOLAR STORAGE

**Among the options for solar storage technologies are batteries and rock bed storage. The latter poses potential problems of fungus growth and invites the use of herbicides and fungicides with their incident problems.**

**The former, i.e. batteries, pose a range of health and environmental concerns. Battery systems include lead/acid batteries now widely used as well as nickel/iron and nickel/zinc types. Among advanced systems under development for longer range applications are sodium/sulfur, lithium/metal sulfide, zinc/chlorine, zinc/air, and iron/air. The production, use and disposal of these batteries will probably entail the use of substantial quantities of lead, nickel, antimony, zinc, and other materials that are persistent, cumulative environmental poisons. Toxic gases can also be released as a result of fires or overheating in the case of accident or failure of battery chargers. ["Hazards of Solar Energy", p72]**

Bossong's preoccupation with small-is-beautiful technologies leads him to neglect the fact that large scale industrial storage systems for thousand megawatt utilities will be equally havoc producing. The technologies involved, such as nitrate salt heat storage, reservoir pumping plants, battery storage, etc. are too diverse to fully examine, but their environmental impact is substantial. One final quote from Bossong on alcohol fuels:

## ALCOHOL FUELS

**Producing alcohol from diseased crops or agricultural wastes is potentially an attractive way way to increase farm income, create new community based businesses . . . There is, however, a danger that agricultural wastes that would otherwise be plowed under to regenerate the soil would instead be used for alcohol production; the result could be a gradual depletion of the soil.**

**There is further the possibility that farm land now used to produce food could instead be converted into alcohol fuel production lands, thereby removing needed crops from the marketplace.**

**. . . alcohol for automotive fuel could result in a "mining of the soil if all the cover is removed" according to the U.S. Department of Agriculture. ["Hazards of Solar Energy", p73]**

It's a pretty gruesome solar world Ken Bossong has painted. Of course, his remarks are taken somewhat out of context and may overstate some of the risks of solar technology. There are two points we wish to make: 1) solar energy has its own set of costs and risks (i.e., it is not free), and 2) distorting the risks solar hazards is just as counterproductive as distorting similar levels of risk for nuclear technology.

If perceived risks are reality, as the opponents of nuclear energy and Yucca Mountain contend, then with a little manipulation of the "facts" concerning solar energy the public might be made to believe solar energy is a blight on civilization. This might especially be so if the pro-solar community could be persuaded to cough up fifty million dollars to create a Nevada Solar Project Office, staffed of course by nuclear industry supporters whose only purpose was to undermine the image of solar energy.

Unfortunately, the groups who have radicalized the Yucca Mountain debate and made solar energy the Holy Grail of the politically correct, seem shocked when their own tactics are used against them. Should the nuclear industry have used scaremongering against solar technologies (as well as character assassination and diversion of funds that we've covered before), the Greens would have protested vehemently.

There are many aspects of solar energy technology that are quite encouraging. Our objective is not to diminish the enthusiasm for solar research, but merely to put it into context. Engineers have a concept known as "engineering tradeoffs" which helps them keep a touch on reality. Lawyers, sociologists, political policy activists and the staffs of most environmental energy think tanks have no such philosophical rule of thumb and no such contact with reality, especially in regard to solar technology. Engineers test their products in the marketplace to see whether they will fly both physically and economically. The environmentalists have no such restraint and conduct solar engineering by law suit and harassment. Whether solar energy becomes the wave of the future, will not be decided in academia, but when investors see a chance for profit in a working technology. Before abandoning nuclear technology for environmental or sociological reasons, it would be nice to see a solar system which actually produces 1000 megawatts reliably.

According to Bossong:

**The public should not be cajoled into thinking that solar is a convenient way to encourage or sustain present wasteful and irresponsible lifestyles. Solar is a solution to problems of pollution, national security, unemployment, etc. when used in conjunction with other policies that respect the environment, individual rights, etc. It is not a total panacea. ["Hazards of Solar Energy", p75]**

We heartily agree. Perhaps solar plus nuclear might solve many of the problems we face, but of course this in part depends on solving the nuclear waste disposal problem and the construction of Yucca Mountain. In the mean time, we should be aware of the potential paradox that solar energy represents. That is, that the most natural and "free" energy source, the sun, may require numerous hidden environmental tradeoffs for its widespread utilization.

# 70. Peak Oil/Coal

In 2011, the price of oil again passed the $100 / barrel and the International Energy Agency (IEA) warned of the 'burden of oil consumption' – that is peak oil. While this proved a statistical blip in the market, how events will play out in oil and coal markets in the coming years is a matter of guesswork. Hydrocarbon fuel uncertainties justify diversification to nuclear energy as a power source which will defy sharp market fluctuations through reliable energy supplies.

Peak oil, an event based on M. King Hubbert's theory, is when the maximum rate of extraction of petroleum is reached, after which it is expected to enter terminal decline. Peak oil theory is based on the observed rise, peak, fall, and depletion of aggregate production rate in oil fields and is often confused with oil depletion. However, peak oil is the point of maximum production, while depletion refers to a period of falling reserves and supply.

The concept of peak oil, or a similar concept of peak coal, is important because it marks the peak production of "cheap" hydrocarbon energy. Recent technological advances such as fracking have allowed a transitioning from conventional crude oil and coal to Natural Gas Liquids (NGLs) or oil derived from other unconventional resources. However this may mask the global problem we face as third world countries race to energize their economies on the back of hydrocarbon fuels. This implies fierce competition for hydrocarbon fuel resources, whether we are indeed close to peak energy production from these resources or not.

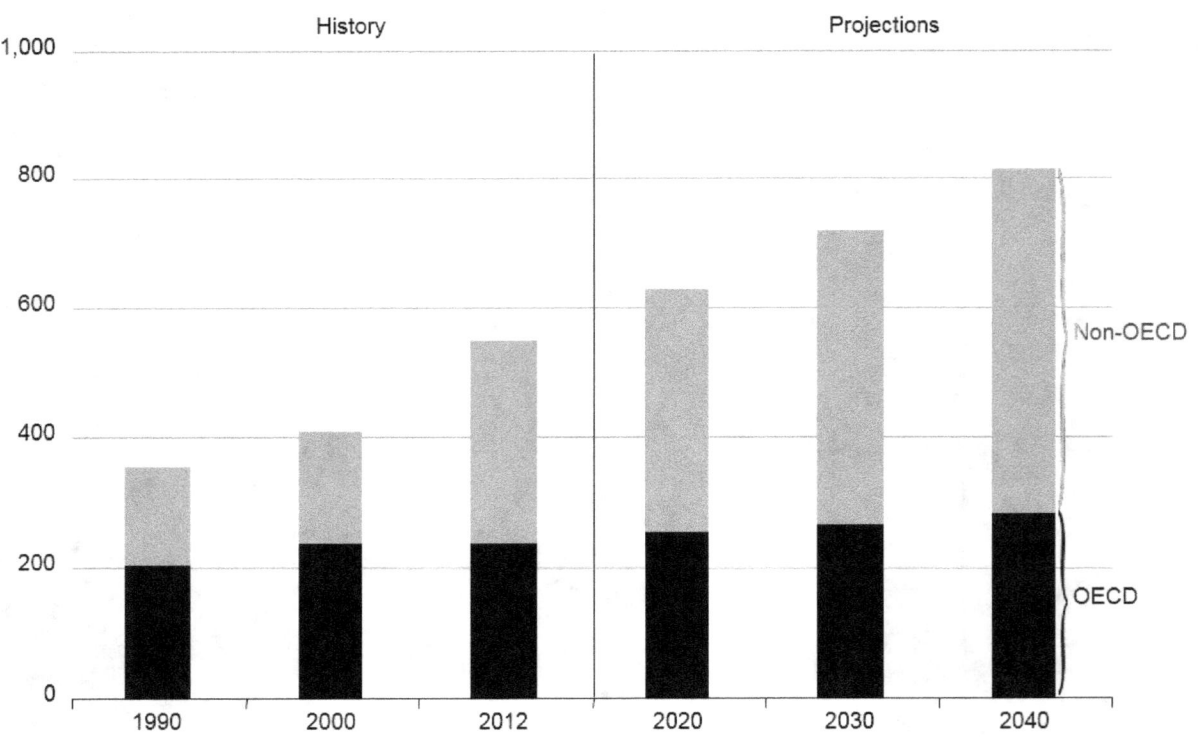

World energy consumption, 1990-2040

The economic argument is that resource constraints will be abated by technology and substitution. If international demand for oil increases, the price will increase to a point at which unconventional production becomes profitable. Indeed, oil and coal production from the tar sands, fracking and other unconventional resources has occurred in response to increasing demand which in turn elevated prices, not because they offered some competitive advantage when compared to conventional crude oil.

Prior to about 2012, "Peak Oil" theory tended to be associated with the oil and gas industry itself. M. King Hubbert, a geologist with Shell, is recognized as the modern inventor of "Peak Oil" theory. Matt Simmons, CEO of Simmons & Co., promoted a "Peak Oil" theory from 2004 until his death in 2010, based on a belief that Saudi Arabian oil fields had already peaked, and Kingdom's reserves were vastly over-reported.

Anti-fossil fuel activist groups began to co-opt "Peak Oil" theory around 2012 in order to discourage investment in oil and gas projects. Miseading "Peak Oil" studies from groups like the Post-Carbon Institute ensued on the pretense that global oil supply was about to "peak", and begin to decline. This drove governments to subsidize the development and deployment of alternative fuels, often leading to boondoggles.

In reality thanks to the development of alternative technologies necessary to produce oil and natural gas from fracking and shale formations previously thought to be permanently inaccessible, the world is now awash in untapped hydrocarbon resources. In 2014 it was revealed that Saudi oil production had not only not "peaked", but they had held back a vast amount of excess production capacity, turned anyone still trying to promote "Peak Oil" theories from the supply side of the equation into immediate laughingstocks.

## PEAK OIL DEMAND VS PEAK SUPPLY

Given international industrialization of hydrocarbon fuels in developing countries, growing America's economy for the next decades on an oil/coal energy base would subject us to non-stop competition for energy resources. But if the recessions of 1973, 1980, and 2008 have taught us anything, it is that high energy prices severely strain the economy.

In a society where economic growth is an essential part of the solution to almost every societal problem, peak oil/coal, whether from declining inventories or increased international demand, presents extraordinary challenges. The growth of the economy requires increasing energy supply, but increased American production of hydrocarbon fuel supplies will likely be siphoned off to developing countries in their growth phases.

The billion dollar question is: at what price of oil does our economy stop growing? Over the past 40 years, when petroleum expenditures as a percent of GDP increased much beyond 5.5%, the economy tended towards recessions. This tendency is due mainly to the fact that oil infiltrates almost every facet of an industrial economy, from personal disposable income, to manufacturing, to service sectors. Therefore higher oil prices restrain growth via declining discretionary consumption as individuals allocate more money towards gasoline and home heating, or as the cost of producing a good increases, etc. The true import of peak oil, therefore, may not be sustained high prices, but economic shrinkage.

Continuing oil shocks are likely, even if the US is lucky enough to escape production problems in the near term. Such shocks are typically associated with recessions, which would imply increased unemployment, surging budget deficits and possibly more pressure on housing prices and the financial sector. In short, long term economic stability may depend strongly on returning to a nuclear energy strategy, in part to allow for the electrification of currently oil dependent transportation.

Thus, in the best case, the world is facing periodic tight oil and coal markets into the indefinite future. In the worst case, our country may be heading into multiple oil and coal shocks and recession as world events fluctuate. Consequently, we benefit to the extent that we can substitute nuclear power for oil based energy production.

# 71. Third World Poverty

The relationship between Yucca Mountain and Third World Poverty is more direct than most realize. If, as some claim, the Amazon is being torched and places like Madagascar are being denuded of wood to fuel cooking fires, it is clear that the Third World cannot overcome its crushing poverty based on a wood energy supply or with primitive slash-and-burn industries. Since the developed nations already consume the majority of petroleum reserves, fast followed by developing nations like India and China, the Third World will not advance to First World status based on fossil fuels.

In a world of limited energy resources, either super energy consumers like America must lower their standards of living (perhaps to subsistence levels) to allow the Third World to pull itself up by its bootstraps, or the Third World must become resigned to its backward status. Those are the optimistic alternatives; pessimistically we already face increasing levels of terrorism and infectious global mini-wars over energy resources. Fortunately, a world of limited energy and material resources is avoidable given a suitable alternative energy reserve, a role for which nuclear energy is particularly suited.

Environmental activists concerned with risks at Yucca Mountain have failed to address the question of what damages their protests might cause outside the North American continent. Without nuclear energy as part of America's energy mix, demand would likely transfer to hydrocarbon fuels, directly competing with the Third World. Many environmentalists believe energy competition is unavoidable and the Third World is doomed to poverty. According to influential environmentalist and ally to the Safe Energy Communication Council Jeremy Rifkin:

**"However, this too must be said: no Third World nation should harbor hopes that it can ever reach the material abundance that has existed in America over the past few decades. To put its faith in Western-style development is a cruel hoax, simply because it is a physical impossibility even if there were a complete redistribution of the world's resources. According to economist Herman Daly:**

**"If it require roughly one-third of the world's annual production of mineral resources to support that 6% of the world's population residing in the U.S. at the standard of consumption to which it is thought that the rest of the world aspires, then it follows that present resource flows would allow the extension of the U.S. standard to at most 18% of the world's population, with nothing left over for the other 82%. Without the services of the poor 82%, the "rich" 18% could not possibly maintain their wealth. A considerable share of world resources must be devoted to maintaining the poor 82% at at least subsistence. Consequently, even the 18% figure is an overestimate.**

**It is thus impossible for the rest of the world to develop as the United States has. In fact, as we have already seen, absolute resource scarcity makes it impossible that even the United States can continue at anything near its present level of energy flow. This is not, however, to dismiss the absolute necessity of fostering economic development in the Third World. The question is: What kind of development is appropriate to poor nations? [Rifkin, Jeremy; Entropy, Plenum, 1984 p193]**

This apocalyptic view of energy resources and the potential of the Third World to improve its condition is typical of the environmental movement. But the notion of impending collapse of our resource base due to "absolute resource scarcity", or that we cannot sustain or improve our "present level of energy flow" is not supported by engineering analysis, unless one neglects the nuclear energy option. Extremists trying to nullify the nuclear energy option by choking waste storage at Yucca Mountain may therefore become the prime catalysts of resource collapse and environmental destruction. Forcing the Third World to artificially compete for non-nuclear energy sources, may ironically create a self-fulfilling prophecy of environmental doom, war and economic collapse. Since the Green movement variously opposes strip mining for coal, oil exploration at sea, nuclear reactors and nearly every energy option, this leaves little room to improve the economic status of the Third World. Again according to Jeremy Rifkin:

Several appropriate models for Third World development already exist. Before Mao's death, the People's Republic of China organized itself in a way that maintained the rural base of the society and favored labor-intensive production. China is not a rich society, but very few people are jobless or homeless. More attention should also be turned to the Gandhian economic model. During the anti-colonial movement led by Gandhi, the symbol of the struggle became the hand operated spinning wheel, a simple piece of appropriate technology that allowed each Indian to have some control over his or her own economic livelihood even in the poorest or most remote village. Gandhian economics favors the country over the city, agriculture over industry, small scale techniques over high-technology. Only this general set of economic priorities can lead to successful Third World development. But once again, it must be said that high-energy-flow nations like the United States must be willing to undertake sacrifices. [Entropy, p192]

Apparently, extremists within the environmental movement like Rifkin believe energy depletion is inevitable and spinning wheels are preferable to Yucca Mountain. Maoist China and Gandhian India are strange economic purgatories to be wished on the Third World and one wonders what kind of utopia environmentalists envision for America? It is interesting to note the evolution of SECC ally, former senator Tim Wirth, who later became undersecretary of State for global affairs. Slated to become Secretary of Energy under the Clinton administration, but replaced by Hazel O'Leary, Wirth was given the role of population control negotiator at the 1994 International Conference on Population Control in Cairo. The belief that the carrying capacity of Mother Earth has been reached (perhaps true without nuclear energy), made coercive population control an undercurrent at the Cairo conference. Wirth applied the limited resources model even to the U.S.:

Q: How well is the United States doing in stabilizing its population?

A: Not as well as we ought to. We have a rapidly growing population in the U.S., the most rapidly growing population of any major industrial country. . . . . One of the questions we have to ask ourselves is, do we want 500 million people in the U.S.? [USA Today, Sept. 6, 1994, p11A]

The energy bounty nuclear technology delivers threatens to rearrange the energy equation towards worldwide abundance. Americans are thus being asked to give up their cars, stereos, consumer goods and even progeny as part of the price of defeating Yucca Mountain and the nuclear industry. The Green's resource scarcity is an artifact of their self imposed nuclear energy scarcity, for thermodynamics tells us resource limitations become moot. with sufficient energy supplies.

Only an energy crisis artificially created through the elimination of nuclear energy (thereby choking fossil fuel supplies) could cause the Third World to look to the spinning wheel as its salvation. America's energy policy, now decided by lawyers, sociologists, political geographers and the environmental elite instead of by scientists and engineers versed, thus becomes a matter of life and death. Indeed, anti-nuclear hysteria may condemn much of the world's population to groveling poverty and starvation.

The University of Nevada Las Vegas Department of Engineering has a large contingent of East Indian graduate students. Their view of the possibility of turning India into a solar nation is pessimistic. The alternative, not only for India but the entire Third World, is to actively promote nuclear power for humanitarian, security and economic reasons.

While the Third World cannot join the First World on the back of fossil fuels, nuclear energy would bridge this energy resource gap. Lessened pressures on fossil fuels would in turn relieve destructive forces in the Middle East and global oil markets, making nuclear energy a vital component to worldwide security. Third World use of nuclear power would also reduce pollution, eliminating hydrocarbon emissions through a modernized infrastructure. Finally, a strong Third World offers new markets for our goods, and relieves immigration pressures.

The catch in using nuclear energy in the Third World is the disposal problem. Ironically, opposing the construction of Yucca Mountain may in effect cut off research and development in the only country with the scientific and financial resources to properly design and build a geologic repository. Whatever solution the U.S. implements, dry cask storage or geologic repository, will no doubt be safe due to strict environmental laws in this country. It is not clear that on-site, above-ground storage in a hundred Third World countries whose research and development programs are nil, will be so adequate.

Thus stopping Yucca Mountain may have the glorifying effect of increasing America's dependence on oil, slowing Third World development, and being the catalyst for numerous wars, but also causing nuclear pollution on a global level from unregulated and poorly engineered repositories worldwide.

# PART IX: BEST INTERESTS

# 72. The Nuclear Energy Option

The fate of nuclear energy is intimately tied to the fate of the Yucca Mountain repository. The legitimate question is whether the political and technical agony over the repository is balanced by the prospective benefits of retaining America's nuclear energy capacity. Nuclear energy does have a number of advantages over other large-scale energy resources that are not generally acknowledged by the public and which seem to be ignored by the media.

## ADVANTAGES:

- Abundant fuel reserves of uranium and thorium exist. With the potential reprocessing of nuclear waste, the energy reserves from nuclear sources are nearly inexhaustible, measured in thousands of years.

- While costs are somewhat higher than natural gas, they are competitive with other energy sources and may drop considerably as regulations delaying the licensing of nuclear plants are streamlined and a new generation of safe and efficient designs are brought on-line.

- Global warming (or even cooling) is not affected by nuclear energy. No carbon dioxide, sulfur oxides, nitrogen oxides, hydrocarbons or chlorofluorocarbons are emitted. Properly employed, nuclear energy is environmentally safe.

- Nuclear waste is compact, isolated and monitored, making it less of an environmental disposal problem than the diffuse gases and ash of the hydrocarbon economy. The entire nuclear waste from one year of operation of a nuclear plant fills a volume the size of two cubic meters.

- Unlike solar energy, nuclear energy is available continuously (it is timescale convenient). It can also be situated close to end-users on small amounts of land (it is geographically convenient).

Despite these advantages, nuclear energy is not the ultimate solution to all of the world's problems. Just as Ken Bossong of Public Citizen was forced to admit in more candid days that solar energy, the great hope of environmentalists, has problems, we should admit nuclear energy's problems as well.

## DISADVANTAGES:

1) Radioactive substances can be extremely dangerous and need to be monitored with care. They can cause cancer and genetic defects.

2) Reactor safety needs to be continuously improved, both through better designs and better management of these sites. Fortunately, safer designs already exist and the main delay in implementing their construction is opposition from the environmentalists to any form of nuclear industry.

3) The federal government should not subsidize nuclear energy, just as it should not subsidize solar energy. Nuclear energy's future should be played out in the marketplace, not in the halls of congress in a battle between industry and environmental lobbyists.

4) Costs must be brought into check. There is a long way to go towards solving the institutional problems inflating the cost of nuclear energy. Equivalent progress needs to be made in trimming costs at the design and construction phase as well.

5) Finally, the end disposal problem still needs to be resolved. While Yucca Mountain has as of this writing been tentatively canceled, it is clear there are no other options on the table and its renewal appears to be a viable long term storage solution.

The problems facing nuclear energy are mostly political, not technical. In a sense, nuclear energy is no longer a cutting edge industry; it has been used successfully for more than fifty years. Implementing nuclear technology no longer depends on quantum scientific breakthroughs and paradigm shifts, we are at least two generations behind in implementing safe technology. However, the industry's future depends on there being a stable and rational political environment in which long term stable investment can be made and nuclear energy developed as part of a mixed energy economy.

One of the ironies of the environmentalist's claims that nuclear energy is too expensive is that costs of nuclear energy are driven by activities of the protest movement itself. In fact, the opposition to Yucca Mountain has specifically attempted to drive up costs by delay tactics and lawsuits. At the national level, Shutdown Strategies proposed by various Naderite affiliates (Public Citizen, Natural Resources Defense Council) seek to cancel reactor siting and the construction of the repository. In Nevada, Bob Loux, Senator Bryan, Senator Reid, Mayor Goodman and others have made it known that they hope to derail the entire repository siting process by litigating transportation routes nationwide, claiming the market values of property along the route will be driven down by perceived risks.

Consequently, the greatest deterrent to the safe use of nuclear energy may be the environmental watchdogs themselves. By forcing the utilities, and manufacturers to divert resources from hard science and plant safety into public relations, the protesters may have delayed the development of truly safe nuclear energy systems by many decades. In place of rational science has come the need to fulfill spurious licensing requirements and fight legal delays. Unlocking the legal and political barriers which now govern the development of nuclear energy would significantly hasten the implementation of safer and/or more economical nuclear technologies now on the threshold of being built.

## THRESHOLD TECHNOLOGIES

A number of technologies can improve the efficiency, safety and economics of nuclear energy, vitalizing our industrial base while saving our environment.

- High Temperature Gas Reactors (more efficient and tolerant of loss-of-coolant accidents).

- Modular mass produced standardized reactors (design and manufacturing error reduction).

- Integral Fast Breeder Reactors (consume significant portion of own waste)

- Linear proton accelerator transmutation (recycling of long-halflife to short-halflife wastes plus energy production).

- Direct reuse of spent fuel after minimum reprocessing (it may be possible to simple grind up used reactor assemblies and recover more energy from second generation assemblies)

- Thorium reactors (Thorium is more abundant than uranium and the nuclear reaction chain burns closer to completion.

Nuclear energy can be cost effective, it does not pollute the atmosphere, it adds less radioactivity to the environment than coal burning processes, and its waste product is concentrated because of the high energy-to-mass ratio of the entire cycle. Thus technological advances leading to cheap nuclear power would allow a national renaissance, both environmental as well as economic.

## THE YUCCA MOUNTAIN IMPACT

The availability of a long-term waste repository obviously affects future utilization of nuclear technology by putting a lock on the development and construction of new reactors. While this is the primary effect of delays in bringing Yucca Mountain online, the evolving design of the repository itself has a profound effect on the future of nuclear energy. The retrievability of spent fuel from the repository obviously is a boundary condition affecting which nuclear reactor technologies evolve. The rising possibility that new processes may be able to recycle spent fuel also add to the motivation to design the repository as a geological monitored retrievable storage facility rather than simply a waste tomb. Serious thought towards incorporating retrievability criteria in the repository and transportation design will likely be repaid with future dividends if and when spent fuel becomes a valuable and recoverable commodity.

The most telling question in this case may be in the design of multipurpose unit canisters capable of being unloaded of their contents at future times with minimal radiation risk. One problem this raises is what constitutes an optimum package size; large canisters of the proposed 125 ton variety will be hard to manipulate. Can such canisters still be handled perhaps two centuries in the future? Physically, such canisters should also be able to withstand occasional rockfalls within the repository. Rockfalls in turn could be somewhat alleviated through rock-bolting of tunnel walls and appropriate concrete reinforcement. Each of these and other choices, however, affect the final repository configuration.

Long term monitored retrievable storage at Yucca Mountain may make sense specifically because of the uncertainties in future technologies and institutions. Any interim aboveground solution to nuclear waste storage must deal with the possibility that economics, political instability, natural disasters, apathy or a host of unforeseen consequences will result in nuclear waste polluting future generations. While on-site or aboveground storage may be more flexible and responsible to future generations, it can be argued that geologic storage is a rational solution even under a number of aborted retrieval and closure scenarios.

## THE NUCLEAR PROMISE

It would be naive to propose that nuclear energy will lead to some Utopian civilization, because every technology requires tradeoffs between benefits, costs and risks. In any event, it is unlikely that nuclear energy is the entire solution to our energy problems, just as solar, hydrocarbon, hydro and other energy technologies have not proven to be perfect solutions. Nevertheless, even in a mixed energy economy, increased utilization of nuclear power could lead a number of positive benefits:

- The reduction of smokestack pollution responsible for acid rain, global warming and health threats.
- The reduction of automobile emissions by bringing on line electric cars, trams, railways and super speed trains.
- A decreased dependence on foreign oil, lessening military commitments and enhancing security.
- Effective subsidization of the Third World by lessening the need of the industrialized nations for limited petroleum imports.
- A general decrease in energy costs, boosting economic growth.

Nuclear energy is not perfect, but it provides options not offered by other energy resources. Nuclear energy, in synergy with other traditional and alternative energy sources, can bring America into a new era. But this can only happen if the industry is unfettered of unnecessary legal and economic constraints and if a solution to the nuclear waste disposal question is reached.

# 73. Unanswered Questions

Oversight of the study of Yucca Mountain is dispersed over multiple levels in a dozen organizations, ranging from the Nuclear Waste Technical Review Board, to the Senate Energy Committee to the Nuclear Waste Project Office at the state level. These entities have a duty to ask tough and even embarrassing questions of the Department of Energy and the scientists studying Yucca Mountain.

In contrast, the anti-repository lobby in Nevada has not been required to justify its positions on Yucca Mountain with the same rigor. Senators Bryan and Reid, Governor Miller, the Sun newspaper, the Nuclear Waste Project Office and the environmental special interests that have invaded Nevada seem to have promoted their political agenda with little concern for whether nuclear waste is in reality safe to transport or store. This has distorted the public's views of the dangers of nuclear waste, inciting mass panic.

Nevadans who have attempted to keep an open mind about the nuclear waste issue have been brutalized by media attacks, hounded by elected officials and had their professional motives ridiculed by outside anti-nuclear agitators. While it is not suggested that those opposing Yucca Mountain should be subjected to these same tactics, the environmental lobby has left unresolved a number of unanswered questions about its logic and motives which it seems fair they be asked to address:

1) Nuclear bombs made from the same types of radioactive materials that would come to the Yucca Mountain waste site and capable of incinerating all of Southern Nevada fly over Las Vegas into Nellis Air Force Base every day. Nuclear annihilation is a terrifying risk, yet none of Nevada's politicians have suggested shutting down Nellis Air Force Base because of this kind of nuclear transport. If the dangers posed by Nellis AFB are so remote as to be inconsequential, why are the smaller dangers of Yucca Mountain viewed as a threat?

2) Las Vegans watched above-ground multi-megaton nuclear explosions at the Nevada Test Site from their backyards 45 years ago. Each of those explosions spread at least as much nuclear waste around the Vegas Valley as could be expected from the worst disasters that can happen at the Yucca Mountain repository. If Nevada's politicians and environmentalists are sincere about their concerns of contamination from the repository site, shouldn't Las Vegas be evacuated to save Nevadans from the much greater surface contamination left over from previous bomb testing?

3) If the dangers of earthquakes are so great at Yucca Mountain that it's too dangerous to store nuclear material there under a thousand feet of rock, then shouldn't Hoover Dam be dismantled? A fault runs under the dam and an earthquake of the magnitude that would disrupt the waste site could conceivably break the dam and send a cascade of water into Southern California. Is it logical to be for the dam but against the waste site, given that earthquake damage to Hoover dam is potentially more life threatening?

4) If there is a major earthquake danger in Southern Nevada, should Nevada have allowed the building of Steve Wynn's Mirage and Wynn casinos, Sheldon Adelson's Venetian, Kirk Kerkorian's City Center and other thirty story highrises that could collapse and kill ten thousand people inside in an instant. The resorts are only a few miles from McCarran Airport; surely a 747 crashing into the MGM Grand would be catastrophic (as was a similar jet crash into an apartment complex in Amsterdam in mid 1992). Rationally, the anti-nuclear coalition should favor closing McCarran International Airport and banning high density highrises.

4) If there is a major volcano risk at Yucca Mountain, shouldn't Las Vegas be abandoned before the city is subjected to the risk of annihilation? Volcanic activity at Yucca Mountain sizable enough to disrupt the repository might be as great as the explosion of Mt. Pinatubo in the Philippines and would likely destroy Las Vegas, only 90 miles away. If the volcano risk at Yucca Mountain makes that site unusable, does this also condemn Las Vegas?

5) If exposure to radiation levels of the sort expected for Nevadans during nuclear waste transport are intolerable, isn't there an obligation to ban air flights into Las Vegas' McCarran Airport? High altitude flight in an airliner exposes humans to measurable cosmic ray radiation. If even slight increases in radiation exposure due to nuclear waste transport are intolerable, then so should be the radiation exposure due to air travel.

6) The potassium in human bones is radioactive and a married couple is exposed to radiation levels similar to that which Southern Nevadans would be exposed to from the waste dump. Should Senators Bryan and Reid introduce a bill in the Senate to ban sex?

7) Should television be banned (the screen gives off X-ray radiation), should microwave ovens be outlawed (microwave radiation), should sun-tanning salons be illegal (high levels of UV radiation), and should post offices be condemned (uranium traces in their granite walls)? These radiation sources present cumulative risks similar in magnitude to Yucca Mountain.

8) Thirty-five percent of adult Nevadans smoke and that smoke has proven high levels of cancer causing radiation (greater than the waste repository will ever expose Nevadans to). Nearly 2000 Nevadans per year die of smoking related disorders and over ten thousand years that represents twenty million Nevadans who will die from cigarettes. Should cigarettes be outlawed in Las Vegas, the smoker's paradise?

9) Among the things that will kill thousands of people over the next ten thousand years of the nuclear repository's existence are: drunk drivers, pool drownings, cigarettes, guns, the common cold, AIDS, smog, icy sidewalks, drugs, airplane crashes, kitchen knives, pit bulls, football, ad infinitum. For example, each year there are about twelve child pool drownings in Southern Nevada and over ten thousand years that comes to 120,000 deaths. Should Nevada introduce a legislative ban on swimming pools?

10) The Pepcon rocket fuel blast in 1989 that rocked the Las Vegas valley and damaged Henderson, Nevada, suggests the nearby Kerr McGee plant is more dangerous than truckloads of inert nuclear spent fuel headed towards a repository. The acid cloud from the Timet titanium metals plant in Henderson, Nevada may well kill more people over ten thousand years than the nuclear waste site. Explosions of tankers of gasoline that travel Nevada's highways and rails will kill many times the number of people that will ever die from the waste repository (e.g., the 1992 Bakersfield train wreck). Should all risks similar to Yucca Mountain be banned?

11) Should lifesaving forms of medical radiation be banned, since they present similar (though vanishingly small) radiation risk levels to the general population as Yucca Mountain. Indeed, Citizen Alert espoused such a view in a position paper.

12) Nuclear power doesn't cause acid rain, it doesn't strip-mine coal fields, it doesn't put radiation in the air (not like burning natural gas and coal with natural radioactive elements does), it doesn't create unhealthy smog, it doesn't affect the ozone layer, it doesn't cause the greenhouse effect. Could the environmental protest against Yucca Mountain lead to the degradation of the environment and the deaths of thousands of people?

These are tough questions. However, the Yucca Mountain opposition seems not overly concerned with logical consistency or answering tough questions. Senators Reid and Bryan, Bob Loux and the vast network of anti-repository organizations all have had a vested interest in keeping the population afraid of nuclear waste. The resulting mix of nuclear hysteria and political ambition is perhaps a nuclear madness more dangerous than radiation from Yucca Mountain will ever be.

# 74. Credentials

Perhaps the real question at Yucca Mountain is not one of science, but of who we should trust and believe.

Analyzing the backgrounds of the key players at Yucca Mountain, one is struck by the fact that this appears to be a battle of the hard sciences versus the social sciences. Proponents of Yucca Mountain tend to be engineers, scientists or technical managers with experience working in high-tech environments (DOE, its subcontractors and those developing $3^{rd}$ and $4^{th}$ generation reactors). In contrast, the opposition is heavily weighted with lawyers, psychologists, history majors, political scientists and what might be called the soft sciences (concentrated in the political establishment, NWPO's socioeconomic studies and the Washington environmental lobby).

The motivations of these two groups are strikingly different. The hard scientists abhor politics and controversy and would like nothing better than to be left alone to do their job -- i.e. building a repository. The social scientists are in contrast primarily concerned with political issues, sometimes to the exclusion of an understanding of the technical issues. This divergence in motivations - one side focusing on the building of a repository while the other emphasizes politics - is so antagonistic that one can question whether the battle over Yucca Mountain is about mitigating the dangers of radiation, or whether it is about a clash between technocentric versus anthropocentric academic world views.

The reason this question is important is because we need desperately to determine who represents the best interests of America and indeed, the world as a whole. Some would argue that both sides of this debate have good intentions and that therefore each side has an equal right to not only be heard, but also to have their advice implemented. Unfortunately, good intentions often have unintended negative consequences. It is the thesis of this book that the unintended consequences of the politicization of science, as especially exemplified by the social scientists and an army of environmental lobbyists brought in by the Nevada Nuclear Waste Policy Office, have been extremely detrimental to the common good.

## EQUITY VS THE ELITE

The "equity" argument voiced by Nevada's politicians and the socioeconomic consultants to NWPO has muddied the question of credentials and politicized the science at Yucca Mountain. It is only equitable, it is claimed, that the least advantaged man exposed to risk should influence technical questions regarding risk. Loosely based on Rawlsian Ethics, the theory that the "least advantaged man" should have controlling input over complex technical issues, irrespective of his expertise, calls into question our ability to safely and economically design and build any complex technologies.

On the surface the equity argument seems plausible, but most people understand that the reason one hires experts to develop projects like Yucca Mountain is because they *are* experts. Moreover, proponents of equity are not necessarily forwarding the interests of the disenfranchised, for they in fact consider themselves to be an expert elite! What seems to have arisen is a pseudo-technical class of lawyers, sociologists, politicians and assorted policy addicts who lack technical expertise but nevertheless want a final say in such matters, generally finding leverage as advocates for some disadvantaged population.

Consequently, protest against Yucca Mountain has become a social and political movement as much as a question of safe technology. The wisdom of allowing social scientists and lawyers (often acting as self-appointed surrogates for the common people) to have veto power over technical projects is a matter of concern. There is the possibility that these social advocates may be more interested in accumulating power for themselves than in representing the popular will. The question this raises is whether society wishes to have its technology designed by non-technologists, as appears to be the intent of some of the protesters against Yucca Mountain, or whether we trust our technologists to conduct sophisticated science free from armchair quarterbacking.

Using equity as the final arbiter of technological issues can paradoxically lead not only to a lessening of the opportunities available to the elite classes, but also to the impoverishment of the "least advantaged" men and women this philosophy is designed to protect. This is because advanced technologies are generally expensive to begin with, and at first only affordable to elites (airplanes, railroads, automobiles), though in the long term the main beneficiaries are the least advantaged. Ironically, elite political advocates may be the only ones who truly gain long term from egalitarianism.

## COMPROMISED CREDENTIALS

Transforming Yucca Mountain from a technological question into a debate over social equity issues has created problems for repository opponents. A lack of technical credentials forced them to overstep their expertise in an attempt to legitimize their political positions. This led to situations in which political geographers, psychologists, lawyers, English and history majors frequently issued statements on nuclear technology though they lacked any training or experience in that field.

While the DOE and its sub-contractors working at Yucca Mountain have many problems of their own regarding their management of the project, there is a qualitative difference between their failures (typically bureaucratic inefficiencies) and the lapses of their opponents (often profound scientific naiveté). DOE and its affiliated scientific support may not always have Nevada's best interests in heart, but they are credentialed to the hilt in the fields in which they conduct science.

Watchdog institutions in academia, government and the media have not questioned credentials sufficiently, denying the public vital feedback. For example, though Nevada's media continually questioned the credibility of DOE and its subcontractors, the expertise of NWPO's consultants, the environmental groups and the State's political establishment were never seriously investigated.

The lack of validation of the credentials of consultants hired by Nevada's Nuclear Waste Project Office led to dubious science. For example, NWPO's funding of a political geographer to discuss the retrieval of nuclear waste from bore holes clearly overstepped proper guidelines. The use of a sociologist to give comments on the radiation deaths and nuclear radiation risks at Three Mile Island (though no fatalities have been recorded there nor expected) on the NWPO sponsored radio show was also suspicious. In general, a high school history teacher as executive director of the Nuclear Waste Project Office, a housewife turned anti-nuclear protester becoming the chief nuclear information officer for the State, a history major acting as their chief transportation consultant, ad infinitum, all contributed to a lack of confidence in the science conducted by NWPO.

The lack of technical sophistication on the part of Nevada's various environmental lobbies was also ignored by the media. For example, the pronouncements of Citizen Alert activists Bob Fulkerson (an English major) and Chris Brown (a religion major) on the engineering validity of subjects ranging from hydrogen-solar generators, to nuclear rockets, to Yucca Mountain were accepted by the media without question.

While everyone has a right to voice an opinion on the subject of Yucca Mountain, not all those opinions should carry the same weight. Technical experience and hands-on work with a technology do count in the real world. Consequently, if sanity is to be brought to the evaluation of the nuclear waste repository at Yucca Mountain, one of the first steps is for the credentials and competence of the competing voices to be judged and verified in government, media, academic and industry spokespersons.

The best way to demonstrate this problem of credentials is to take inventory of the prime actors on both sides of the Yucca Mountain controversy. What becomes quickly clear is that there is an obvious difference in the backgrounds of those on opposite sides of the fence. Pro-nuclear forces tend to be either scientific experts in fields related to nuclear energy, business leaders or project managers who have built successful enterprises and relied on engineering analysis in the past. In contrast, the anti-nuclear forces as a whole have rarely engaged in the building of any enterprise (even solar enterprises!). What we have is a battle between doers and critics.

To demonstrate these differences, we have assembled a players program covering the actors on the Yucca Mountain stage. Things which should be noted:

- Only a small minority of the opposition have experience working for industry (much less the nuclear industry) and consequently they possess little hands-on-feel for technology.

- The opposition is top heavy with lawyers, politicians, social scientists and the soft sciences who tend to view the entire world from a political perspective.

- Few of the opposition appear to have been actively engaged in the design or construction of any major engineering structures.

While there are those who will argue that lawyers, politicians and social scientists play an irreplaceable part in the technical study of Yucca Mountain, this argument does not carry much force. One could equally suggest that lawyers, politicians and social scientists should participate when someone's car is fixed (it is certain they would find themselves a niche in the garage), however no one is insane enough to suggest that their own car be fixed in this way. Allowing a national nuclear waste repository to be designed by politically motivated technical dilettantes carries significantly greater risks.

# 75. Nevada's Best Interests

Nevadans have been caught in a vise between the unbridled ambitions of their politicians, the revolutionary environmentalism of the Green movement, the machine politics of Washington and the hardball negotiating tactics of the nuclear power industry. Nevadans deserve better than to be crushed between these dinosaurs.

Nevada's options may not be as limited as they seem. The essential dismantlement of the Nuclear Waste Project Office started with the unseemly departure of Bob Loux was helpful but not enough. The 2016 elections and coming elections in 2018 will change the political landscape in Nevada and nationally. and have a major impact on the state's willingness to negotiate for benefits. While the Yucca Mountain repository may be opposed by a majority of Nevadans, many also believe that negotiating for benefits in compensation for a project they view as inevitable is a rational course their legislators should pursue.

Given the paradox that the likelihood that the nuclear waste repository will be constructed at Yucca Mountain despite the opposition of Nevada's powerful former Senator Reid, it is time for Nevada to begin some hard introspection. Nevada's window of opportunity in regard to negotiations for benefits may rapidly close, given the risks from nuclear waste transportation and storage in the state are extremely close to zero. If Nevada's political opposition is locked in cement, it may not be worth the trouble to compensate Nevada. The disaster at Fukushima may actually increase the need to resolve the storage issue quickly, given the amounts of nuclear material stored in vulnerable pools and on-site storage.

Yet, Yucca Mountain is a major project with substantial impacts on the infrastructure and economy of Nevada and the state should not be required to receive nuclear waste without benefits. Nevadans have coexisted with the Nevada Test Site for sixty-five years though in environmental terms the NTS is potentially a worse neighbor than the Yucca Mountain repository. There is no reason such a symbiotic relationship cannot be built with the repository. Nevada doesn't need to pursue the nuclear waste site, but it does need to actively pursue ample compensation for receiving any waste.

This compensation might include, but not be limited to, the following:

1) 100 million dollars per year in fees, indexed to inflation and retroactive to the 1987 Nuclear Waste Amendment Act. This represents more 10% of a state budget now in heavy deficit.

2) Guarantees that a railroad spur will be built to Yucca Mountain that avoids Las Vegas completely and limits the highway traffic of radioactive waste. Highway upgrades to ensure the safety of nuclear waste transportation that does come by road.

3) The funding of a nuclear studies institute within the university system to attract the brightest minds in nuclear and high-energy research to Nevada. Las Vegas should become a center not only for nuclear waste disposal, but also for the study of new reactor designs, transmutation and fuel recycling.

4) Nevada should ask for limited-liability deed to the waste. Nuclear waste will be recyclable into usable products in the not so distant future and represents a monetary gold mine for the state for centuries (if not millennia) to come.

5) Nevada should receive sizable subsidies for general education, at the grade school through university levels to compensate for the influx of workers.

6) Water resources are a problem for the entire Silver State, and especially Southern Nevada. In exchange for accepting Yucca Mountain, tradeoffs should be made for water rights, especially access to a greater allotment of Colorado River water. California must be willing to trade water allotments for access to the repository.

7) General infrastructure upgrades are needed. This includes local street and highway upgrades, as well as possible widening of the I-15 corridor to Los Angeles.

8) The facilities for construction of canister overpacks and other components should be required to maintain facilities in Nevada.

9) First consideration for other federal projects. Especially of interest would be:

(a) a new mag-lev train between Las Vegas and Los Angeles.

(b) a nuclear transmutation facility.

(c) a solar research facility at the Nevada Test Site

10) Ecological improvements or Endangered Species Act abatements and funding (desert tortoise, Devil's Hole pupfish)

11). Land swaps. More than 80% of Nevada is now in federal hands.

Nevadans need to understand that the Yucca Mountain repository is likely to be built without any of the above compensation and without needed oversight if the present political course is continued. Opposition by local politicians has not diminished the odds that Yucca Mountain will be built, as evidenced by the votes in Congress against efforts by senator Reid to terminally derail the process. Similarly, the environmental movement has only proven capable of delaying but not defeating the repository.

Before Nevada can regain control over the Yucca Mountain issue, much less begin to negotiate for benefits, a political housecleaning will be necessary. To ensure proper oversight of Yucca Mountain, the Nevada Nuclear Waste Project Office will need to be reengineered:

1) The Nuclear Waste Project Office needs to be gutted of its present personnel and replaced with people who know something about the nuclear waste industry. Robert Loux nor his successor Bob Halstead were not technically qualified to run this agency. Typically, an MBA with an engineering bachelor's background and experience in the nuclear field should be employed. Such qualified individuals are not hard to come by; in fact that is why the presence of a secondary-education major in this position is such a travesty.

2) The politically motivated Nevada Nuclear Waste Task Force needs to be replaced with a neutral information service run out of the Nevada university system. This would depoliticize the state's informational efforts and more importantly, offer the citizens of Nevada an educational organization capable of understanding and explaining the real social, economic and technical risks of Yucca Mountain, not just the perceived risks!

3) All contracts awarded by NWPO to the level of individual grants need to be reviewed. A majority of these contracts went out-of-state or to groups with political agendas which exclude objective science. The university system of Nevada and local engineering and consulting firms should be relied on and Requests For Proposals must be widely disseminated within the state. (In fact, most NWPO contracts should be limited to in-state resources to ensure Nevada is represented in this process and not replaced by disinterested academics.)

4) Competitive bids should be required on all contracts over $50,000 and should be open for public scrutiny. The Code of Federal Regulations should be followed in this regard (10- CFR 600-438)

5) All reports done for the state, and all data collected, should be easily accessible for use by residents of the state. NWPO has hidden or squelched much of the socioeconomic data, making its collection pointless.

The 2018 elections may change the dynamics of the Yucca Mountain debate. The replacement of a political establishment which sees little need to negotiate for the best interests of Nevadans may be the first step towards breaking the political logjam over Yucca Mountain. Representatives capable of understanding technical issues like Yucca Mountain, mining, the Nevada Test Site, water rights, etc. seem to be a priority for Nevada's future.

Gaming rules the State of Nevada, mining and ranching are its most important industries and the federal government's impact on the state is huge. After Senator Reid's retirement and loss of political leverage in the U.S. Senate on Yucca Mountain, Nevadans at some level understand that this makes their junior senators poor advocates for their state. Bryan and Reid failed to influence the rise of Indian gaming, riverboat gambling and other gaming venues outside the State, Heller and Cortez Mastos have even less clout. They have not been particularly effective in protecting Nevada's interests in regard to mining, cattle ranching or the Nevada Test Site

Since Nevadans cannot count on federal funding for alternative energy schemes to be a magic bullet, reality may be close to setting in. Developing a compensation strategy for Nevada is thus a high priority as Nevada faces a decision point. On the one hand lies Yucca Mountain and the prospects of a wide range of benefits, on the other Yucca Mountain and no benefits. The choices do not seem too difficult.

# 76. Best Interests of America

Yucca Mountain is a critical fork in the road for America. The nuclear waste repository is a necessary cornerstone of our hopes for energy self sufficiency, securing the availability of 20% of our energy production now derived from nuclear reactors and allowing us to move to 4th Generation Nuclear Reactor technology. This non-polluting energy can allow us to continue development of our natural resources without the environmental burdens traditionally associated with fossil fuels and the resource limitations of solar systems.

Arrayed against Yucca Mountain and nuclear energy are forces whose philosophy is to politicize every aspect of our technology and environment. Unfortunately, the alternative worldview they offer is one of uneconomic solar technologies that are unlikely to survive without massive federal subsidy. Solar utopianism is very different from solar realism, which views energy from the sun as a useful alternative energy option but not a political end in itself. Nuclear energy in parallel with solar energy and hydrocarbons offers America options while solar utopianism offers the paralysis of our society and possibly the destruction of the environment we wish to preserve.

Yucca Mountain is also about whether we will base our technological progress on hard science, or on psychological fantasies and popular myths about risks. If we choose the latter course, allowing polls interpreted by sociologists to take the place of sound engineering in our decision making process, we will have to admit that the scaly monsters that hid beneath our beds and plagued our childhood dreams are indeed real. There is a better course.

Humans depend on energy (literally, the ability to do work) to survive and prosper. Just as surely, the environment also depends on our access to abundant energy because humans can avoid resource depletion and destruction of the ecology through energy substitution. For example, hydrogen has great promise as a fuel because of its low air pollution (it burns to form water and limited NOx). Instead of using depletable oil reserves, hydrogen might someday be substituted as a fuel in many systems, but only if there is a primary energy source to generate that hydrogen. With abundant energy, we can become a hydrogen economy (or some other appropriate mix) while keeping the cars, air-conditioning, airplanes and telecommunications that have created modern civilization. Without abundant energy sources, conservation becomes our only option, an option that dangerously limits our future.

Finding a dependable, cheap and abundant energy source is not just America's problem, it is the major problem facing both Third World and First World countries alike. Places like India, the Middle East, China, Pakistan and other Third World countries cannot frivolously reject nuclear energy because of irrational fears generated by environmental lawyer/activists. Forcing the Third World to depend on fossil fuels and strip their rain forests for the fuel to bring them into the modern era is not a rational economic or environmental solution.

The greatest danger America faces may now be economic rather than from nuclear weapons. Two of our major long term economic competitors, Japan and India, have already opted for nuclear energy as part of their energy solution. China has already significantly tapped its coal reserves and must eventually also go down a nuclear path if it is to preserve its GDP growth. Forcing America to compete with these economies without nuclear energy is a non-solution. Consequently, what is at stake at Yucca Mountain is both America's economic and environmental future. We offer the following suggestions for America:

## DEPOLITICIZATION OF AMERICAN TECHNOLOGY

The central lesson of Yucca Mountain is that America cannot allow its technology to be politicized. Scientific truth *does* exist; the question of whether 2 + 2 = 4 is not a political question subject to popular polls. Nuclear power *is* a vital part of our energy future.

Unfortunately, our nation is now in the midst of a philosophical battle over who controls our science and technology. The new trend appears to be to socialize and therefore politicize our nation's scientific decisions. However, perceived risks are not a sufficient basis on which to judge energy policy or technology in general. Risk analysis, based on scientific data and expert technical judgment is the only approach to evaluating complex technologies which is self-correcting and avoids paralysis. Political processes, once introduced into scientific investigations, by their very nature replace truth with perceived truths whose progressive accumulation over time can only be swept away at great expense. Some will claim that science operates in a "milieu" of competing interests, but the fact is that at the end of the day, 2 + 2 really does equal 4.

The muddying of the technology debate, in nuclear issues as well as other sciences (genetic engineering, environmental remediation, etc.) is a direct result of what we will call the "stakeholder" problem.

## STAKEHOLDER ISSUES

At Yucca Mountain, the DOE has attempted to involve "the people" in some of the decision making processes regarding characterization of the repository through "stakeholder" meetings. While this is theoretically a commendable effort because of its egalitarian motivations, we've seen that such efforts can lead paradoxically to counterproductive and unintended consequences.

According to Dr. Daniel Dreyfus, former director of the Office of Civilian Radioactive Waste management, the word "stakeholder" is defined to include anyone and everyone interested in Yucca Mountain. Unfortunately, this leads to an infinite chain of stakeholders whose stake in the outcome is successively more tenuous. In fact, not everyone in America is at equal risk from the transport and disposal of nuclear waste at Yucca Mountain, so while everyone may be a Yucca Mountain stakeholder in a loose sense, there is certainly a hierarchy of stakeholders. Indian tribes, the state of Nevada, nuclear ratepayers, the nuclear industry, etc., all hold distinct positions within the stakeholder hierarchy as defined by the Nuclear Waste Policy Act. At the end of the stakeholder chain are the national environmental public interest groups, who while well meaning have no natural legal "standing" in the debate (either through exposure to risk or monetary involvement). Yet, national environmental activists have exercised near veto power over the development of nuclear reactors and the Yucca Mountain repository through their manipulation of the stakeholder input process.

In the future, Congress must be ready to make ground rules defining who has standing in technological decision making process and what level of influence this standing represents. Despite the opposition of environmentalists this must lead to a resurgence in the construction of nuclear reactors freed from entangling red tape. America's greatness is clearly based on its utilitarian traditions, not on Rawlsian egalitarianism. Recognizing that allowing populist movements to veto technological projects may be non-elitist, egalitarian and fair, but not wise, is thus crucial to bringing sanity back to the development of Yucca Mountain and technology in general.

Present legislative efforts to make the Environmental Protection Agency subject to utilitarian cost/benefit analysis is an encouraging move towards making a hierarchy of stakeholders based on costs (risks) and benefits (monetary gain and property rights). A similar clarifying of the stakeholder hierarchy surrounding Yucca Mountain is also called for.

## FINAL RECOMMENDATIONS: AMENDING THE NUCLEAR WASTE POLICY ACT

Nevada must be recognized as a primary Yucca Mountain stakeholder with substantial rights as a risk taker, property owner and because of the potential impact on the local gaming industry. However, a national cost/benefit analysis does not suggest the state should be given ultimate veto power over the repository. Normally, this would lead to a bargaining process in which state political leaders would negotiate to determine appropriate compensation to alleviate any inequities. Since this does not seem possible given the present political climate, it appears certain that when the Nuclear Waste Policy Act is revisited, the state's political establishment must be sidestepped in favor of trying to reach a wise, if not fair, solution.

Given a solution will need to be manufactured without Nevada political input, two issues which will need to be addressed are compensation to the state of Nevada and restructuring of the state's oversight duties. Regarding compensation, it may be appropriate to return to Sen. Bennett Johnston's rejected 1987 amendment to the Nuclear Waste Policy Act which proposed levels in the area of $100 million per year. Grassroot Nevada citizen's groups and prominent citizens are working to develop a benefits structure that is rational and meets Nevada's economic needs.

Also important in the amending of the Nuclear Waste Policy Act will be to make sure the Nevada Nuclear Waste Project Office becomes a constructive rather than destructive element in the repository study. Rather than strangling necessary state oversight, some simple options may prevent the runaway hysteria that NWPO has caused:

1) Enforcement and upgrading of the laws on competitive bidding (10-CFR 600-436) . Also require Requests For Proposals on all substantial contracts.

2) Require regular full audits of the Nevada Nuclear Waste Project Office.

3) Stiffen the professional competency requirements for agency staff to prevent them from becoming political footballs.

4) Enforce the utilization of local expertise to provide local feedback. Requiring local residence for all employee, contractor personnel, and National Laboratory scientists who spend the majority of their time working at the site.

Involve stakeholders in the process of selecting external peer reviewers.

Favor local industries and firms as sources for supplying goods and services to the program.

Aim to design a repository system whose predictable performance exceeds by a substantial margin the standards set up by regulators.

Develop contingency plans should Yucca Mountain prove unsuitable for a repository.

*******************************

Finally, we would like to leave you with the thought that America must not search out the fair solutions to its technical problems, but the wise solutions. There is no question that Yucca Mountain is not fair to Nevada. However, neither would it be fair to allow nuclear spent fuel to remain above ground at hundreds of sites around the continent for future generations to deal with, simply because we do not have the courage to choose truth over politics.

To the best of my ability, I have provided over the past pages the background behind the philosophies driving the debate over Yucca Mountain and nuclear energy. Since most of the readers of this book will be intimately involved in the siting process, it is now up to readers to use this information to forward the process of responsibly disposing of this nation's nuclear waste.

# Index

www.ingramcontent.com/pod-product-compliance
Lightning Source LLC
Chambersburg PA
CBHW081106170526
45165CB00008B/2346